Specially invited lectures presented at the

XIIth and XIIIth MICROSYMPOSIA ON MACROMOLECULES

INTERNATIONAL UNION OF
PURE AND APPLIED CHEMISTRY
MACROMOLECULAR DIVISION
in conjunction with

CZECHOSLOVAK ACADEMY OF SCIENCES
and
CZECHOSLOVAK CHEMICAL SOCIETY

MACROMOLECULAR MICROSYMPOSIA—XII and XIII

Specially invited lectures presented at the

XIIth and XIIIth MICROSYMPOSIA ON MACROMOLECULES

at Prague, Czechoslovakia
20–23 and 27–30 August 1973

Symposium Editor
B. Sedláček

LONDON
BUTTERWORTHS

ENGLAND: BUTTERWORTH & CO. (PUBLISHERS) LTD.
LONDON: 88 Kingsway, WC2B 6AB

AUSTRALIA: BUTTERWORTHS PTY. LTD.
SYDNEY: 586 Pacific Highway, Chatswood, NSW 2067
MELBOURNE: 343 Little Collins Street, 3000
BRISBANE: 240 Queen Street, 4000

CANADA: BUTTERWORTH & CO. (CANADA) LTD.
TORONTO: 2265 Midland Avenue, Scarborough, M1P 4SL

NEW ZEALAND: BUTTERWORTHS OF NEW ZEALAND LTD.
WELLINGTON: 26–28 Waring Taylor Street, 1

SOUTH AFRICA: BUTTERWORTH & CO. (SOUTH AFRICA) (PTY) LTD.
DURBAN: 152–154 Gale Street

The contents of this book appear in

Pure and Applied Chemistry, Vol. 38, Nos. 1–2 (1974)

Suggested U.D.C. *number* 541 · 64 (063)

ISBN 0 408 70639 2

Printed in Great Britain by Page Bros (Norwich) Ltd., Norwich

CONTENTS

SCIENTIFIC AND ORGANIZING COMMITTEES

XIIth Microsymposium

M. BOHDANECKÝ
Chairman, Programme Committee

J. BIROŠ, K. DUŠEK, P. KRATOCHVÍL
Members, Programme Committee

B. SEDLÁČEK
Microsymposium Editor

S. NEŠPŮREK
Technical Committee

XIIIth Microsymposium

J. ŠTAMBERG
Chairman, Programme Committee

J. ČOUPEK, K. DUŠEK, J. KÁLAL
Members, Programme Committee

B. SEDLÁČEK
Microsymposium Editor

S. NEŠPŮREK
Technical Committee

ORGANIZED STRUCTURES IN BLOCK COPOLYMERS: STABILITY DOMAIN AND STRUCTURAL STUDY

BERNARD GALLOT

Centre de Biophysique Moléculaire, CNRS, Avenue de la Recherche Scientifique, 45045, Orleans, France

ABSTRACT

In this paper we describe a safe method for the determination of the structure of block copolymers in concentrated solution and in the dry state. This method involves the use of both low angle x-ray diffraction and electron microscopy. Using these two techniques, we have established the existence of five types of structures for block copolymers and shown that when the composition of the copolymer in the insoluble blocks increases one observes successively: a centred cubic structure, an hexagonal structure, a lamellar structure, an inverted hexagonal structure and an inverted centred cubic structure. We also describe the effect of the concentration of the solvent, the temperature and the polymerization of the solvent.

1 INTRODUCTION

It is now well known that, in the presence of suitable solvents, amorphous block copolymers exhibit liquid crystalline phases[1,2]. The structural elements of these liquid crystalline phases are in the order of some hundred Ångstroms. The study of structures on this scale obviously calls for low angle x-ray scattering and electron microscopy.

X-ray diffraction easily provides the structural type (lamellar, hexagonal or cubic) and the main lattice parameter (thickness of the sheets for a lamellar structure, distance between the axes of the cylinders for a hexagonal structure, distance of the centres of the spheres for a cubic structure). But the detailed determination of the structure and the calculation of all its parameters require an hypothesis on the respective disposition of the different blocks and the solvent.

The use of electron microscopy avoids such an hypothesis[3–5]. However, the state of viscous liquid of the mesomorphic gels presents important difficulties for electron microscopic studies. To overcome these difficulties it is necessary to solidify the copolymer–solvent system by a post polymerization of the solvent[3,4]. Another difficulty with electron microscopy is the necessity for the staining of one block.

Nevertheless, the use of both x-ray diffraction and electron microscopy is the safest method to determine the structure of block copolymers in

PAC—38—2—B

solution and in the dry state. We shall justify this assertion in the present paper.

2 STRUCTURAL DETERMINATION

2.1 Principle of a safe structural determination[4, 5]

In a practical point of view, how have we to proceed to perform a safe determination of the structure of block copolymers?

At first we prepare mesomorphic gels by dissolution of the copolymer in a monomer which is a preferential solvent of one block. The principal monomers used are: styrene, MMA, BMA, VAcO, dichloropropene, acrylic acid. The principal copolymers studied are: polystyrene–polybutadiene (SB)[4, 6], polybutadiene–polystyrene–polybutadiene (BSB)[5, 7], polystyrene–polybutadiene–polystyrene (SBS)[8, 9], polystyrene–polyisoprene (SI)[1, 10, 11], polyisoprene–polyvinyl-2- or polyvinyl-4-pyridine (IVP)[12—14], polyisoprene–polymethylmethacrylate (I.MMA)[15], polybutadiene–polyvinylnaphthalene (B.VN)[16], polybutadiene–polymethylstyrene (B.MS)[17], polybutadiene–polybenzylglutamate (BG)[18].

Then, using well focussed and strictly monochromatic x-rays, we resolve the structure of the mesomorphic gel by low angle x-ray scattering. We fully polymerize the monomer by u.v. light or with a peroxide. We verify, by low angle x-ray scattering, that the periodic structure has not been destroyed by polymerization and we measure its new parameters.

After studying the solid sample by x-ray diffraction, we cut it with an ultramicrotome, stain the diene block by fixation of osmium tetroxide on the double bonds of the polydiene, and study the ultrathin sections with an electron microscope. Resulting from the method of staining used, the diene blocks appear black and the other blocks light on electron micrographs.

2.2 Determination and description of the structures

The study of x-ray diffraction patterns obtained with both mesomorphic gels (concentrated solution of the copolymer in a preferential solvent of one block) and organized copolymers (solid samples obtained by polymerization of a monomer used as solvent) prepared from block copolymers containing between 10 and 90 per cent of one block shows that:

(i) all x-ray patterns obtained can be classified in three families, each corresponding to a well determined structural type: lamellar, cylindrical or cubic;

(ii) x-ray patterns obtained before and after polymerization of the solvent belong to the same family.

Therefore, the structural type is not modified by polymerization and the results, provided by electron microscopy on solid samples, can be extrapolated to mesomorphic gels without any risk of error.

Using the results of both x-rays and electron microscopy we have established the existence of one type of lamellar structure, two of hexagonal structure and two of cubic structure. We shall describe these structures and give their characteristic features in x-ray diffraction and electron microscopy.

2.2.1 Lamellar structure

The first family of x-ray diffraction patterns is characterized by the presence of a series of five or seven sharp lines in their central region; their Bragg spacings are in the ratio: 1, 2, 3, 4, 5, 6, 7 . . . typical of a layered structure. A typical electron micrograph provided by an ultrathin section of a solid sample giving a diffraction pattern characteristic of this layered structure is shown in *Figure 1*.

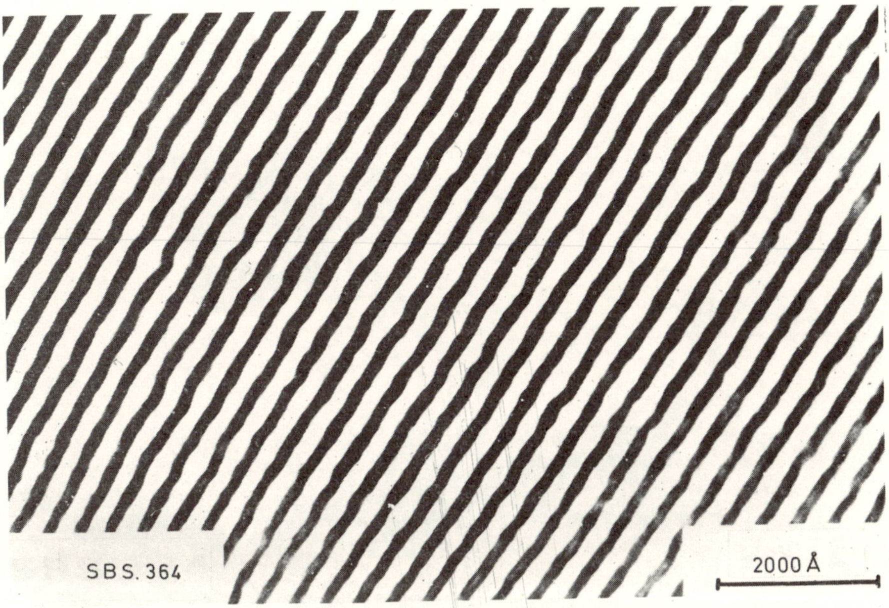

Figure 1. Electron micrograph of the copolymer SBS.364 swelled with 30% MMA

In this figure (corresponding to the copolymer SBS.364 swelled in 30 per cent MMA) one can see a striated structure formed by parallel stripes alternately black (containing the polybutadiene blocks coloured by osmium tetroxide) and white (containing the polystyrene blocks swelled by the polymerized MMA). This striated structure results from the section of the lamellar structure by a plane perpendicular to the planes of the sheets.

Therefore we can describe the lamellar structure pictured in *Figure 2* as a set of plane parallel equidistant sheets; each sheet results from the superposition of two layers, one formed by insoluble blocks B, the other by the solution in the preferential solvent S of the soluble blocks A.

The total thickness $d = d_A + d_B$ of a sheet is given directly by the Bragg spacing of the x-ray pattern, or directly measured on the electron micrographs (to have an accurate value of d by electron microscopy it is necessary to use electron micrographs provided only by sections perpendicular to the planes of the sheets[5].

The thicknesses d_A and d_B of the two layers are directly measured on the micrographs or calculated from the intersheet spacing d given by x-rays, using a formula based on simple geometry.

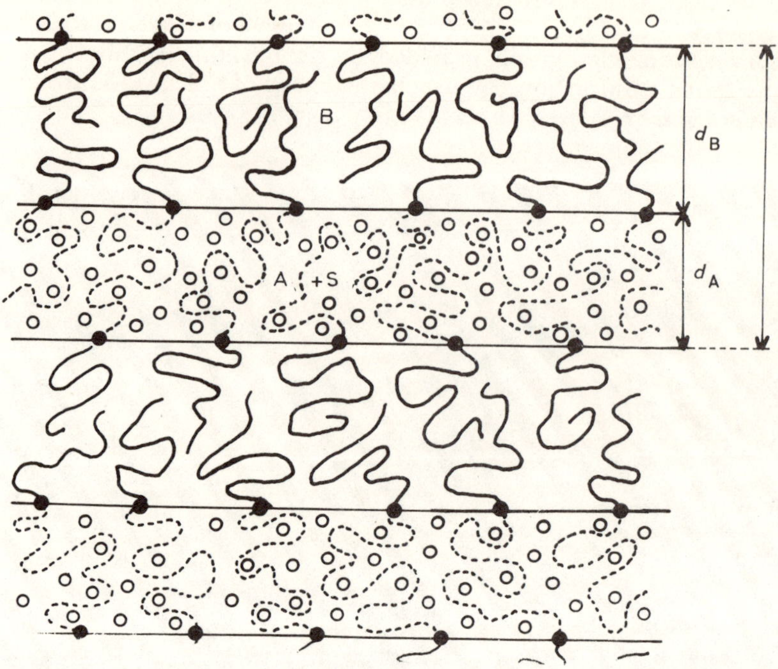

Figure 2. Schematic representation of the lamellar structure of a BAB copolymer

$$d_A = d\left\{1 + \frac{C(1 - X_A)\overline{V}_B + (1 - C)\varphi_B\overline{V}_S}{CX_A\overline{V}_A + (1 - C)\varphi_A\overline{V}_S}\right\}^{-1} \tag{1}$$

where C is polymer concentration in the solution, X_A is concentration of the A block in the copolymer, $\overline{V}_A$ and $\overline{V}_B$ are specific volume of the A and B blocks, $\overline{V}_S$ is specific volume of the solvent, and φ_A and φ_B are partition coefficients of the solvent ($\varphi_A + \varphi_B = 1$).

If the solvent S dissolves only one block (A for instance), $\varphi_A = 1$ and $\varphi_B = 0$, then:

$$d_A = d\left[1 + \frac{C(1 - X_A)\overline{V}_B}{CX_A\overline{V}_A + (1 - C)\overline{V}_S}\right]^{-1} \tag{2}$$

2.2.2 *Hexagonal structures*

The second family of x-ray diffraction patterns is also characterized by the presence of five sharp lines in the central region. But their reciprocal spacings are in the ratio: 1, $\sqrt{3}$, $\sqrt{4}$, $\sqrt{7}$, $\sqrt{9}$, which is characteristic of a bidimensional hexagonal array.

Electron micrographs of solid samples providing such x-ray diagrams have allowed us to distinguish two different structures corresponding to the same family of x-ray patterns. *Figures 3* and *5* are examples of such electron micrographs. They are of two types: *Figure 3* is characterized by black spots on a white background, while *Figure 5* is characterized by white spots on a black background.

2.2.2.1 Hexagonal structure—In *Figure 3*, one can see electron micrographs of sections in two perpendicular directions (the insert is at a slightly higher scale than the main figure) of the copolymer **BSB**.374 (31.5 per cent polybutadiene).

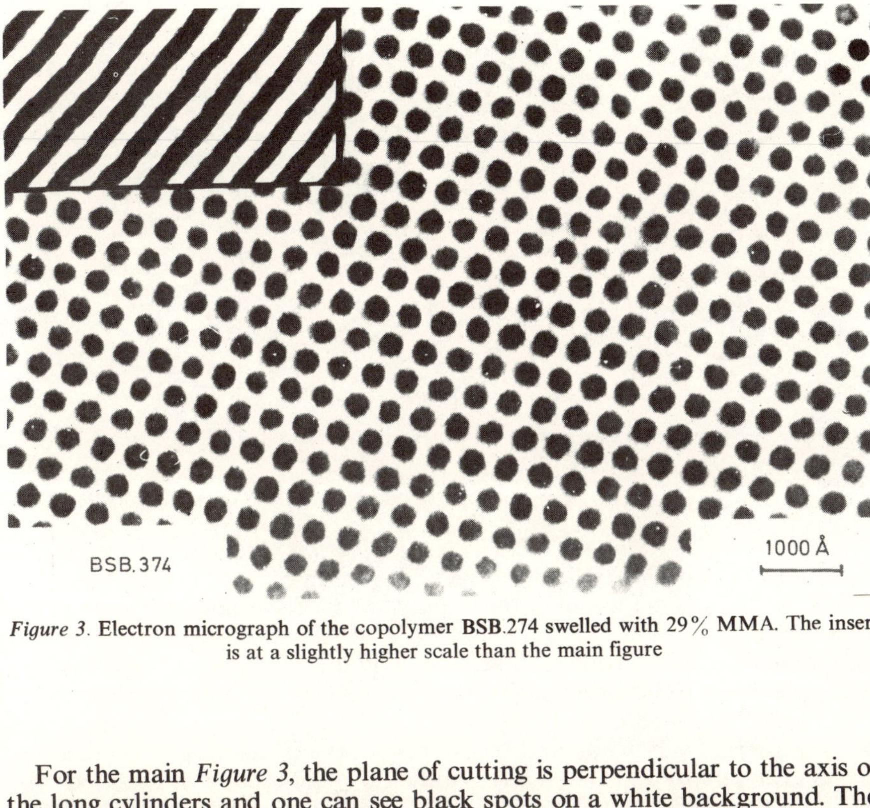

Figure 3. Electron micrograph of the copolymer **BSB**.274 swelled with 29% MMA. The insert is at a slightly higher scale than the main figure

For the main *Figure 3*, the plane of cutting is perpendicular to the axis of the long cylinders and one can see black spots on a white background. The spots have the shape of circles which are distinct and isolated: they are arranged in an hexagonal array in agreement with x-ray results. The insert is a section of the same copolymer by a plane perpendicular to the plane of cutting of the main figure. One can see a striated structure in which the black stripes are sections of polybutadiene cylinders by a plane nearly parallel to the direction of their axes. So we are in the presence of cylinders and not of spherical particles.

Therefore the cylindrical hexagonal structure (pictured in *Figure 4*) consists of indefinitely long cylinders arranged in a regular hexagonal two dimensional array; the long cylinders containing the insoluble blocks B are separated from one another by the solution of the soluble blocks A.

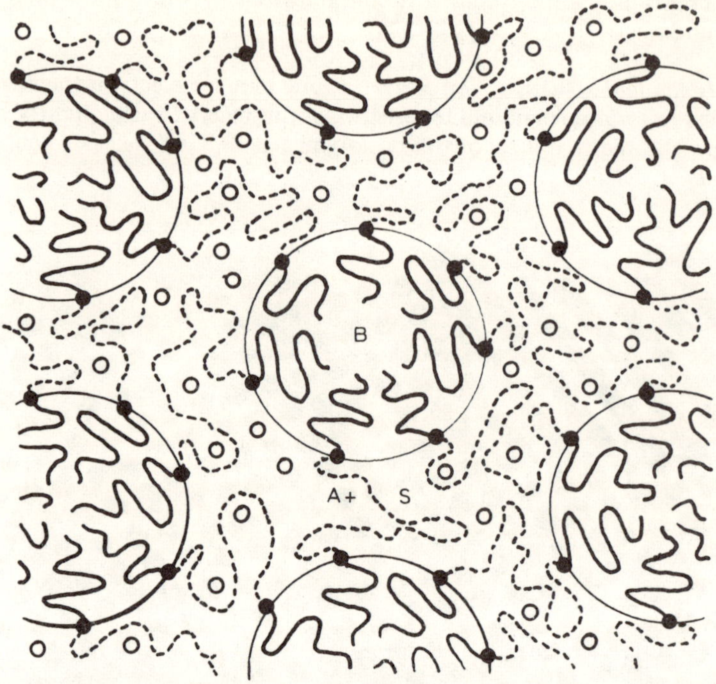

Figure 4. Schematic representation of the hexagonal structure of a BAB copolymer

2.2.2.2 Inverted hexagonal structure—In *Figure 5*, one can see micrographs of sections in two perpendicular directions (the insert is at a slightly higher scale than the main Figure) of the copolymer BSB.421 (74.8 per cent polybutadiene).

In the main figure, one can see, on a black background, white spots arranged in an hexagonal array: they are sections of long cylinders by a plane nearly perpendicular to the direction of their axes. In the insert. one can see alternatively white and black stripes: the white stripes are sections of the cylinders by a plane nearly parallel to the direction of their axes. In the main *Figure 5*, the circles are white, therefore the cylinders are filled with the polystyrene blocks A in solution in the solvent S as pictured in *Figure 6*.

From the micrographs of *Figures 3* and *5*, one can deduce the existence of two types of cylindrical structures: one with cylinders filled by insoluble polybutadiene blocks, the other with cylinders filled by polystyrene in solution in the solvent. We call these hexagonal and inverted hexagonal structures respectively. They can be distinguished from one another only by electron microscopy.

2.2.2.3 Characteristic parameters of the two cylindrical structures—The characteristic parameters of the two types of hexagonal structures are the distance D between the axes of two neighbouring cylinders and the diameter

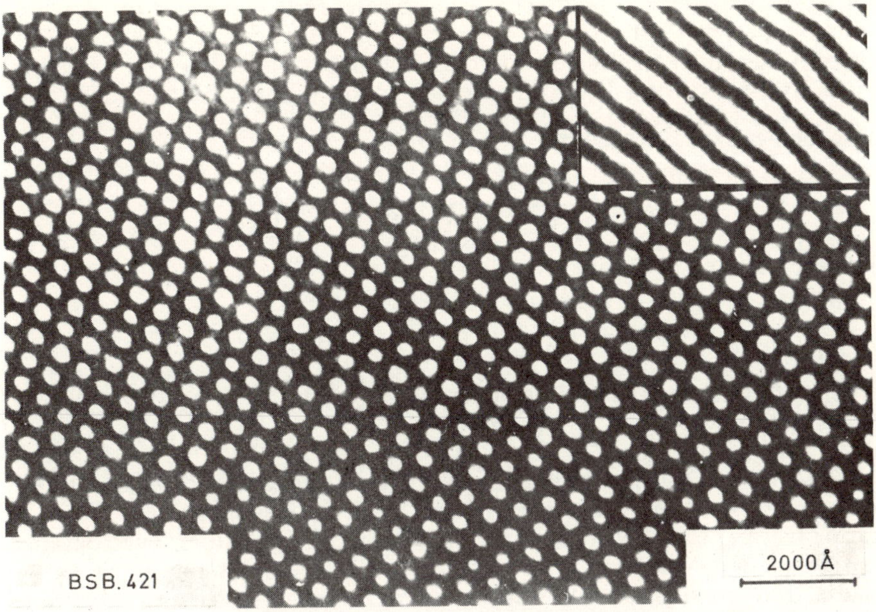

Figure 5. Electron micrographs of the copolymer BSB.421 swelled with 30% MMA. The insert is at a slightly higher scale than the main figure

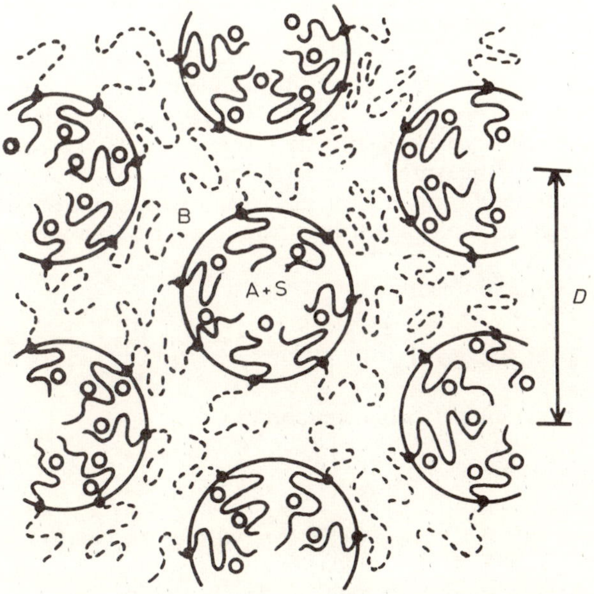

Figure 6. Schematic representation of the inverted hexagonal structure of an AB copolymer

$2R$ of the cylinders. D is directly given by the Bragg spacings of x-ray diffraction patterns, or directly measured on electron micrographs. $2R$ is directly measured on electron micrographs[5] or calculated from the value of D given by x-rays, using formulae based on simple geometrical considerations.

For the hexagonal structure:

$$R^2 = \frac{D^2\sqrt{3}}{2\pi}\left[1 + \frac{1 - CX_B}{CX_B} \cdot \frac{\bar{V}_{AS}}{\bar{V}_B}\right]^{-1} \tag{3}$$

For the inverted hexagonal structure:

$$R^2 = \frac{D^2\sqrt{3}}{2\pi}\left[1 + \frac{CX_B}{1 - CX_B} \cdot \frac{\bar{V}_B}{\bar{V}_{AS}}\right]^{-1} \tag{4}$$

where $\bar{V}_{AS}$ is the specific volume of the solution of the block A in the solvent S.

2.2.3 Cubic structures

The third family of x-ray diffraction patterns is characterized by the presence of four sharp lines in their central region. Their reciprocal spacings are in the ratio: 1, $\sqrt{2}$, $\sqrt{3}$, $\sqrt{4}$ which is characteristic of a cubic structure; but from such patterns it is very difficult to know if the cubic lattice is simple or centred.

Electron micrographs of solid samples providing such x-ray diagrams have allowed us to distinguish two different structures corresponding to the same family of x-ray patterns[19]. One type of electron micrograph is characterized by black spots on a white background (SB copolymers < *ca.* 18 per cent polybutadiene) and the other type by white spots on a black background (SB copolymers > *ca.* 80 per cent polybutadiene)[19]. For the two types of micrograph the circular spots are arranged in an hexagonal lattice (plane 111) or in a square lattice (plane 100)[20].

Therefore we can deduce the existence of two types of cubic centred structure: the cubic centred structure and the inverted cubic centred structure,

The characteristic parameters of the two types of cubic structures are: (a) the side a of the cubic cell directly given by the Bragg spacings of x-ray patterns or directly measured on electron micrographs; (b) the diameter $2R$ of the spheres directly measured on electron micrographs or calculated from the value of a given by x-rays using formulae based on simple geometrical considerations.

For the centred cubic structure:

$$\bar{R}_{cs} = a\left(\frac{3}{4\pi}\right)^{\frac{1}{3}}\left[1 + \frac{1 - CX_B}{CX_B} \cdot \frac{\bar{V}_{AS}}{\bar{V}_B}\right]^{-\frac{1}{3}} \tag{5}$$

For the inverted centred cubic structure:

$$\bar{R}_{cs} = a\left(\frac{3}{4\pi}\right)^{\frac{1}{3}}\left[1 + \frac{CX_B}{1 - CX_B} \cdot \frac{V_B}{\bar{V}_{AS}}\right]^{-\frac{1}{3}} \tag{6}$$

2.2.4 Comparison of x-ray and electron micrograph results

The values of the structural parameters found by x-rays and electron microscopy are in good agreement as illustrated by *Tables 1, 2* and *3*, for SB, BSB and SBS copolymers.

Table 1. Parameters of the lamellar structure

Copolymer	Gel of departure %	d, Å XR	d, Å EM	d_A, Å XR	d_A, Å EM	d_B, Å XR	d_B, Å EM
BSB.343	29 MMA	404	425	238	225	166	200
BSB.375	29 MMA	425	405	280	245	145	160
SBS.364	30 Styr	369	380	200	205	169	175
SBS.332	37 AcV	442	420	279	250	163	170
SB.33	30 MMA	365	375	247	240	118	135

d is inter-sheet spacing, d_A is thickness of the polystyrene layer, d_B is thickness of the polybutadiene layer, XR is x-ray diffraction, and EM is electron microscopy.

Table 2. Parameters of the hexagonal structure

Copolymer	Gel of departure %	D, Å XR	D, Å EM	$2R$, Å XR	$2R$, Å EM
BSB.372	29 MMA	348	320	155	140
BSB.373	29 MMA	403	365	200	185
BSB.374	30 Styr.	448	420	236	220
SBS.333	30 MMA	475	460	280	260
SBS.334	25 MMA	480	460	270	240
SB.32	25 Styr.	391	370	212	190

D is distance between the axes of two neighbouring cylinders, $2R$ is diameter of the cylinders, XR is x-ray diffraction, and EM is electron microscopy

Table 3. Parameters of the inverted hexagonal structure

Copolymer	Gel of departure %	D, Å XR	D, Å EM	$2R$, Å XR	$2R$, Å EM
BSB.421	30 MMA	413	400	280	240
SBS.365	35 Styr.	545	590	415	400
SB.35	35 MMA	568	530	442	420
SB.36	30 MMA	663	710	460	450
SB.36	30 Styr.	670	655	460	440

D is distance between the axes of two neighbouring cylinders, $2R$ is diameter of the cylinders. XR is x-ray diffraction, and EM is electron microscopy.

For the intersheet spacing d of the lamellar structure and the distance D between the axis of two neighbouring cylinders of the hexagonal and the inverted hexagonal structure the agreement is very good (the measurements on electron micrographs being made on a great number of sheets or cylinders). The average difference is about 5 per cent and never exceeds 10 per cent.

For d_A (thickness of the polystyrene layer), d_B (thickness of the polybutadiene layer), and $2R$ (diameter of the cylinders) the agreement is not so good but nevertheless it is satisfactory.

2.2.5 *Localization of the solvent*

Two methods can be used for studying the localization of the solvent in the mesomorphic structures.

2.2.5.1 *Study of the glass transitions by differential scanning calorimetry (DSC)*—For AB and ABA block copolymers of styrene and butadiene the T_g of the polybutadiene 1,2 blocks is in the vicinity of 0°C. So the use of a DSC equipped with tight cells has allowed us to evaluate the partition coefficient of the solvent between the two types of blocks by measuring the T_g of the polybutadiene 1,2 blocks in the dry copolymer and in mesomorphic gels of different concentrations in different solvents[21,22]. We have seen that, in the limits of the experimental error, the T_g of the polybutadiene 1.2 blocks is the same in the dry copolymer and in the mesomorphic gels. Therefore the solvent (MEK, MMA, styrene) does not enter the polybutadiene domains or, if it eventually enters them, it is in very small amounts (less than 10 per cent).

On the contrary, for polyisoprene–polyvinylpyridine copolymers with a microstructure of the polyisoprene block of about half-and-half 3.4 and 1.2), the T_g of the polyisoprene blocks changes with the amount of solvent in the mesomorphic gels; therefore there is an important partition coefficient for the solvent in such copolymers[23].

In the case of polystyrene–polyisoprene 1,4 *cis*, it has been impossible to use the DSC technique, because the T_g of the polyisoprene 1,4 *cis* blocks is near −80°C, that is to say in the vicinity of the melting point of the solvent used.

2.2.5.2 *Study of the structure factor*—It is well known that the respective intensities of the different diffraction orders on x-ray patterns change with the respective thicknesses of the two layers for the lamellar structure, with the diameter of the cylinders for the hexagonal structure, and with the diameter of the spheres for the cubic structure. Thus the study of the relative intensities of the reflections allows us to evaluate these structural parameters and then to calculate the partition coefficient.

To illustrate this procedure we shall take the example of the lamellar structure.

The structure factor of a lamella is given by[11]:

$$F(S) = \int_{\text{lamellar}} \rho(z) \cos 2\pi Sz \, dz \tag{7}$$

where Oz is the direction perpendicular to the plane of the sheet and $\rho(z)$ is the distribution of the electron density sketched in *Figure 7*. By integration of the equation 7, taking account of the symmetry of the structure, one obtains:

$$F(d_A/d) = K(\nu_A - \nu_B)\frac{\sin \pi n d_A/d}{\pi n d_A/d} \tag{8}$$

where n is the diffraction order and ν_A and ν_B are the number of electrons per Ångstrom cube for the layers of thickness d_A and d_B.

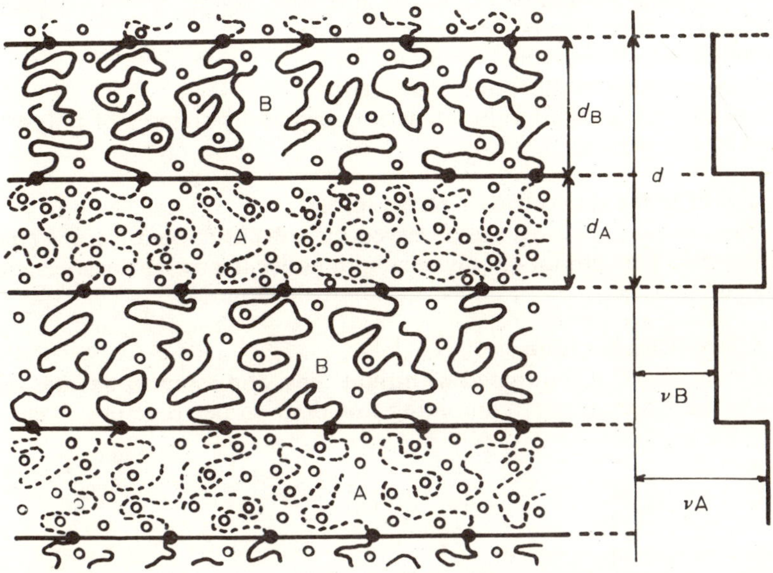

Figure 7. Schematic representation of the distribution of the electron density for a lamellar structure[11]

The intensity of the reflections is proportional to F^2 and the comparison of the relative intensities of the different diffraction orders gives the value of d_A/d. Then, by using formula 1, one can calculate the value of the partition coefficient φ_A.

We have found for toluene and SI 1,4 *cis* copolymers, $\varphi_A = \varphi_{PS} = 0.6$; for MMA and SB 1,2 copolymers, $\varphi_A = \varphi_{PS} > 0.9$, in agreement with the DSC results.

3 DOMAINS OF STABILITY OF THE DIFFERENT STRUCTURES

The principal factors governing the stability of the mesomorphic structures are the composition of the copolymer, the concentration of the solvent and the temperature.

3.1 Effect of the composition of the copolymer

For SB, BSB and SBS copolymers for instance, the type of structure adopted by the copolymer is determined by its composition. When the proportion of polybutadiene increases from 10 to 90 per cent one observes successively:

(a) a cubic centred structure formed by spheres of polybutadiene till about 18 per cent;

(b) an hexagonal structure formed by cylinders of polybutadiene till about 35 per cent;

(c) a lamellar structure formed by alternate layers of polystyrene and polybutadiene till about 65 per cent;

(d) an inverted hexagonal structure formed by cylinders of polystyrene swelled by the solvent till about 80 per cent;

(e) an inverted centred cubic structure formed by spheres of polystyrene swelled by the solvent.

3.2. Effect of the concentration of the solvent

Mesomorphic structures are observed for solvent concentrations smaller than about 45 per cent for copolymers with molecular weight between about 20000 and 200000.

3.3 Effect of temperature

The study, by DSC equipped with tight cells, of mesomorphic gels of SB, BSB and SBS copolymers containing less than 45 per cent of solvent, has

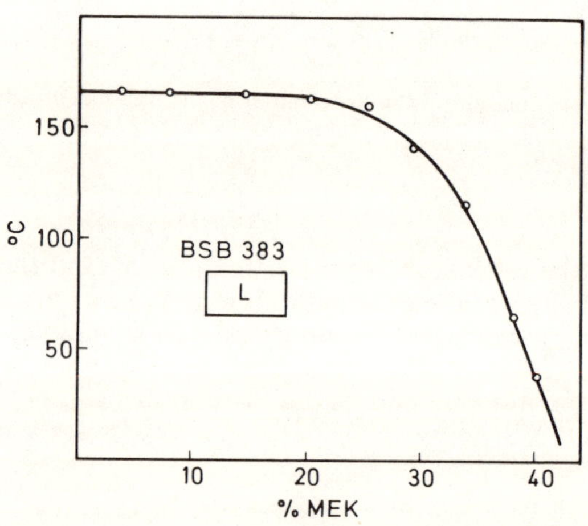

Figure 8. Phase diagram of the system BSB.383/MEK. Lamellar structure

shown that such copolymers exhibit only one mesophase as a function of temperature. The structure is lamellar (*Figure 8*), hexagonal (*Figure 9*), inverted hexagonal (*Figure 10*), centred cubic or inverted centred cubic (not represented) depending on the composition of the copolymer[5,9].

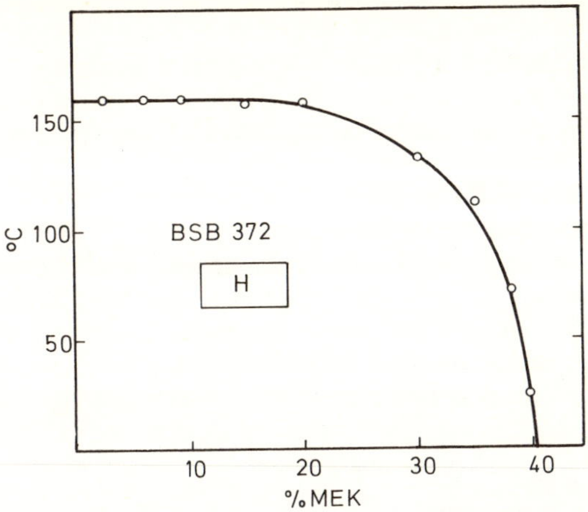

Figure 9. Phase diagram of the system BSB.372/MEK. Hexagonal structure

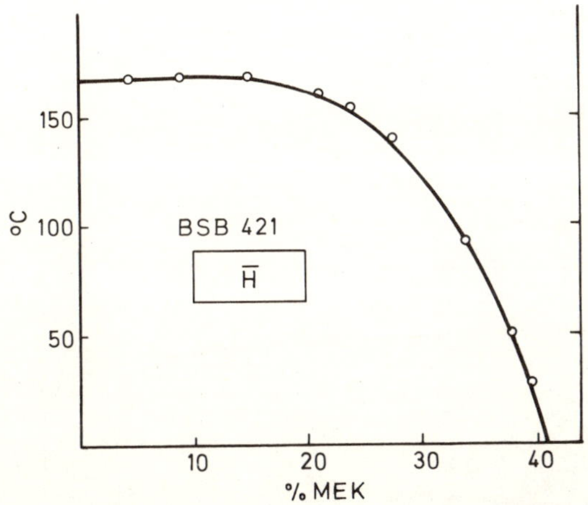

Figure 10. Phase diagram of the system BSB.421/MEK. Inverted hexagonal structure

4 FACTORS GOVERNING THE STRUCTURAL PARAMETERS

Among the factors governing the geometrical parameters of the mesomorphic structures, we shall only describe here the effect of the solvent concentration, the molecular weight of the copolymer and the polymerization of the solvent.

4.1 Influence of the solvent concentration

We have to consider two cases. The first consists of systems copolymer–solvent with $\varphi_A \geqslant 0.9$, that is to say, the solvent practically does not enter the domains formed by the B blocks. Such systems are generally characterized by a decrease of the lattice parameter (intersheet spacing for the lamellar structure, distance between cylinders or spheres for the hexagonal or cubic structures) when the solvent concentration increases. Copolymers SB, SBS and BSB with polybutadiene 1,2 blocks, put in solution in MEK, MMA or styrene, are representative of this first class of systems. The second case is formed by systems copolymer–solvent with $0{\cdot}4 < \varphi_A < 0.6$, where the solvent swells the two blocks. Such systems are generally characterized by the fact that the lattice parameter increases with solvent concentration. A typical example is polystyrene–polyisoprene 1,4 *cis* copolymers in solution in toluene or styrene.

4.1.1 AB and ABA polystyrene–polybutadiene 1,2 copolymers

We shall describe the influence of the solvent concentration upon the three following types of structure: lamellar, hexagonal and inverted hexagonal.

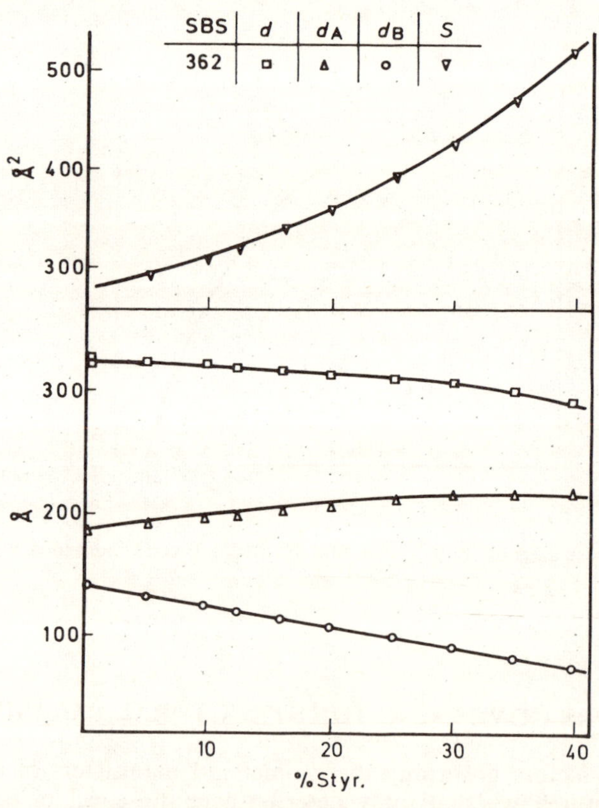

Figure 11. Variation of the parameters of the lamellar structure with solvent concentration. Copolymer SBS.362[9]

4.1.1.1 *Lamellar structure*—The effect of the solvent concentration on the geometrical parameters of the lamellar structure is illustrated by *Figure 11* where we have plotted the structural parameters of the copolymer SBS.362 versus styrene concentration. One can see that when the solvent concentration increases: (a) the intersheet spacing d and the thickness d_B of the insoluble polybutadiene layer both decrease; (b) the thickness d_A of the soluble polystyrene layer and the average surface S available to a molecule both increase.

4.1.1.2 *Hexagonal and inverted hexagonal structures*—In order to illustrate the effect of the solvent concentration on the two types of hexagonal structures, we have plotted the variation of the structural parameters of the hexagonal and the inverted hexagonal structures versus solvent concentration in *Figure 12* and *Figure 13* respectively.

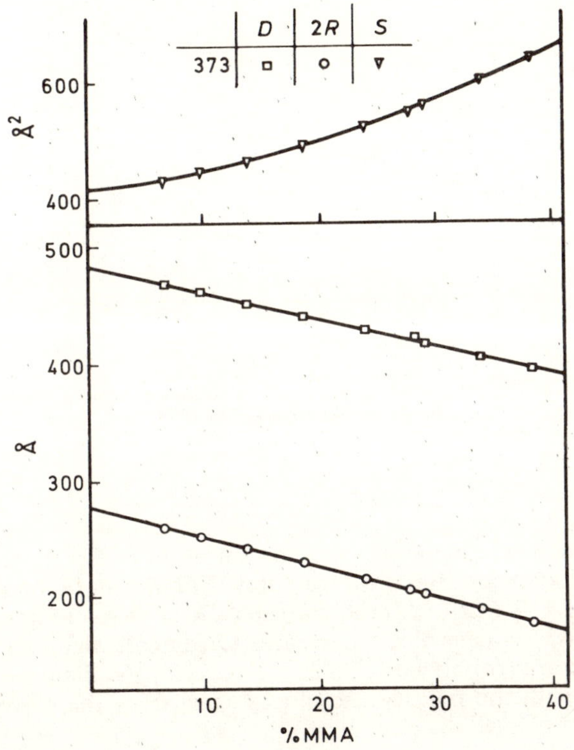

Figure 12. Variation of the parameters of the hexagonal structure with solvent concentration. Copolymer BSB.373[5]

One can see that for both the hexagonal and the inverted hexagonal structures, when the solvent concentration increases, the distance D between the axes of two neighbouring cylinders decreases and the average surface S

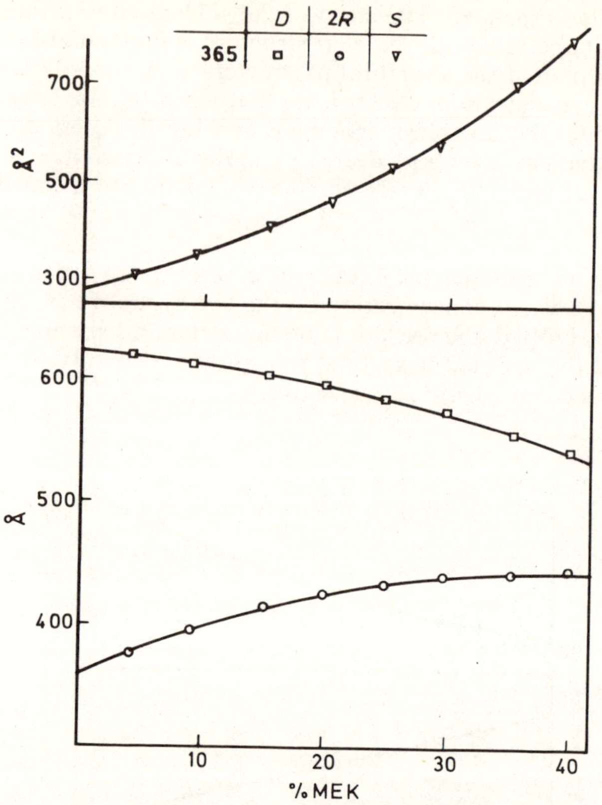

Figure 13. Variation of the parameters of the inverted hexagonal structure with solvent concentration. Copolymer SBS.365[9]

available to a molecule increases. But the variation of the diameter of the cylinders is different for the two structures. For the hexagonal structure (copolymer BSB.373, *Figure 12*) the diameter $2R$ of the cylinders filled with the insoluble polybutadiene blocks decreases. On the contrary, for the inverted hexagonal structure (copolymer SBS.365, *Figure 13*) the diameter $2R$ of the cylinders filled with the polystyrene blocks swelled by the solvent (MEK) increases.

4.1.2 *Copolymers polystyrene–polyisoprene 1,4* cis

We shall only describe the behaviour of the lamellar structure. We shall take as example the copolymer polystyrene–polyisoprene 1,4 *cis* SI.822 in toluene solution (*Figure 14*). For this system the partition coefficient $\varphi_A = \varphi_{PS} = 0.6$[11]. One can see on the figure that the structural parameters (the intersheet spacing d, the thickness d_A of the polystyrene layer and the thickness d_B of the polyisoprene layer) all increase with toluene concentration. We have also plotted in dotted lines the variation of d_A and d_B calculated some years ago[10] with the hypothesis $\varphi_A = 1$, to show the influence of the partition coefficient on the values of d_A and d_B.

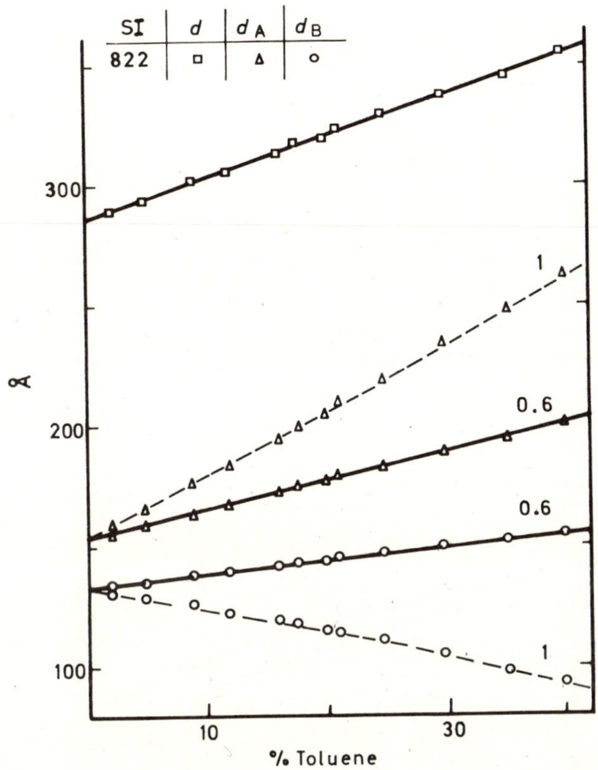

Figure 14. Variation of the parameters of the lamellar structure with solvent concentration. Copolymer SI.822. Partition coefficient $\varphi_A = 0.6$

4.2 Influence of the molecular weight of the copolymer

Among the different copolymers studied it is the lamellar structure of BSB copolymers which has the largest domain of stability. So we shall illustrate the influence of the molecular weight of the copolymer on the structural parameters with the results obtained for the lamellar structure of BSB copolymers. As shown in *Figure 15*, the intersheet spacing d increases monotonically with the molecular weight of the copolymer.

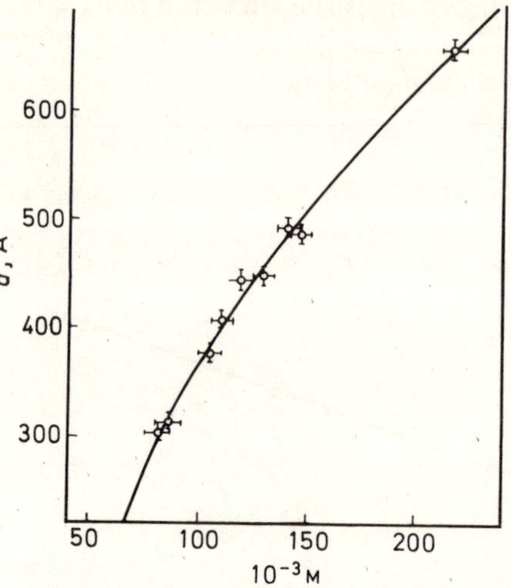

Figure 15. Variation of the thickness (extrapolated to zero solvent concentration) of the sheets of the lamellar structure of BSB copolymers versus the molecular weight of the copolymer[5]

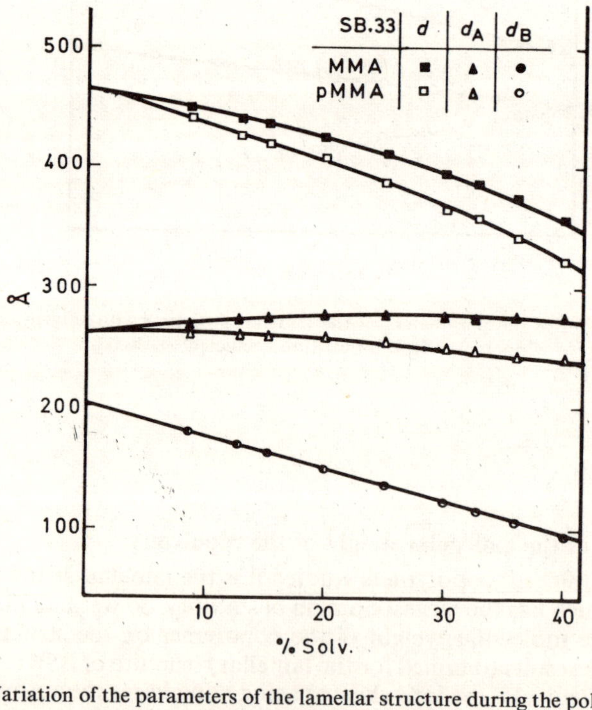

Figure 16. Variation of the parameters of the lamellar structure during the polymerization of the solvent. System SB.33/MMA[4]

4.3 Influence of the polymerization of the solvent

Mesomorphic gels prepared by swelling of the copolymer in a monomer (e.g. styrene, vinyl acetate, methyl methacrylate) have been converted to organized copolymers by polymerization of the monomer by u.v. light[4].

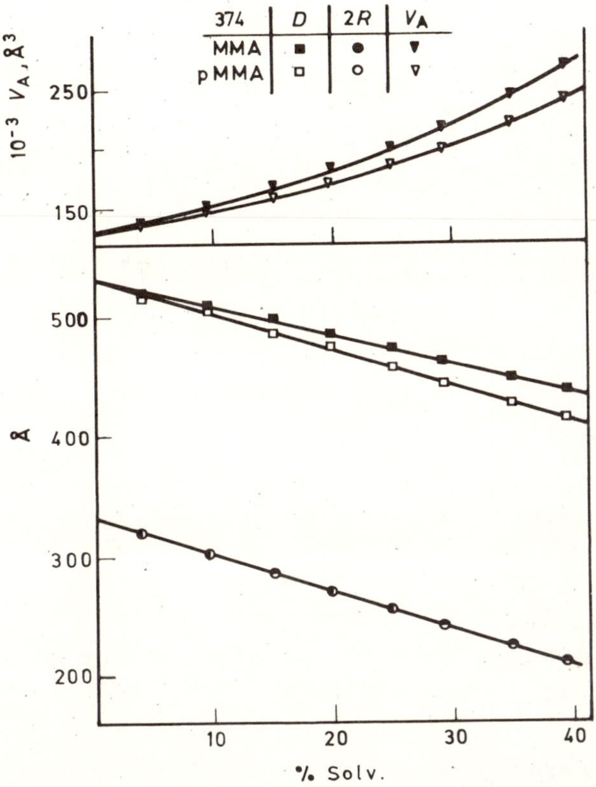

Figure 17. Variation of the parameters of the hexagonal structure during the polymerization of the solvent. System BSB.374/MMA[5]

For all systems studied x-ray diffraction gives the same structure before and after polymerization of the solvent. The *Figures 16* (SB.33), *17* (BSB.374) and *18* (SBS.365) illustrate the variation of the structural parameters of each structural type during polymerization. The examination of these figures shows that the polymerization of the monomer involves a contraction of the characteristic parameter of the soluble polystyrene blocks: d_A for the lamellar structure (*Figure 16*), D-$2R$ for the hexagonal structure (*Figure 17*) and $2R$ for the inverted hexagonal structure (*Figure 18*).

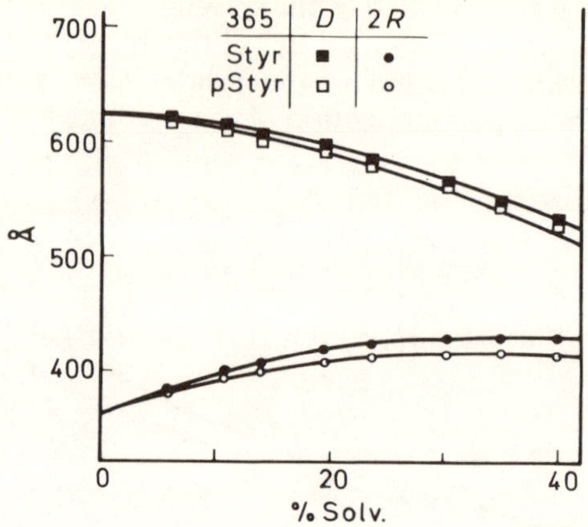

Figure 18. Variation of the parameters of the inverted hexagonal structure during the polymerization of the solvent. System SBS.365/Styrene[9]

5 STRUCTURAL STUDY OF DRY COPOLYMERS

The determination of the structure of dry copolymers has also been performed by using both low angle x-ray diffraction and electron microscopy. Nevertheless, a regular structure is obtained with dry copolymers only after submitting the copolymer to annealing, extrusion or shearing, depending upon the structural type of the polymer.

5.1. Cubic structure

Copolymers containing less than about 18 per cent of one block reveal a cubic structure after annealing at high temperature ($T > 150°C$) during more than 30 days[20].

5.2 Hexagonal structure

Copolymers with composition in the shortest blocks between about 18 and 35 per cent reveal a well ordered hexagonal structure after extrusion at high temperature in an apparatus of the type described by Keller[24–26].

5.3 Lamellar structure

Terris has obtained macrocrystals of lamellar structure with SI copolymers, by applying to the copolymer heated at high temperature a gradient of shear of about 0.07 s^{-1} (ref. 33).

A similar apparatus built in Orleans consists of an external cylinder made of two parts (perfectly adjustable) to allow an easy withdrawing of the oriented sample from the mould and a rotating inner cylinder. The diameter of the external cylinder is $R_1 = 52$ mm and the diameter of the inner cylinder is $R_2 = 50$ mm. The height of the cylinders is 50 mm. The apparatus operates under high vacuum. Eight grams of copolymer are enough to perform an experiment.

6 THEORIES OF THE MORPHOLOGY OF BLOCK COPOLYMERS

All theories are based on equilibrium considerations[27–32]. As with all problems involving the equilibrium between phases, the most favourable configuration corresponds to that which minimizes the total free energy. The problem of microphase segregation involves an evaluation of the change in the contribution to the free energy, when a certain dispersed phase is formed from some reference state, which generally corresponds to perfect mixing of the phases.

As the free energy change is:

$$\Delta G = \Delta H - T\Delta S$$

the problem will be solved if changes in the enthalpy (ΔH) and in the entropy (ΔS) can be calculated.

To simplify we shall only consider Meier's theory[29]. In Meier's theory, particular attention is given to the uniform filling of space by the chain segments to achieve a uniform density and an accurate evaluation of the configurational entropy constraints imposed by the particular domain morphology is made.

To make an analytical approach possible, he assumes that random flight statistics is applicable enabling the general diffusion equation to be used. Then he calculates the probability distribution of chains within a domain as a function of its size. The stable domain size is that corresponding to a uniform density. Thus he finds: for spheres, radius $R = 1.33\alpha\bar{K}\,M^{\frac{1}{2}}$; for cylinders, radius $R = 1.00\alpha\bar{K}\,M^{\frac{1}{2}}$; and for lamellae, thickness $d = 1.40\alpha\bar{K}\,M^{\frac{1}{2}}$ where $\bar{K}$ is a constant relating unperturbed end-to-end chain dimensions to molecular weight, M is the molecular weight of chains in dispersed phase, and α is the chain expansion parameter.

Thus, in order to achieve a uniform density of chain segments within a domain, the chains must have end-to-end dimensions which are different from those of a random coil.

Then he evaluates the chain density in the matrix and employs a domain separation which leads to a matrix possessing a uniform density. When such a condition is satisfied, the chain expansion parameters in the A and B phases are related by the following expressions:

$$\alpha_B \simeq \alpha_A \qquad \text{for spheres}$$

$$\alpha_B \simeq \alpha_A \left(\frac{M_B}{M_A}\right)^{\frac{1}{6}} \qquad \text{for cylinders}$$

$$\alpha_B \simeq \alpha_A \left(\frac{M_B}{M_A}\right)^{\frac{1}{2}} \qquad \text{for lamellae}$$

The evaluation of α_A and α_B in order to predict domain dimensions is carried out by minimizing the free energy consisting of the three following contributions:

(a) the elastic energy due to the deformation of the A and B random coils,
(b) the surface free energy due to the interface between A and B segments, and

(c) the constraint free energy due to the fact that A and B segments are confined to restricted volumes.

Thus Meier calculates the relative free energies for each domain shape and finds that morphological changes occur at determinate ratios of block lengths: expressing his results in per cent weight fraction of the shortest component, he finds 33 for the sphere–cylinder and 37.5 for the cylinder–lamellae transitions.

Experimentally we have found 18 $\pm$ 2 for the sphere–cylinder and 35 $\pm$ 5 for the cylinder–lamellae transitions.

The domain of stability predicted by Meier's theory is too large for spheres and too small for cylinders.

7 CONCLUSION

In this paper, we have shown that the safe determination of the structure of block copolymers in concentrated solution and in the dry state involves the use of both low angle x-ray diffraction and electron microscopy. Using these two techniques systematically we have established the existence of five types of structure and determined their domains of stability. The description of these five structures can be summarized as follows:

(i) The cubic structure results from the arrangement in a centred cubic lattice of spheres filled by the insoluble blocks of the copolymer.

(ii) The hexagonal structure consists of long cylinders filled with the insoluble blocks and arranged in an hexagonal array.

(iii) The lamellar structure consists of equidistant, plane and parallel sheets: each sheet is formed by the superposition of two layers: one containing the insoluble blocks the other containing the soluble block and the solvent.

(iv) The inverted hexagonal structure consists of long cylinders filled with the soluble blocks swelled by the solvent and arranged in an hexagonal array.

(v) The inverted cubic structure consists of spheres filled with the soluble blocks swelled by the solvent and arranged in a centred cubic lattice.

Studying the localization of the solvent in these structures we have shown that block copolymers can be classified in two categories. The first consists of copolymers where the solvent does not enter the microdomains formed by one type of blocks; such copolymers exhibit the five types of structures already described. The second consists of copolymers where the solvent swells the two types of blocks; for such copolymers it is not necessary to distinguish between the cubic and the inverted cubic structures, and between the hexagonal and the inverted hexagonal structures. For both categories, it is the composition of the copolymer, or more precisely the volume of the phases formed by the two types of blocks, which determines the structure adopted by the copolymer. Theoretical treatments predict such a behaviour at least qualitatively[14, 29].

REFERENCES

[1] B. Gallot, R. Mayer and Ch. Sadron, *C.R. Acad. Sci. Paris C.* **263**, 42 (1966).
[2] B. Gallot, Second International Liquid Conference, Kent, Ohio, August 1968.
[3] B. Gallot, Third International Liquid Conference, Berlin, August 1970.
[4] A. Douy and B. Gallot, *Mol. Cryst. Liq. Cryst.* **14**, 191 (1971).
[5] A. Douy and B. Gallot, *Makromol. Chem.* **156**, 81 (1972).
[6] A. Douy and B. Gallot, *C.R. Acad. Sci. Paris, C.* **272**, 440 (1971).
[7] A. Douy and B. Gallot, *C.R. Acad. Sci. Paris, C.* **272**, 1478 (1971).
[8] A. Douy and B. Gallot, *C.R. Acad. Sci. Paris, C.* **274**, 498 (1972).
[9] A. Douy and B. Gallot, *Makromol. Chem.* **165**, 297 (1973).
[10] A. Douy, R. Mayer, J. Rossi and B. Gallot, *Mol Cryst. Liq. Cryst.* **7**, 103 (1969).
[11] M. Gervais, A. Douy and B. Gallot, *C.R. Acad. Sci. Paris, C.* **276**, 391 (1973).
[12] J. Rossi, B. Gallot and Ch. Sadron, *C.R. Acad. Sci. Paris, C.* **271**, 920 (1970).
[13] B. Gallot and Ch. Sadron, *Macromolecules* **4**, 514 (1971).
[14] Ch. Sadron and B. Gallot, *Makromol. Chem.* **164** 301 (1973).
[15] J. Rossi and B. Gallot, unpublished results.
[16] A. Douy and B. Gallot, unpublished results
[17] A. Douy and B. Gallot, unpublished results.
[18] A. Douy, B. Perly and B. Gallot, in press.
[19] A. Douy and B. Gallot, in press.
[20] E. Pedemonte, A.Turturro and U. Bianchi, *Polymer* **14**, 145 (1973)
[21] A. Douy, M. Gervais and B. Gallot, *C.R. Acad. Sci. Paris, C.* **270**, 1646 (1970).
[22] M. Gervais, A. Douy and B. Gallot, *Mol. Cryst. Liq. Cryst.* **13**, 289 (1971).
[23] M. Gervais and B. Gallot, unpublished results.
[24] A. Keller, E. Pedemonte and F. Willmouth, *Kolloid-Z.* **238**, 385 (1970).
[25] J. Dlugosz, A. Keller and E. Pedemonte, *Kolloid-Z.* **242**, 1125 (1970).
[26] A. J. Folkes *et al.*, *Kolloid-Z.* **251**, 1 (1973).
[27] U. Bianchi, E. Pedemonte and A. Turturro, *Polymer* **11**, 268 (1970).
[28] D. Leary and M. Williams, *J. Polymer Sci. B.* **2** 335 (1970).
[29] D. J. Meier, *J. Polymer Sci. C.* **26**, 81 (1969).
[30] T. Inoue, T. Soen, T. Hashimoto and H. Kawai, *J. Polymer Sci. A.* **2–7**, 1283 (1969).
[31] S. Krause, *J. Polymer Sci. A.* **2–7**, 249 (1969).
[32] S. Krause, *Macromolecules* **3**, 84 (1970).
[33] J. Terris, *Thèse*, Strasbourg (July 1973).

THE OPTICAL PROPERTIES OF ANISOTROPICALLY ORDERED SOLUTES IN CHOLESTERIC LIQUID CRYSTALLINE MESOPHASES

F. D. SAEVA

Xerox Corporation, Rochester Research Center, 800 Phillips Road, Webster, New York 14580, USA

ABSTRACT

Liquid crystalline mesophases may be either thermally (thermotropic) or solvent (lyotropic) induced. The anisotropic ordering of molecules within each of these classes may then be of the smectic, nematic or cholesteric type.

The present work deals solely with liquid crystalline mesophases of the cholesteric type where one of the many requirements for the formation of cholesteric mesophases is the existence of chiral single molecules. The chiral molecules may be derivatives of cholesterol or certain polypeptides, e.g., poly-γ-benzyl-L-glutamate (PBLG) in helix supporting solvents. We have previously observed that achiral solutes display circular dichroism (CD) in thermotropic cholesteric liquid crystalline mesophases. This extrinsic liquid crystal induced circular dichroism (LCICD) has been more recently observed in lyotropic cholesteric mesophases and employed to probe the internal molecular organization of both classes of cholesteric mesophases, as well as to provide spectroscopic and conformational information concerning the interaction of solutes in liquid crystalline mesophases in general.

Distinct differences have been observed between the LCICD of anthracene in thermotropic cholesteric mesophases composed of cholesteryl derivatives and lyotropic (PBLG) systems. The results of these LCICD studies will be discussed in terms of the following mechanisms: (a) helical organization of solute molecules, and (b) solute molecules exposed to a helical organization of liquid crystal molecules.

INTRODUCTION

In recent years there has been a resurgence of interest in liquid crystals since the early studies by Friedel in the 1920s. Presumably, this recent interest is a result of the development of new instrumental techniques for characterizing the properties of these intriguing materials, the realization of their importance in biological systems, as well as their use in novel device applications.

The purpose of this contribution is to compare the optical properties of achiral solutes in thermotropic and lyotropic cholesteric mesophases and to indicate the information that is provided both about the solute as well as the internal structure of these mesophases from circular dichroism (CD) studies. A brief review of liquid crystal classes and mesophase types is provided along with the known spectroscopic applications of liquid crystals with the object of placing the subject matter of this contribution into the proper perspective.

Brief description of liquid crystals

Liquid crystals, commonly referred to as the fourth state of matter, exhibit strong birefringence as a result of the anisotropic organization of associated single molecules. Liquid crystals may be divided into the following two main classes: (1) thermotropic (thermally induced), and (2) lyotropic (solvent induced).

Thermotropic liquid crystals are the most widely studied and normally consist of organic molecules which are fairly rigid and rod-like in structure and possess at least one polarizable group. Lyotropic mesophases, on the other hand, generally contain both hydrophobic and hydrophilic groups within a single molecule. A variety of substances, e.g. 9-bromo-phenanthrene-3-sulphonic acid[1], long chained fatty acids[2], certain ionic dyes[3] as well as concentrated solutions of a variety of synthetic polypeptides such as poly-γ-benzyl-L-glutamate[4], exhibit lyotropic liquid crystalline behaviour in various solvent systems.

Each of the liquid crystal types may be further characterized as being either smectic, nematic or cholesteric. For an up-to-date review of the

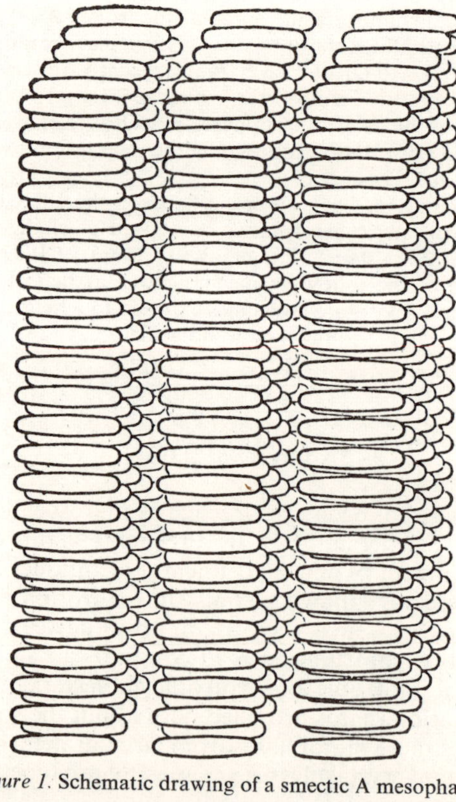

Figure 1. Schematic drawing of a smectic A mesophase

structure and physical properties of liquid crystals see the review article by Brown. Doane and Neff[5].

The smetic mesophase is the most highly structured of the three classes. Molecules in smectic mesophases are stratified in layers where there is a preferred directionality of the long molecular axis. As a consequence of the two dimensional order found in smectics they tend to be the most highly

Figure 2. Schematic drawing of a nematic mesophase

viscous of the three classes. Various types of smectic mesophases have been observed each with a slightly different molecular arrangement. See *Figure 1* for a schematic drawing of the smectic A mesophase type.

Nematics, on the other hand, contain molecules with uniaxial molecular alignment (one dimensional order), described schematically in *Figure 2*, within randomly distributed groups of molecules and behave optically like uniaxial crystals. Nematic mesophases are normally of lower viscosity than smectic mesophases reflecting the weaker intermolecular interactions between molecules.

Cholestic mesophases are a special type of liquid crystal in that they require the existence of a centre of chirality within the single molecules of which the mesophase is composed[6]. Cholesteric compositions may also be obtained by the addition of chiral solutes. which may not be liquid crystalline themselves, to nematic structures[7]. Resolution into enantiomers of a nematogenic compound possessing a racemic centre also has been shown to result in a cholesteric material[8]. The molecular organization within this phase is such that there is a unidirectional alignment of molecules within single layers. The single layers are stacked so that the direction of the long axes of the molecules in one layer is displaced slightly from the direction in the adjacent layer tracing out a helical structure as shown schematically in *Figure 3*. The helical structure may be either left- or right-handed depending on nature of the cholesteric material. The pitch (p) of the cholesteric helix can be determined from the wavelength (λ_0) of the selective reflection of circular polarized light (CPL) either from absorption or circular dichroism (CD) measurements by means of the following expression $p = \lambda_0/n$ if n, the

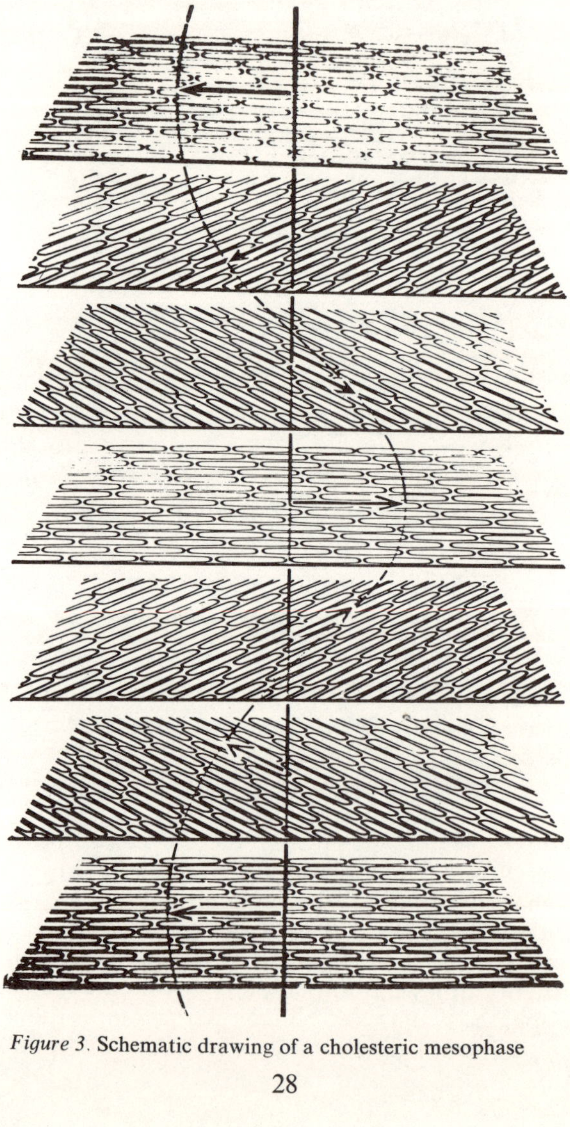

Figure 3. Schematic drawing of a cholesteric mesophase

refractive index of the mesophase, is known. Cholesteric mesophases are best known for the beautiful iridescent colours they exhibit as a result of the selective reflection of visible light when pitch values lie within a certain range.

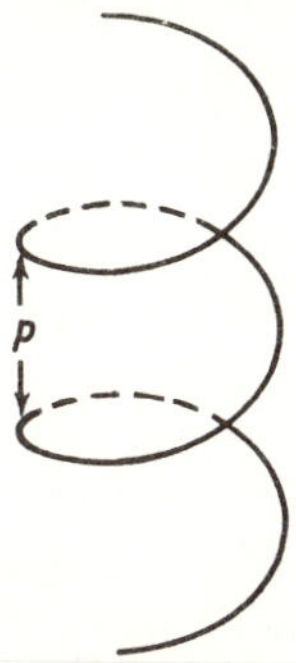

Spectroscopic applications of liquid crystals

Polarization data for electronic transitions have been obtained by orienting solute molecules in aligned nematic[9] and cholesteric mesophases[10]. The observed polarization data are related to the extinction coefficients along molecular fixed axes, ε_x, ε_y, ε_z through the following expression:

$$\varepsilon_{\parallel} - \varepsilon_{\perp} = S_x\varepsilon_x + S_y\varepsilon_y + S_z\varepsilon_z$$

The quantities $\varepsilon_{\parallel}$ and $\varepsilon_{\perp}$ are the extinction coefficients measured with the electric vector of light parallel and perpendicular, respectively, to the direction of alignment of the long axis of the liquid crystal molecules. S_i is a parameter which describes the orientation of the solute in the liquid crystal matrix and is defined as

$$S_i = (3\cos^2\theta_i - 1)/2$$

where θ_i is the angle which the ith molecular axis makes with the long axis of the liquid crystal molecules, and may be derived from liquid crystal n.m.r. studies[11–13].

From polarization and n.m.r. studies in liquid crystals it was concluded that solute molecules align in the best packing arrangement from steric considerations, e.g. planar molecules orient their long axis parallel to the long axis of the liquid crystal molecules.

A positive S-value indicates that the corresponding axis is preferentially oriented parallel to the magnetic field, which is the same as the molecular axis of the liquid crystal molecules of positive dimagnetic anisotropy. A maximum value of $+1$ would correspond to perfect alignment of axis, while a negative S-value (maximum-$\frac{1}{2}$) indicates that the corresponding axis is preferentially oriented perpendicular to the magnetic field direction. S_x for anthracene, for example, normally has a positive value.

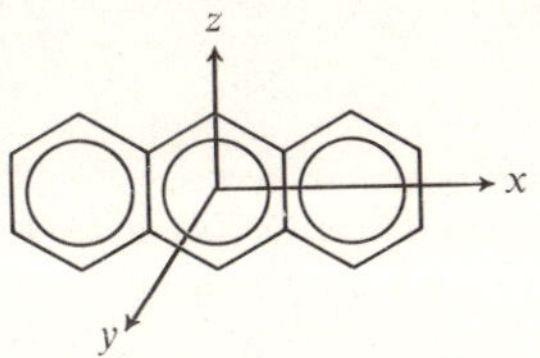

The optical properties of solutes in liquid crystals provide both spectroscopic information about the solute as well as preferred conformation of the solute in the liquid crystal. These studies, however, provide little information concerning the internal molecular structure in the liquid crystal matrix.

CHOLESTERIC LIQUID CRYSTAL INDUCED CIRCULAR DICHROISM (LCICD)

CICD in Thermotropic cholesterics

Recently achiral molecules dissolved in thermotropic cholesteric liquid crystals have been observed to exhibit enormous CD in the region of their absorption bands whose sign is dependent on the sense (chirality) of the cholesteric helix[14]. *Figure 4* shows the CD and absorption spectrum of a left- and right-handed helical cholesteric mesophase in which *N*-(*p*-methoxybenzylidene)-*p*-butylaniline (I) has been dissolved. Selective reflection of circular polarized light from the cholesteric matrix[15] occurs between 500 and 600 nm while (I) absorbs between 230 and 390 nm. The helical structure of the cholesteric mesophase was found essential for the observation of extrinsic CD. This was shown by the loss of liquid crystal induced circular dichroism by the conversion of the helicoidal cholesteric mesophase into a unaxial nematic, i.e. unwinding the cholesteric helix, by means of an electric field[14, 16].

Further CD studies have shown the sign of the LCICD to be dependent on the polarization of the electronic transition within the solute as well as the

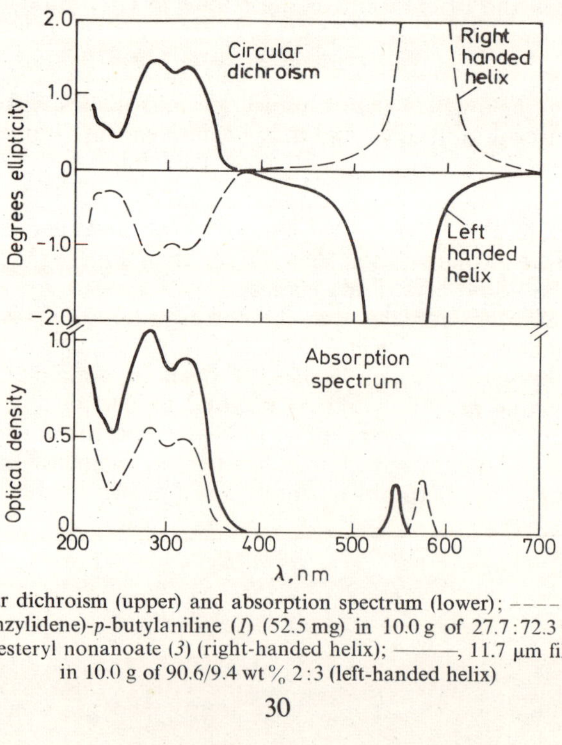

Figure 4. Circular dichroism (upper) and absorption spectrum (lower); -----, 6.3 μm film of *N*-(*p*-methoxybenzylidene)-*p*-butylaniline (*1*) (52.5 mg) in 10.0 g of 27.7:72.3 wt % cholesteryl chloride (*2*)–cholesteryl nonanoate (*3*) (right-handed helix); ———, 11.7 μm film of 1 (63.2 mg) in 10.0 g of 90.6/9.4 wt % 2:3 (left-handed helix)

chirality of the cholesteric helix[17]. *Figure 5* presents the LCICD and absorption spectrum of pyrene which is known to possess $\pi \rightarrow \pi^*$ electronic transition moments at wavelengths > 200 nm, polarized along different molecular axes[18, 19].

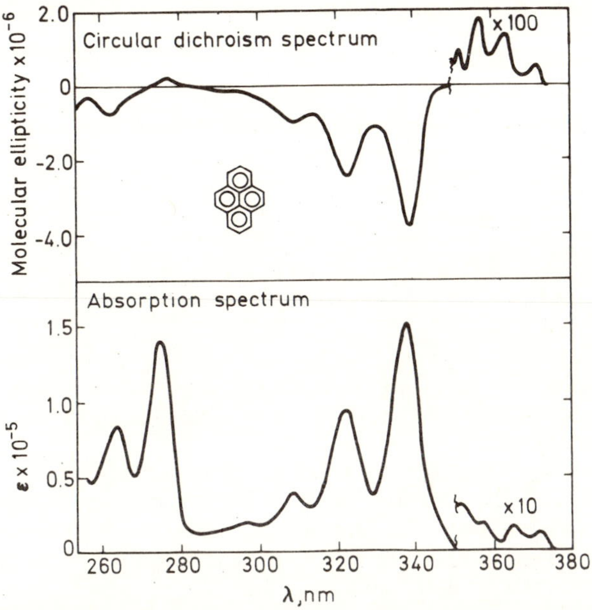

Figure 5. Circular dichroism and absorption spectrum of pyrene in 70/30 wt % cholesteryl chloride/cholesteryl nonanoate (λ_0 = 630 nm, T = 29°C)

The LCICD spectrum of pyrene (*Figure 5*) presents some spectroscopically significant information concerning the polarizations of the electronic transitions. The O–O bands for the ^1L_b, ^1L_a and ^1B_b pyrene transitions occur at 372, 339 and 277 nm, respectively. The transversely polarized (short axis) in-plane O–O band of the ^1L_b and ^1B_b transitions show positive CD (i.e. $\varepsilon_L > \varepsilon_R$) in a right-handed helicoidal cholestric mesophase consisting of 70/30 wt % cholesteryl nonanoate/cholesteryl chloride. All the vibrational bands within the ^1L_b transition appear to be of the same polarization in contrast to the ^1B_b transition which appears to be of mixed polarization[20, 21] or may contain overlapping transitions which are of opposite CD signs as indicated by the lack of match between the absorption and LCICD spectra. The ^1L_a transition, on the other hand, is longitudinally polarized (long axis in-plane) and shows negative CD.

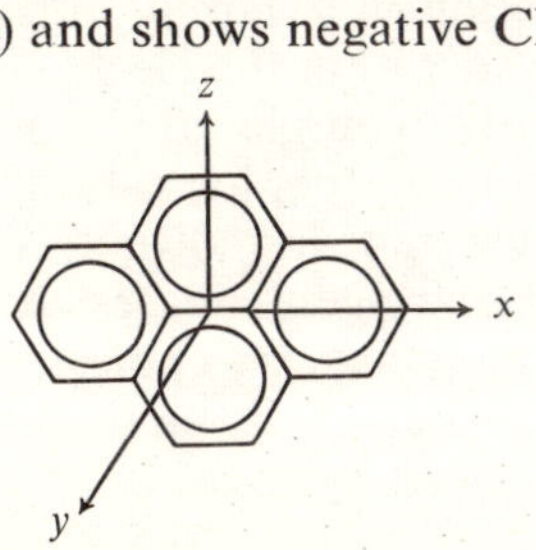

O–O transition (nm)	Polarization
372	z
339	x
277	z

More recently the LCICD sign has been observed to be dependent on the position of λ_0 of the cholesteric pitch band relative to the wavelength of the absorption band[22]. For right-handed cholesteric mesophases composed of cholesteryl chloride–cholesteryl nonanoate mixtures the LCICD sign for a transition moment with a preferred orientation parallel to the long axis of the liquid crystal molecules is summarized in *Table 1*.

Table 1. Sign of LCICD as function of λ_0

Helix sense	λ_0/λ_{ab}*	LCICD sign
Right-handed	>1	$-$
	<1	$+$
Left-handed	>1	$+$
	<1	$-$

* λ_{ab} denotes wavelength of absorption.

The intensity of solute LCICD is a linear function of solute concentration over a limited range[22] and of sample thickness, and non-linear functions of temperature and pitch.

Experimentally, then, the LCICD sign for ordered solutes is dependent on the following: (*i*) the chirality of the cholesteric helix; (*ii*) polarization of absorption bands within a solute; (*iii*) preferred conformation of the solute with respect to the long molecular axis of the liquid crystal molecules, and (*iv*) the position of λ_{ab} relative to λ_0.

The LCICD intensity, on the other hand, is a function of: (i) pitch of the cholesteric helix, and (ii) temperature.

The preceding experimental observations for low concentration of solute agree with recent theoretical studies[23, 24] that extend the theory of electromagnetic radiation in non-absorbing cholesteric liquid crystal to the absorbing situation by adding a frequency dependent complex distribution to the spiralling dielectric tensor of the liquid crystal itself. The experimental sign of the CD depends upon the direction of polarization of the absorption band with respect to the background birefringence. Related studies including infrared CD investigations of the thermotropic cholesteric liquid crystals themselves have been recently reported[24–27].

LCICD in lyotropic cholesterics

Certain synthetic polypeptides, e.g. poly-γ-benzyl-L-glutamate (PBLG), are known to exist in an α-helical conformation in a variety of solvents[28]. Extrinsic circular dichroism has been observed within the electronic transitions of certain dyes[29, 30], such as acridine orange, complexed to isotropically oriented polypeptide molecules while in an α-helical conformation. Concentrated solutions of these polypeptides, in helix supporting solvents, have been found by Elliott and Ambrose[31] to form birefringent phases. Robinson[4] subsequently characterized these birefringent phases as cholesteric liquid crystalline mesophases, where the helical polypeptide is analogous to the cholesteryl derivative in thermotropic cholesteric systems.

Achiral molecules, such as anthracene and pyrene, which do not exhibit induced circular dichroism in dilute or concentrated isotropic solutions of PBLG, do indeed exhibit LCICD in the anisotropic birefringent lyotropic cholesteric mesophases formed by PBLG in helix supporting solvents such as chloroform, methylene chloride and dioxane.

LCICD has been observed for a number of achiral molecules dissolved in birefringent concentrated solutions of PBLG in helix supporting solvents. We believe this induced effect to be quite general and independent of chemical structure of the solute in contrast to rigid requirements for solutes that complex to isotropically oriented helical polypeptides[29, 30]. The extrinsic CD of the achiral solute is associated with the formation of a birefringent cholesteric mesophase and disappears when the α-helical molecules of PBLB become randomly oriented by means of small changes in concentration of PBLG while the relative concentration of anthracene to PBLG is held constant. In other words, the extrinsic CD of anthracene in PBLG–dioxane mixtures is associated only with the lyotropic cholesteric mesophase.

The observed LCICD for anthracene in lyotropic cholesteric mesophases formed by PBLG in dioxane are distinctly different from that observed in thermotropic cholesteryl mesophases composed of cholesteryl derivatives[22]. *Figure 6* presents the LCICD and electronic spectra of anthracene dissolved in a thermotropic and a lyotropic cholesteric mesophase. The LCICD spectrum of anthracene in 18/82 wt% PBLG/dioxane follows its absorption spectrum quite closely, and the CD for the pitch band and anthracene absorption bands are both of negative sign. The chirality of the cholesteric helix is then left-handed, i.e. of opposite chirality† to the helicity of the polypeptide[33, 34]. The LCICD spectrum of anthracene, on the other hand, in a thermotropic cholesteric mesophase composed of 60/40 wt% cholesteryl chloride/cholesteryl nonanoate shows CD bands of both positive and negative sign whose relative intensity does not follow its absorption spectrum.

The variation in the LCICD of anthracene between the two cholesteric systems is attributed to the difference in ability of the mesophases to physically order the solute. In the thermotropic system the most preferred conformation of the anthracene involves alignment of its long axis with the long axis of the liquid crystal molecules. In the lyotropic system, however, alignment of the long or short axes of anthracene perpendicular to the long axis of the liquid crystal molecules seems to be equally preferred. This conclusion is consistent with the relative sizes of the solute anthracene (~10 Å) and PBLG molecules (~850 Å for MW = 125000)[35]. The ordering of solute molecules by the liquid crystal would be most efficient for solutes of comparable size and shape to the liquid crystal molecules. The small variation in pitch between the two cholesteric mesophases was found not to be the cause of the difference in the LCICD spectra for anthracene presented in *Figure 6*.

Enantiomeric lyotropic cholesteric mesophases are produced by PBLG and PBDG in dioxane solvent as indicated by the LCICD spectra for anthracene in *Figure 7*. The chirality of the cholesteric helix in the lyotropic mesophase can be established by determining the sign of the CD in the region of the reflective wavelength of the cholesteric pitch band†. Once a correlation

† See Ref. 32.

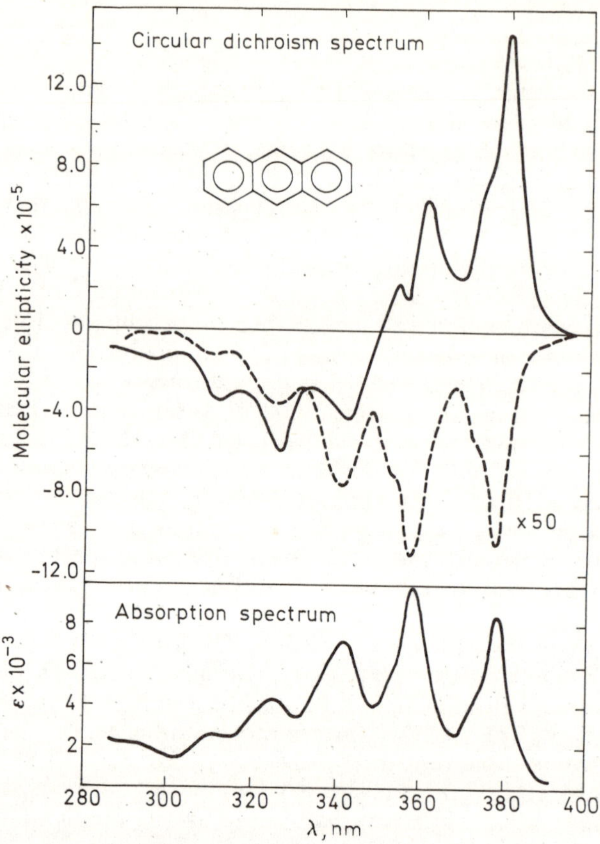

Figure 6. Circular dichroism and absorption spectrum of anthracene in
(———) 1. A thermotropic cholesteric mesophase composed of 60/40 (wt %) cholesteryl chloride/cholesteryl nonanoate (pitch ≃ 9 μm)
(-----) 2. A lyotropic cholesteric mesophase composed of 18/82 (wt %) PBLG/dioxane (MW PBLG = 125000, pitch ≃ 18 μm)

between the sign of the CD for the pitch band and a LCICD band has been made the chirality of the cholesteric helix may be determined simply from the sign of LCICD[33]. The chirality of the cholesteric helix formed by PBLG and PBDG in dioxane (described in *Figure* 7) are left- and right-handed respectively. The difference in LCICD intensity of anthracene in the PBLG and PBDG cholesteric mesophase is attributed to the variation in pitch between the two systems.

Molecular ellipticity values for anthracene in PBLG of differing molecular weight, i.e. 46000, 125000 were *identical* within experimental error for samples of virtually *identical* pitch. The lower MW sample relaxed from the 'Grandjean' to the focal conic texture in one or two hours while the higher MW samples took 24–30 hours to relax.

LCICD has also been observed within the electronic transitions of the polypeptide itself. Rotational strengths for the polypeptide absorption bands

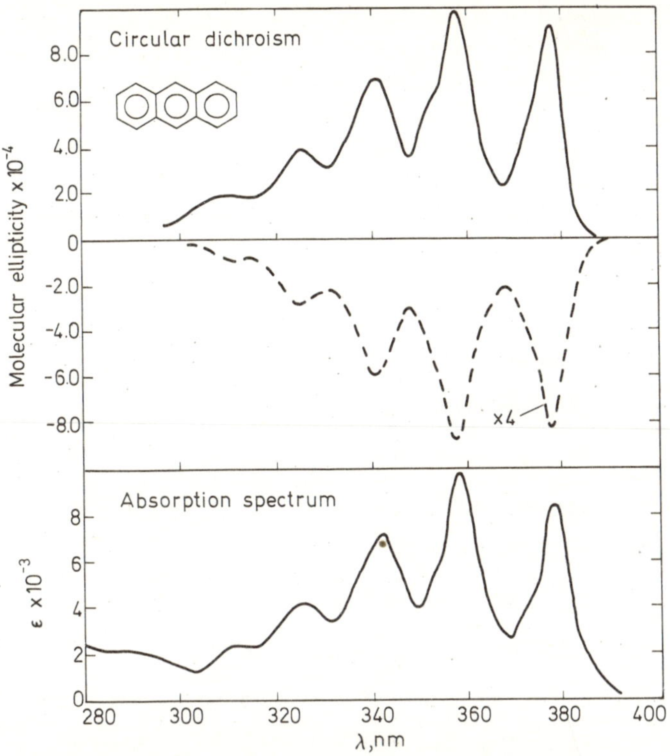

Figure 7. Circular dichroism and absorption spectrum of anthracene in
(-----) 1. A lyotropic cholesteric mesophase composed of 18/82 (wt%) PBLG/dioxane (MW PBLG = 125000, pitch ≃ 18 μm)
(———) 2. A lyotropic cholesteric mesophase composed of 20/80 (wt%) PBDG/dioxane (MW PBDG = 120000, pitch ≃ 21 μm)

are approximately 10^2–10^3 larger in the cholesteric mesophase than in isotropic solution. The ^1L_b electronic transition of the benzyl group in PBLG shows large positive CD for a left-handed cholesteric mesophase.

CONCLUSION

Liquid crystal induced circular dichroism has been observed in achiral noncomplexing solutes in lyotropic cholesteric mesophases. The sign of the LCICD in both lyotropic and thermotropic cholesterics provides: (i) The chirality of the cholesteric helix (provided the region of λ_0 is known), and (ii) the preferred orientation of the solute with respect to the long molecular axis of the liquid crystal molecules.

The intensity of LCICD provides information about the pitch of the cholesteric mesophase.

In lyotropic cholesteric systems small molecules such as anthracene are not ordered to the extent they are in thermotropic systems composed of cholesteryl derivatives.

It is concluded that LCICD arises primarily from the helical organization of the solute alone (mechanism a). It is suggested that LCICD will also exist for randomly distributed solute molecules (mechanism b).

ACKNOWLEDGEMENT

Stimulating discussions with Drs. W. H. H. Gunther, G. Johnson, J. E. Kuder and H. A. Scheraga and technical assistance by G. R. Olin are gratefully acknowledged.

REFERENCES

1. H. Sandquist, *Ber. Dtsch. Chem. Ges.* **48**, 2054 (1915).
2. J. W. Bain, *Colloid Chemistry*, Chemical Catalog Co. Inc., N.Y., Vol. I, p. 138 (1926), (Editor, J. Alexander).
3. J. F. Dreyer, *J. Phys. Colloid Chem.* **52**, 808 (1948).
4. (a) C. Robinson, *Trans. Faraday Soc.* **52**, 571 (1956). (b) C. Robinson, J. C. Ward and R. B. Beevers, *Disc. Faraday Soc.* **25**, 29 (1958). (c) C. Robinson, *Tetrahedron*, **13**, 219 (1961). (d) C. Robinson, *Mol. Cryst.* **1**, 467 (1966) and references cited therein.
5. G. H. Brown, J. W. Doane and V. D. Neff, *CRL Critical Reviews in Solid State Sciences*. 303 (1970).
7. G. Friedel, *Ann. Phys.* (Paris) 273 (1922). •
7. F. D. Saeva, *Mol. Cryst. and Liq. Cryst.* **23**, 171 (1973) and references cited therein.
8. D. Dolphin, Z. Muljiana, J. Cheng and R. B. Meyer, *J. Chem. Phys.* **58**, 413 (1973) and references cited therein.
9. G. P. Ceaser and H. B. Gray, *J. Amer. Chem. Soc.* **91**, 191 (1969).
10. E. Sackmann, *J. Amer. Chem. Soc.* **90**, 3569 (1968).
11. A. D. Buckingham and K. A. McLauchlan, *Progress in Nuclear Magnetic Resonance Spectroscopy*, Vol. 2, Pergamon Press, Oxford, 1967.
12. G. Englert and A. Saupe, *Proceedings of the Intern. Conf. on Liq. Cryst.* 183 (1965).
13. A. Saupe, *ibid.* 207 (1965).
14. F. D. Saeva and J. J. Wysocki, *J. Amer. Chem. Soc.* **93**, 5928 (1971).
15. J. L. Fergason, *Mol. Cryst.* **1**, 293 (1966).
16. J. J. Wysocki, J. Adams and W. Haas, *Phys. Rev. Lett.* **20**, 1024 (1968).
17. F. D. Saeva, *J. Amer. Chem. Soc.* **94**, 5135 (1972).
18. R. M. Hochstrasser, *J. Chem. Phys.* **33**, 459 (1960).
19. V. H. Zimmermann and N. Joop, *Z. Electrochem.* **64**, 138 (1960).
20. O. E. Weigang, Jr., *J. Chem. Phys.* **43**, 3609 (1965).
21. J. Horwitz, E. H. Strickland and C. Billups, *J. Amer. Chem. Soc.* **91**, 184 (1969).
22. F. D. Saeva, P. E. Sharpe and G. R. Olin, *J. Amer. Chem. Soc.* **95**, 7656, 7660, 7882 (1973).
23. G. Holzwarth and N. A. W. Holzwarth, *J. Opt. Soc. Amer.* **63**, 324 (1973).
24. E. Sackmann and J. Voss, *Chem. Phys. Lett.* **14** 528 (1972).
25. K. J. Mainusch and H. Stegemeyer, *Z. Phys. Chem.* N.F. **77**, 210 (1972).
26. B. Schrader and E. Korte, *Angew. Chem. internat. Edit.* **11**, 226 (1972).
27. I. Chabay, *Chem. Phys. Lett.* **17**, 283 (1972).
28. P. Doty, A. M. Holtzer, V. H. Bradbury and E. R. Blout, *J. Amer. Chem. Soc.* **76**, 4493 (1954).
29. E. R. Blout and L. Stryer, *Proc. Nat. Acad. Sci. Wash.* **45**, 1591 (1959).
30. L. Stryer and E. R. Blout, *J. Amer. Chem. Soc.* **83**, 1411 (1961).
31. A. Elliott and E. J. Ambrose *Disc. Faraday Soc.* **9**, 246 (1950).
32. The chirality of the cholesteric mesophase is that indicated by the handedness of circular polarized light transmitted in the region of the pitch band (i.e., a cholesteric mesophase that selectively transmits left-handed circular polarized light in the region of the pitch band is a left-handed helix).
33. F. D. Saeva, *Mol. Cryst. and Liq. Cryst.* **18**, 375 (1972).
34. T. Ooi, R. A. Scott, G. Vanderkooi and H. A. Scheraga, *J. Chem. Phys.* **46**, 4410 (1967).
35. A. Elliott, R. D. B. Fraser and T. D. MacRae, *J. Mol. Biol.* **11**, 821 (1965).

THERMODYNAMICS AND DYNAMICS OF POLYPEPTIDE LIQUID CRYSTALS

WILMER G. MILLER, CHIA CHUAN WU, ELIZABETH L. WEE, GARY L. SANTEE, JUEY H. RAI and KENNETH G. GOEBEL

Department of Chemistry, University of Minnesota, Minneapolis, Minnesota, USA

ABSTRACT

The prediction that molecular asymmetry alone is sufficient to force a phase transition from a disordered to an ordered phase in a system of rodlike particles has been investigated using α-helical synthetic polypeptides as the source of rodlike particles. To this end the temperature–composition phase diagrams for the two component systems polybenzylglutamate (PBLG)–dimethylformamide (DMF) and polycarbobenzoxylysine (PCBL)–dimethylformamide are compared with the Flory lattice model for rigid impenetrable rods. The experimental systems exhibit and the theory predicts three distinct regions in the phase diagram: a narrow biphasic region in which isotropic and liquid crystal phases differing only slightly in composition coexist, a transition region over which solvent is increasingly excluded from the coexisting ordered phase and rods excluded from the coexisting isotropic phase, and a region where almost pure solvent coexists with a highly concentrated liquid crystal phase. A detailed comparison of the theoretical phase diagram for rigid, impenetrable rods with the experimental ones reveals discrepancies which can be attributed to the facts that the experimental rods are neither completely rigid nor impenetrable. Analysis of thermal data on PBLG–DMF indicates that the latent heat for the isotropic to liquid crystal phase transition is small and endothermic. We conclude that molecular asymmetry alone is sufficient to produce a phase transition to an ordered phase.

Dynamical data show that the bulk viscosity may be considerably lower in the ordered than in the disordered phase. Although rod motion in the liquid crystal phase is correlated, electron spin resonance data suggest that individual rod motion about its mean lattice position is greater than in the isotropic solution of equivalent concentration. Additional electron spin resonance studies show that the motion of small rods and of the polymer side-chains is little affected by the presence of a high concentration of long rods, whereas the tumbling of long rods is dramatically influenced by the presence of other long rods.

INTRODUCTION

The thermodynamics and dynamics of flexible linear polymer chains have been thoroughly investigated and are well understood. The flexibility of polymeric chains is highly variable. and depends on chemical constitution and secondary bonding. Stiff chain polymers form a small but interesting set. They may have properties which are dramatically different from and not so well understood as their flexible relatives. Solutions of stiff chain polymers are known[1–6] to organize into an ordered liquid (liquid crystalline) phase at sufficiently high polymer concentration. Theoretical considerations[7–10] indicate that molecular asymmetry alone may be sufficient to explain the stability of the ordered phase relative to the disordered (isotropic) phase.

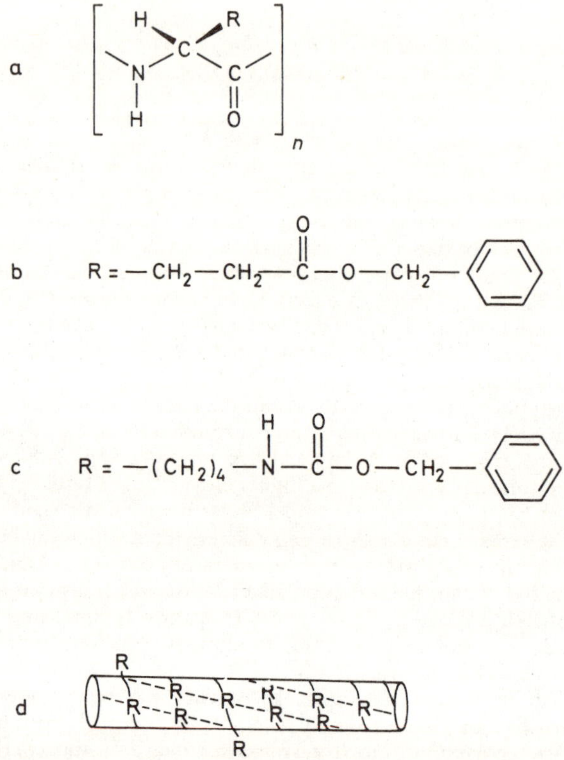

Figure 1. (a) monomeric unit of an α,L-polyamino acid; (b) γ-benzylglutamate side chain; (c) ε-carbobenzoxylysine side chain; (d) portion of an α-helix with attached side chains

The lattice theory of Flory[10] also predicts that as the polymer–solvent interaction becomes increasingly unfavourable there is a sudden and precipitous increase in order and in rejection of solvent from the liquid crystal phase.

In aqueous systems the polymers which have been studied are polyelectrolytes. Addition of small electrolyte destroys the ordered phase indicating that electrostatic interaction is important in stabilizing the ordered phase. With these systems it is thus difficult to establish whether or not molecular asymmetry alone is sufficient to stabilize an ordered phase. The early observations with nonionic, helical polypeptides[5,6] were carried out with

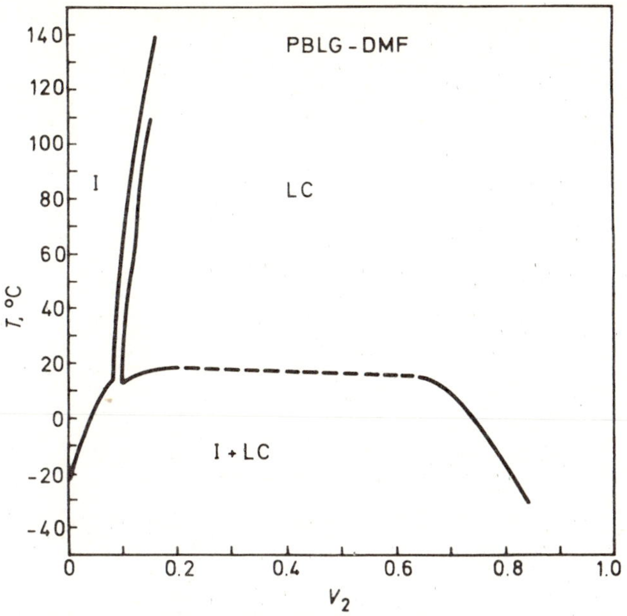

Figure 2. Temperature–composition phase diagram for PBLG (310 000) in DMF[16, 17]. Dashed line indicates area of insufficient data

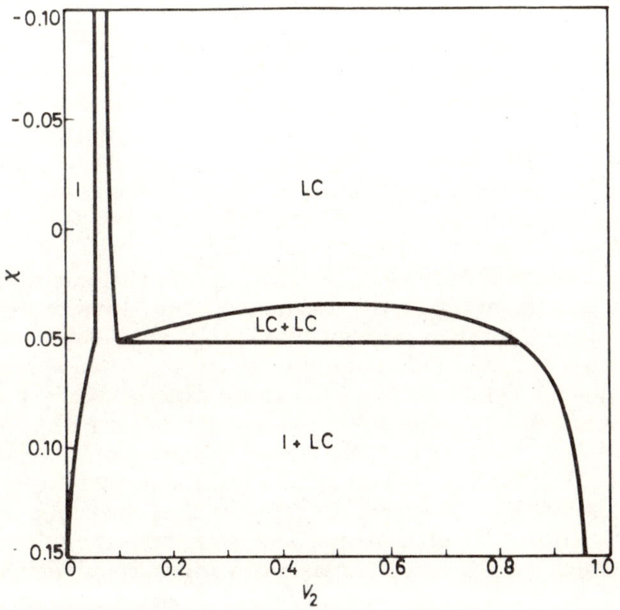

Figure 3. Lattice model phase diagram for rigid, impenetrable rods[10] of axial ratio 150. Phases are isotropic (I) or liquid crystalline (LC)

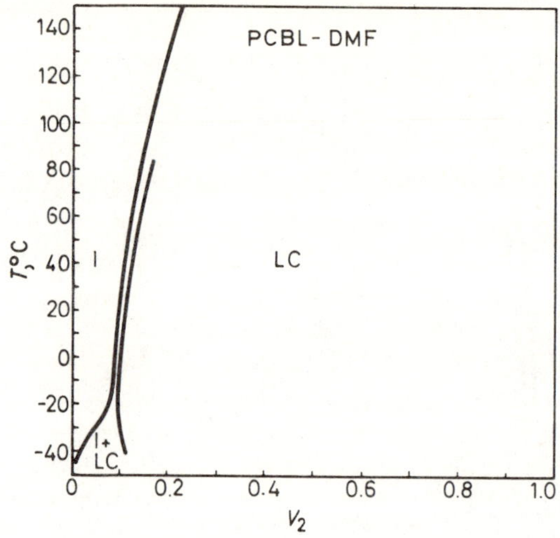

Figure 4. Temperature–composition phase diagram for PCBL (690 000) in DMF[18]

polymer–solvent systems wherein the polymers are now known to have a tendency to associate even at low concentration[11] suggesting that in these systems too, factors other than molecular asymmetry are important. The peptide bond polymers (*Figure 1a*) seem attractive, however, as the side chain may be varied widely while the hydrogen bonded helical, rodlike conformation remains intact. Inasmuch as several helical polypeptides show little or no tendency to associate in dimethylformamide (DMF)[11–13] and that DMF remains liquid over a 250 Centigrade degree temperature range, we felt that the helical polypeptide–DMF system might be attractive for polymer liquid crystal investigation. Studies on the three component system polybenzylglutamate (PBLG)–DMF–nonsolvent indicate that the general features of the Flory phase diagram are observable[14]. The finding[15] that the dilute solution osmotic second virial coefficient for PBLG in DMF had a substantial temperature dependence and became zero near room temperature made the two component PBLG–DMF system particularly attractive. It suggested a change from a thermodynamically good to poor solvent in an accessible temperature range, perhaps making a determination of the temperature–composition phase diagram and of the importance of molecular asymmetry versus polymer–solvent interaction experimentally possible.

A partial temperature–composition phase diagram for PBLG–DMF was published in 1971[16], and has since been extended[17] to cover an approximately 200 Centigrade degree range of temperature and the entire range of polymer composition. A variety of measurements—nuclear magnetic resonance, polarizing microscope, hydrodynamic and isopiestic—were necessary in order to determine the phase boundaries. The various techniques yield phase boundaries in substantial agreement. *Figure 2* shows our current best effort for the PBLG–DMF phase diagram for molecular weight 310 000 (weight average) having a rigid rod axial ratio of about 135. *Figure 3* shows the theoretical rigid, impenetrable rod phase diagram[10] for an axial ratio of 150. *Figure 4* shows a partial phase diagram[18] for the system polycarbobenzoxylysine (PCBL)–DMF for 690 000 molecular weight PCBL with a rigid rod axial ratio of about 190. The similarities are striking. The gross features of the Flory model are certainly evident.

RIGID, IMPENETRABLE ROD THERMODYNAMICS

It is of interest to pursue in more detail the various features of the Flory lattice model. The isotropic–liquid crystal phase equilibria for rigid, impenetrable rods was determined from equations 1–3.

$$\ln(1 - v_2) + (1 - 1/x)v_2 + \chi^{\mathrm{I}} v_2^2 = \ln(1 - v_2^*) + (y - 1)v_2^*/x + 2/y + \chi^{\mathrm{LC}} v_2^{*2} \quad (1)$$

$$\ln v_2 + (x - 1)v_2 - \ln x^2 + \chi^{\mathrm{I}} x(1 - v_2)^2 = \ln v_2^* + (y - 1)v_2^* + 2 - \ln y^2 + \chi^{\mathrm{LC}} x(1 - v_2^*)^2 \quad (2)$$

$$v_2^* = [x/(x - y)][1 - \exp(-2/y)] \quad (3)$$

where x is the rod axial ratio, y the equilibrium degree of disorientation in the liquid crystal phase, χ the polymer–solvent interaction parameter, and v_2 and v_2^* the equilibrium volume fraction of polymer in the isotropic and liquid crystalline phases, respectively. Equations 1 and 2 were obtained by equating chemical potentials, and equation 3 by minimization of the free energy of the liquid crystal phase with respect to rod alignment[10]. The phase equilibrium may be described as consisting of three regions: a narrow isotropic–liquid crystal biphasic region when χ is near zero or negative, a transition region when χ becomes positive where the coexisting ordered phase becomes increasingly rich and the isotropic phase increasingly poor in polymer, and a region where essentially pure solvent is in equilibrium with a highly concentrated polymer solution when χ becomes sufficiently positive. In addition to isotropic–anisotropic equilibria there is a small region where equilibria between two liquid crystal phases minimizes the free energy of the system.

If the ordered phase is perfectly aligned the isotropic-liquid crystal phase equilibria are given[10] by equations 4 and 5. The phase diagram which was

$$\ln(1 - v_2) + (1 - 1/x)v_2 + \chi^{\mathrm{I}} v_2^2 = \ln(1 - v_2^*) - \ln(1 - v_2^* + v_2^*/x) + \chi^{\mathrm{LC}} v_2^{*2} \quad (4)$$

$$\ln v_2 + (x - 1)v_2 - \ln x^2 + \chi^{\mathrm{I}} x(1 - v_2)^2 = \ln v_2^* - \ln(1 - v_2^* + v_2^*/x) + \chi^{\mathrm{LC}} x(1 - v_2^*)^2 \quad (5)$$

calculated from these equations is shown in *Figure 5*, where the ordered phase has been forced to remain totally aligned at all concentrations. There is a qualitative difference between the shape of the phase diagrams in *Figures 3* and *5*. Although it is difficult to quantitatively connect the theoretical χ parameter with the corresponding experimental variable, temperature, it should be a monotonically varying function of temperature. The general correspondence of *Figure 3* and lack of correspondence of *Figure 5* with

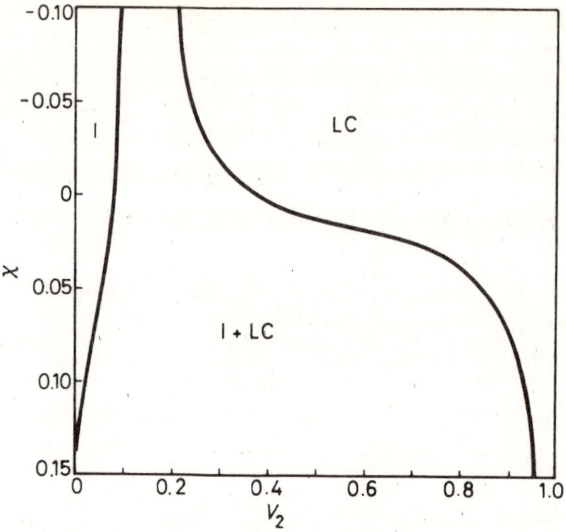

Figure 5. Lattice model phase diagram for rigid, impenetrable rods[10] of axial ratio 150 in which the ordered phase is completely aligned

the experimental phase diagrams indicates that the disorientation entropy makes a significant contribution to the thermodynamics of the dilute, ordered phase and is largely responsible for the prediction of a narrow biphasic region that is effectively χ (or T) invariant over a wide range of χ values. The disorientation index y as a function of composition is shown in *Figure 6*, calculated from equation 3. In the narrow biphasic region the composition of the liquid crystal phase is 0.07–0.10, a region where y is extremely concentration dependent (axial ratio 150). Yet the calculated and experimental curves are in substantial though not complete (see below) agreement.

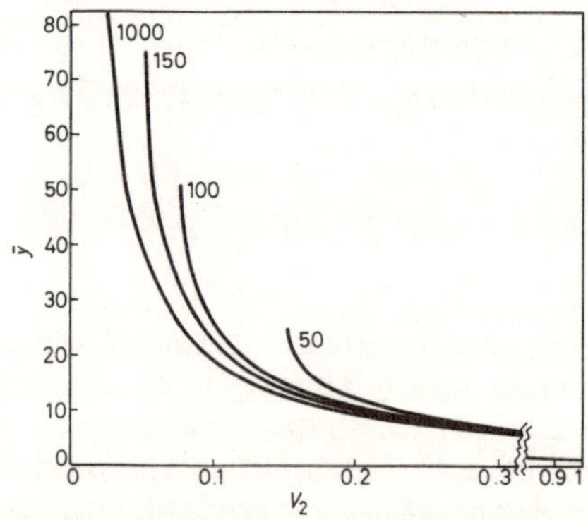

Figure 6. Index of disorientation as a function of composition for the rigid, impenetrable rod model[10]. Rod axial ratios are as indicated

Therefore we conclude that the lattice model treatment of the disorientation is a rather good description in spite of its approximate[10] and concentration sensitive nature. However, this choice of representing the phase equilibria is not unique in giving a phase diagram similar to *Figure 3*[17].

The theoretical phase diagrams were calculated assuming χ^I was equal to χ^{LC}. If we consider χ to be entirely enthalpic and not an excess free energy the equality is equivalent to stating that the conversion of the isotropic phase to the liquid crystal phase at constant composition has no latent heat. It is of considerable importance to determine the existence and magnitude of the latent heat in the experimental systems, and its effect on the theoretical phase diagrams.

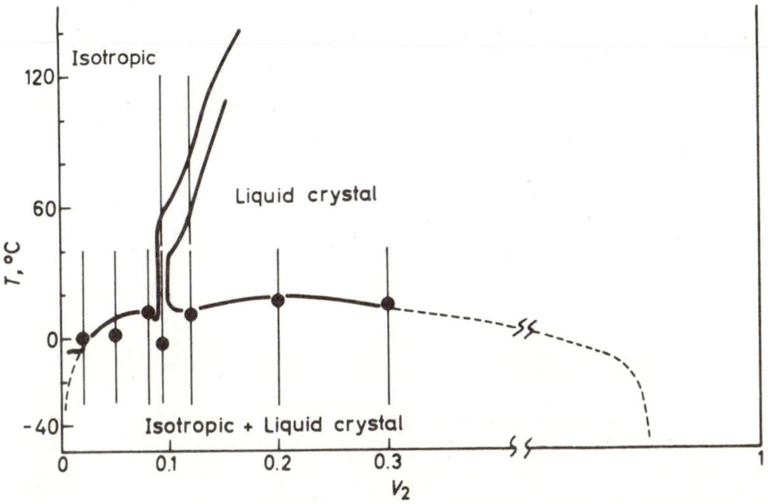

Figure 7. Schematic representation of the differential scanning calorimetry experiments[19]. Vertical lines indicate range of temperature scan. Circles indicate onset of endothermicity when scanning from low to high temperature

Differential scanning calorimetry (DSC) experiments on a 12 volume % solution whereby a PBLG liquid crystal solution was converted to an isotropic solution during the temperature scan gave no evidence of a latent heat[19]. However, DSC experiments were able to detect a thermal transition when a biphasic system at −30°C, composed of nearly pure solvent and highly concentrated polymer phases in equilibrium, was heated to give either an isotropic or liquid crystalline phase at +40°C[19]. The experiments, shown schematically in *Figure* 7, yield heats which can be identified as the heat of mixing two widely different compositions. Similar experiments with PCBL–DMF are difficult due to the necessity of operating at temperatures near the solvent freezing point, and were inconclusive. Also direct heat of mixing measurements have not been successful due to the high viscosity of the concentrated polymeric solutions.

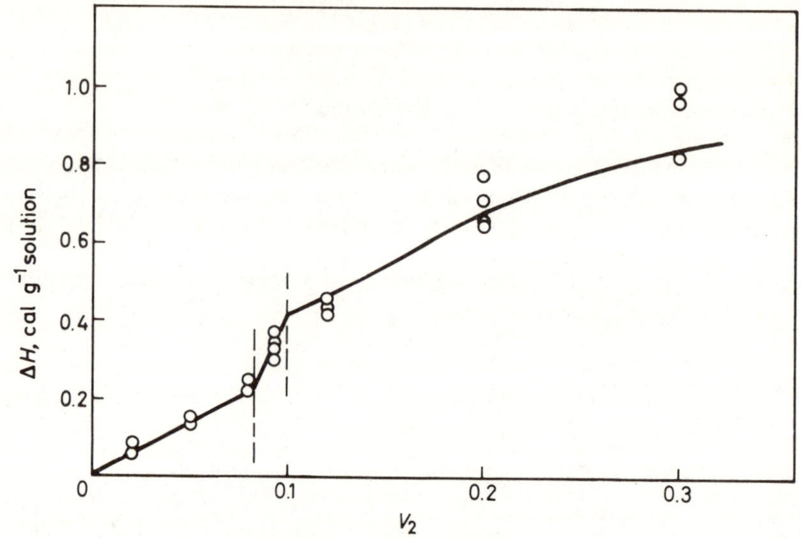

Figure 8. Heat of mixing as a function of PBLG (310000) concentration. Open circles—experimental results[19]; solid line—calculated from equation 6 with $\chi^{\mathrm{I}} = 0.50$, $\chi^{\mathrm{LC}} = 0.70$. $T = 300$ K. $v_2^{\mathrm{I}} = 0.001$, $v_2^{\mathrm{LC}} = 0.84$; dashed vertical lines—phase boundaries at 15–20°C in the narrow biphasic region

The observed DSC heats, shown in *Figure 8*, are small and endothermic which suggests that a van Laar treatment of the heat of mixing is appropriate. If we assume that the χ parameter may be phase dependent but concentration independent within a phase, the heat of mixing in van Laar approximation will be given by equation 6

$$\Delta H_{\mathrm{mix}} = RT(\chi^{\mathrm{I\,or\,LC}} n_1 v_2 - \chi^{\mathrm{I}} n_1^{\mathrm{I}} v_2^{\mathrm{I}} - \chi^{\mathrm{LC}} n_1^{\mathrm{LC}} v_2^{\mathrm{LC}}) \tag{6}$$

where there are n_1 moles of solvent in a solution of composition v_2 in the final state of the system, n_1^{I} and v_2^{I} refer to the initial isotropic phase, and n_1^{LC} and v_2^{LC} to the initial liquid crystal phase. The values of v_2^{I} and v_2^{LC} are fixed and obtainable from the phase diagram, and n_1^{I} and n_1^{LC} are fixed by material balance. Either χ^{I} or χ^{LC} is used in the first term depending on whether the final solution is isotropic or liquid crystalline, respectively. Although there are two unknown quantities in equation 6, χ^{I} and χ^{LC}, the experimental data can be fit with only a narrow range of χ values, typified by $\chi^{\mathrm{I}} = 0.50$ and $\chi^{\mathrm{LC}} = 0.70$. A theoretical line generated from equation 6 using these values is shown in *Figure 8*. Further refinement, such as a concentration dependent χ, is unwarranted.

Equation 6 adequately describes the data and furthermore gives unequal χ values. A latent heat exists and may be calculated in van Laar approximation at any composition from equation 7.

$$\Delta H_{\mathrm{LC}\leftarrow\mathrm{I}} = RT(\chi^{\mathrm{LC}} - \chi^{\mathrm{I}}) n_1 v_2 \tag{7}$$

The latent heat is slightly larger than that previously[19] estimated when the low temperature phase boundaries were of unknown composition. For a 12 volume% solution equation 7 yields a value of 0.16 cal g^{-1} of solution or about 2 millicalories for our sample size, which should be detectable. However, we have a two component system and the phase transition must pass through a biphasic region which for the 12% solution is spread over 30 Centigrade degrees. We feel that the broad transition is probably responsible

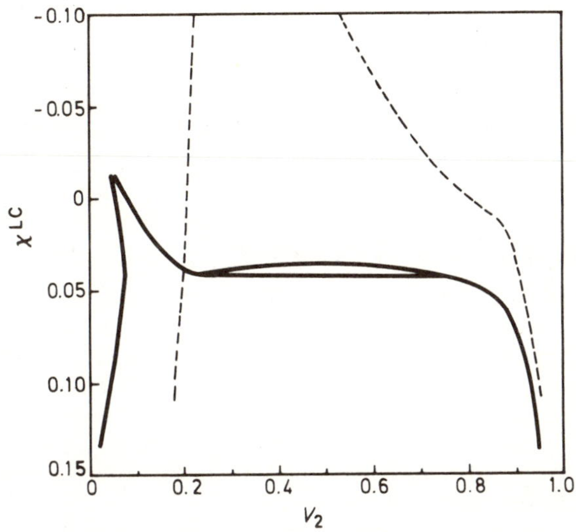

Figure 9. Rigid, impenetrable rod ($x = 150$) phase equilibria with unequal χ. Solid lines, $\chi^{LC} = 1.4\chi^{I}$; dashed lines, $\chi^{LC} = (210/T) - 0.68$, $\chi^{I} = (150/T) - 0.68$

for our inability to observe a latent heat directly, or else equation 7 overestimates it for unknown reasons.

We conclude that in the PBLG–DMF system a small latent heat exists. and is endothermic for conversion of the isotropic to the liquid crystal phase. The enthalpy thus serves to destabilize the liquid crystal phase. Consequently it must be stabilized by entropic considerations, which are dominated presumably by molecular asymmetry. The generality of this observation is unknown. The situation may be similar to the well-known entropy dominated process of rubber elasticity where the associated small energy changes may be positive or negative depending on the rubber.

We turn now to the effect of unequal χ on the rigid. impenetrable rod phase equilibria. The ratio 0.70/0.50 is equal to 1.4 so we consider first the phase equilibria calculated from equations 1–3 with $\chi^{LC} = 1.4\chi^{I}$. The phase diagram is shown in *Figure 9*. In the negative χ region no solutions exist other than the ones shown. In the narrow biphasic region the calculated and experimental values are not even in qualitative agreement. We next consider that χ should be considered as an excess free energy, and not as purely enthalpic. We further recognize that vapour sorption measurements indicate that there is a contribution to the solution thermodynamics from the entropy of solvent–flexible side chain mixing[20–22]. Thus there is experimental justification for considering χ to have an entropic component. Shown in *Figure 9* is a phase diagram calculated with χ^{LC} given by $210/T - 0{\cdot}68$ and χ^{I} by $150/T - 0.68$. By this assignment the enthalpic components of χ retain a ratio of 1.4 and are equal to 0.70 and 0.50 at 300 K, and the entropic components are equal and chosen so that χ^{LC} becomes zero slightly above room temperature. This choice has completely wiped out any resemblance to the experimental phase diagram. With unequal χ the phase diagram is extremely sensitive to the manner in which χ^{I} and χ^{LC} vary. Excess entropy contributions must be carefully chosen in order to produce a narrow biphasic region resembling that in *Figure 3*. At this point they would have to be chosen arbitrarily.

SEMIFLEXIBLE, IMPENETRABLE ROD THERMODYNAMICS

We have emphasized up to now the similarities between the gross features of the experimental and theoretical phase equilibria. A closer inspection reveals that they have in fact considerable differences. In the theoretical diagram (*Figure 3*) the equilibrium concentrations in the narrow biphasic region are almost independent of χ, becoming slightly closer together as χ becomes more negative. Of the two lines the one bounding the isotropic phase is especially independent of χ. With both experimental systems the equilibrium concentrations move to larger values as the temperature increases though their difference remains almost constant. We believe the origin of this discrepancy lies primarily in the fact that the experimental rods are not rigid. Considerable experimental evidence has been accumulated which indicates that at room temperature neither PBLG nor PCBL behaves as a perfectly rigid rod[23–25]. The initial conclusion that PBLG was a rigid helix at room temperature seems likely to be due to an error in molecular weight calibration[24]. And with each polymer the intrinsic viscosity drops considerably with increasing temperature[17, 18].

We can either consider a lattice model of semiflexible rods[9], or a rigid rod model with a temperature dependent axial ratio. We prefer[17] the latter and consider the axial ratio dependence of the rigid, impenetrable rod model. Inasmuch as the equilibrium concentrations are insensitive to χ in the narrow biphasic region it is convenient to consider their dependence on the axial ratio for the athermal case ($\chi = 0$). This is illustrated in *Figure 10*. The difference between v_2 and v_2^* increases as x decreases. In addition the difference between the theoretical v_2 and v_2^* is greater than the experimental values for all x equal or less than the rigid rod value of each polymer. It would take extremely large, negative χ values to bring the theoretical and experimental values into exact agreement. We can though take the experimental v_2 and v_2^* and from *Figure 10* determine to what apparent axial ratio they correspond. Results of such an approach are shown in *Figure 11*. Each of the values calculated from v_2 and v_2^* lies below the rigid rod value for the respective

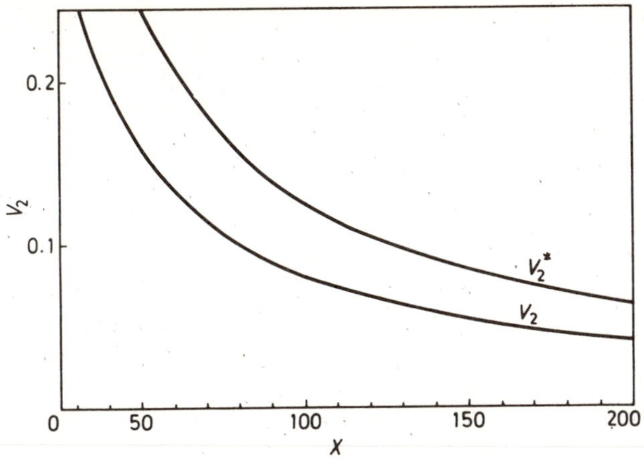

Figure 10. Equilibrium concentrations[10] for rigid, impenetrable rods with $\chi = 0$. Lower curve—line bounding the isotropic phase (v_2); upper curve—line bounding ordered phase (v_2^*).

polymer, and the two sets of apparent axial ratios have a common temperature dependence. The temperature dependence is also the same for the two polymers which would not seem unreasonable as their helical backbones as well as their side chains through the γ-carbon are the same. The temperature dependence of the intrinsic viscosity is nearly the same for the two polymers[17,18]. The apparent axial ratios for PCBL fall in the range deduced

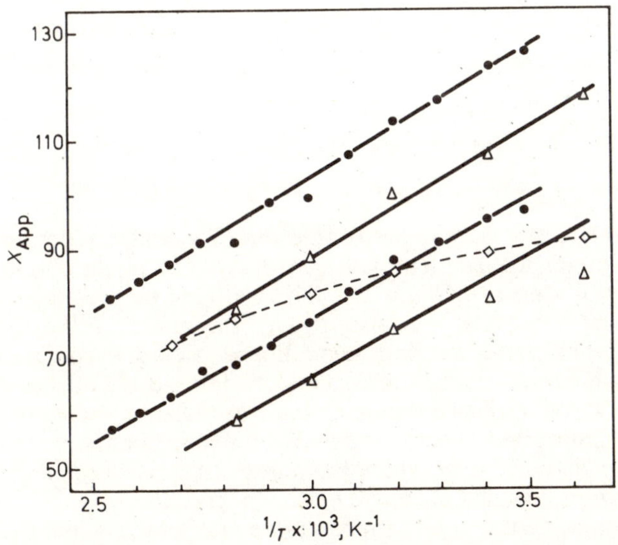

Figure 11. The apparent axial ratio as a function of temperature for PBLG (circles) and PCBL (triangles). Lower two sets—determined from v_2 (*Figure 10*); upper two sets—determined from v_2^* (*Figure 10*); diamonds—axial ratio of PCBL determined from intrinsic viscosity[18]

from intrinsic viscosity measurements although the temperature dependence is not the same. Though the agreement is not exact it is the similarity which leads us to conclude that the shift of the narrow biphasic region to higher concentration at increasing temperature is predominantly a function of increasing rod flexibility.

RIGID, PENETRABLE ROD THERMODYNAMICS

We have seen that the phase equilibria for the experimental systems depart from that predicted for rigid, impenetrable rods because the experimental rods are not rigid. We now consider the fact that the experimental rods are also not impenetrable. The side chains of PBLG (*Figure 1b*) and of PCBL (*Figure 1c*) constitute a sizeable fraction (75 and 79 wt.%, respectively) of the total cross section of the rod. Much evidence has accumulated[20–22, 26] which indicates that unless the side chain is flexible and hence solvent penetrable the rod-like polymers will not dissolve, and the liquid crystal phase cannot be realized. The side chain flexibility introduces three factors: first, there is an entropy associated with flexible side chain–solvent mixing which contributes significantly to the solution thermodynamics; second, the polymer–solvent interaction is predominantly solvent interacting with flexible side chains, the main chain or hard core rod being shielded from contact with solvent; third, the rigid rod cross section in the previous discussions taken as the cross section with side chains of mean extension may not be appropriate for comparison with rigid rod theory. An accounting of the effects of flexible side chains has been attempted[27] patterned after the rigid, impenetrable rod approach. The phase equilibria for this model can be calculated from equations 8–10,

$$\ln(1 - v_2) + [1 - 1/x(1 + mc)]v_2 + [mc/(1 + mc)]\chi^{\mathrm{I}}v_2^2 = \ln(1 - v_2^*)$$
$$+ (y - 1)v_2^*/x(1 + mc) + 2/y + [mc/(1 + mc)]v_2^* + [mc/(1 + mc)]\chi^{\mathrm{LC}}v_2^{*2} \quad (8)$$

$$\ln v_2 + [x(1 + mc) - 1]v_2 - \ln x^2 + \chi^{\mathrm{I}}mcx(1 - v_2)^2 = \ln v_2^* + (y - 1$$
$$+ mcx)v_2^* + 2 - \ln y^2 + 2mcx/y + \chi^{\mathrm{LC}}mcx(1 - v_2^*)^2 \quad (9)$$

$$v_2^* = [x(1 + mc)/(x - y)][1 - \exp(-2/y)] \quad (10)$$

where m is related to the number of side chains of length c per molecule and the other symbols are as identified previously. The phase diagram with mc equal to 0.15 is shown in *Figure 12*. This value of mc was chosen not only to be illustrative but also because it gives the best fit to vapour sorption isotherms[21]. The phase diagram is qualitatively similar to *Figure 3* except that the small liquid crystal–liquid crystal biphasic region has been eliminated. The narrow biphasic region extends to higher χ values because polymer–solvent interaction occurs only with the side chain component of the polymer. The distance over which the narrow biphasic region runs into the positive χ region depends of course on the magnitude of mc.

In common with the rigid, impenetrable rod model when the polymer–solvent interaction becomes increasingly unfavourable there is a point where the system separates into a solvent rich isotropic phase and a solvent poor ordered phase. The ordered phase has a polymer composition of 85–95% by the impenetrable rod model, but a much lower concentration by the penetrable rod model. In the PBLG–DMF system the ordered phase goes from 70–80% polymer during the transition region. Allowing the rod to be semi-flexible but impenetrable, as in the previous section, will not bring the theoretical values into agreement with the experimental ones. The introduction of flexible side chains which alone interact with the solvent softens the exclusion of the solvent. We believe the flexible side chains contribute significantly to the transition region phase equilibria.

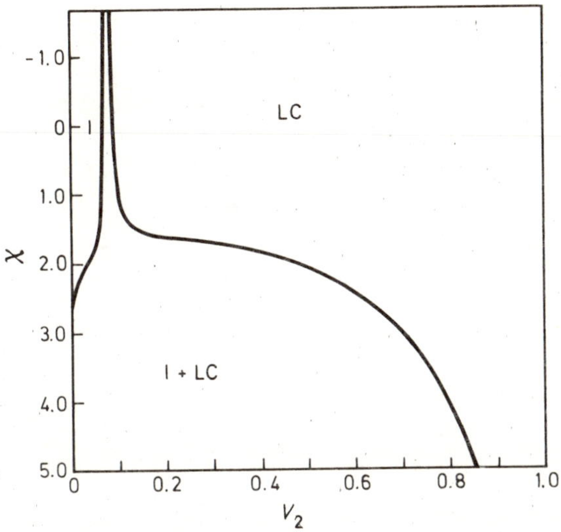

Figure 12. Lattice model phase diagram for rigid rods with flexible, permeable side chains[27]. Axial ratio is 150 and *mc* is 0.15

The flexible side chains also permit a more unfavourable polymer–solvent interaction before the transition region is reached. Our dilute solution osmotic pressure[15], vapour sorption[20,21] and DSC results[19] in PBLG–DMF each indicates a substantial positive heat of mixing at room temperature where a single phase isotropic solution is stable to several volume % polymer, and where a single ordered phase is stable to several volume % solvent. Whereas it is not easy to rationalize these observations by the impenetrable rod model, it is easy to do so with the flexible side chain model. With PCBL only vapour sorption measurements have been made. In contrast to PBLG the conclusion with PCBL is that the polymer–solvent mixing is essentially athermal at room temperature[20]. Since the transition region for PBLG lies just below room temperature we might infer that the transition region for PCBL should lie at lower temperatures. This turns out to be realized, as the transition region is 30–40°C lower for PCBL. We would also predict on the basis of vapour sorption measurements[20] that polycarbobenzoxyornithine–DMF should have a transition region below that of PBLG.

Finally we turn to the narrow biphasic region. As with the impenetrable rod model the line bounding the isotropic phase is effectively independent of χ, and that bounding the ordered phase nearly so. As with the rigid, impenetrable rod model we set $\chi = 0$ and investigate the apparent axial ratio dependence of the equilibrium concentration. We find that they lie close together and are in better agreement than those in *Figure 11*. However, the temperature dependence is similar to the impenetrable rod forced fit. The flexible side chain model affects the disorientation index, and hence would be expected to have a small effect on the narrow biphasic region.

LIQUID CRYSTAL–LIQUID CRYSTAL EQUILIBRIA

The rigid, impenetrable rod model predicts a small region where the minimum free energy of the system is represented by two liquid crystal phases

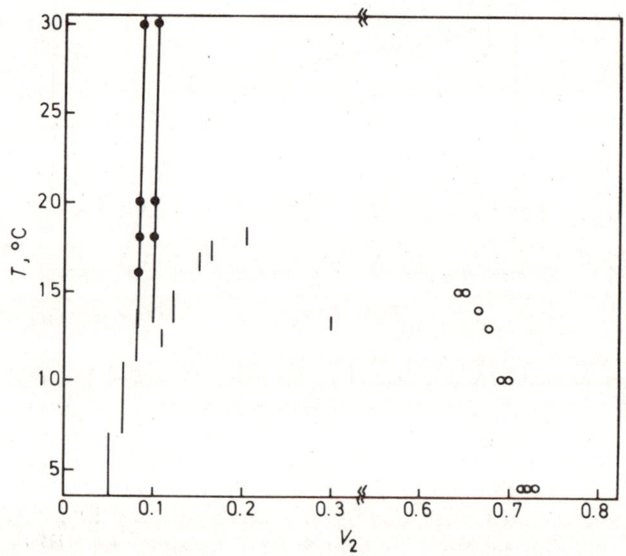

Figure 13. Expanded view of PBLG (310 000)–DMF phase boundary measurements relevant to the existence of an upper critical solution temperature

in equilibrium. Flory thought this biphasic region might disappear in a refined treatment[10]. This biphasic region has an upper critical solution temperature which occurs at about 50 volume %. The critical composition is almost molecular weight independent and thus behaves very differently from its counterpart in random coil solutions.

In *Figure 13* is an expanded scale view of pertinent phase boundary measurements for PBLG. Although there is need for a good method to determine phase boundaries in the 25–60 volume % region, the shape of the regions we have measured requires that the liquid crystal–liquid crystal region must exist. Consequently an upper critical solution temperature must exist for the PBLG–DMF system, which gives rise to the possibility of critical opalescence. On occasion solutions of 15–30 volume% which have been standing at temperatures slightly below room temperature are visually white and opaque, yet when viewed under a microscope show no precipitation. The significance of these observations awaits a quantitative study.

NATURE OF THE PHASE TRANSITION

In our discussions we have assumed that the phase transition was first order, which is also assumed by the Flory lattice model. However, many order–disorder transitions are thermodynamically second, and not first order. It is of interest to consider the evidence relative to the order of the transition.

In the PBLG–DMF system we have provided indirect evidence that there is a small latent heat. In neither system has a latent heat been observed directly. We can be more definite, however, concerning a density difference at least in the narrow biphasic region of the PCBL–DMF system. Particularly in the 10–15 volume% region a sharp meniscus develops with the ordered phase being more dense. As such a biphasic solution is warmed the meniscus is displaced towards the bottom and disappears coincident with crossing the phase boundary shown in *Figure 4*. Even when the visually sharp meniscus is present it is not unusual to find, upon microscopic examination, spherical droplets of liquid crystal in the isotropic phase. With PBLG we have never observed a meniscus by eye in the narrow biphasic region which must indicate that the density difference is very small. In a polarizing microscope one can see that the ordered phase does have a tendency to settle. The ordered phase which is dispersed in the isotropic phase takes on a spherical shape indicating a difference in surface tension. In neither system have we observed a meniscus when a solution enters or is in the wide biphasic region. This may be a kinetic problem (see below).

The similarity of the theoretical and experimental phase diagrams, the existence of a distinct meniscus in one system and the indirect observation of a latent heat in the other give good evidence that we are working with first order transitions in the sense of discontinuous first derivatives of the free energy.

DYNAMICS—BULK FLOW THROUGH THE NARROW BIPHASIC REGION

The viscosity of solutions of PBLG is strongly concentration and shear rate dependent[6, 14, 28–30]. Shown in *Figure 14* is the concentration dependence for several temperatures taken at a shear rate of 19.4 s^{-1}. The phase boundaries are clearly evident. The viscosity drops by a factor of 4 in crossing from the isotropic to the liquid crystal phase. Increasing the shear rate diminishes this effect. At high shear rates the distinction between the viscosity of isotropic and liquid crystal solutions disappears[30] probably because the high shear rate simply orders the isotropic phase. Conversely extrapolation to

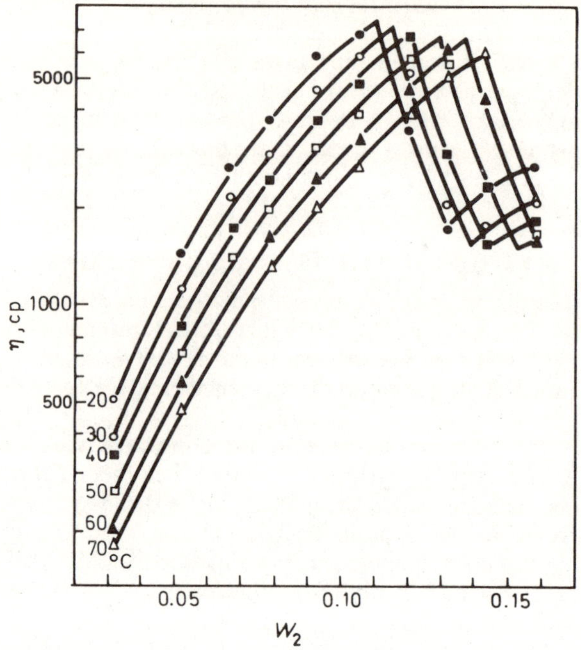

Figure 14. Concentration dependence of the viscosity of PBLG (310 000)–DMF solutions. The shear rate was $19.4\,s^{-1}$

zero shear rate heightens the difference. In the isotropic phase the different temperatures produce a family of parallel lines which may be shifted to give a common curve[27]. At a fixed temperature and composition the shear stress and shear rate are linearly related on a log–log plot in both phases indicating that a 'power law' is obeyed. As one might expect, the liquid crystal phase shows a much greater deviation from Newtonian flow. The only point we wish to make here is that ordering the rods into a mean parallel arrangement leads to considerably easier bulk flow. This effect is so large that it can be visually observed.

DYNAMICS—BULK FLOW NEAR THE TRANSITION REGION

The temperature dependence of the viscosity of 9.2 and 14.4 wt % PBLG–DMF solutions at fixed shear rate is shown in *Figure 15*. With the 9.2% solution, isotropic at room temperature, the viscosity shows a normal temperature dependence until about 17°C where it begins to show a very high temperature coefficient. At low temperature the behaviour of the 14.4% solution, which is liquid crystalline at room temperature, is similar. It is the low temperature behaviour which concerns us at the moment. The sudden increase in viscosity as the temperature is lowered occurs with all concentrations and is independent of whether one is cooling an isotropic or a liquid

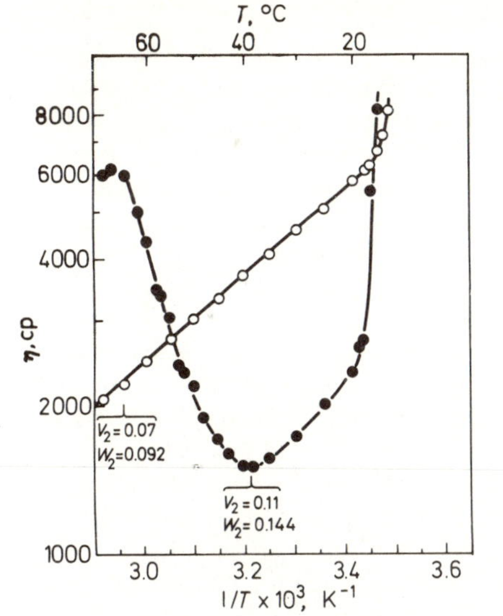

Figure 15. Temperature dependence of the viscosity of PBLG (310 000)–DMF solutions. The shear rate was 19.4 s^{-1}

crystalline phase. The viscosity quickly rises above 10 000 centipoise, the maximum we could measure, and becomes time dependent. The solution eventually behaves as if it were a gel even in solutions containing less than 1 % polymer.

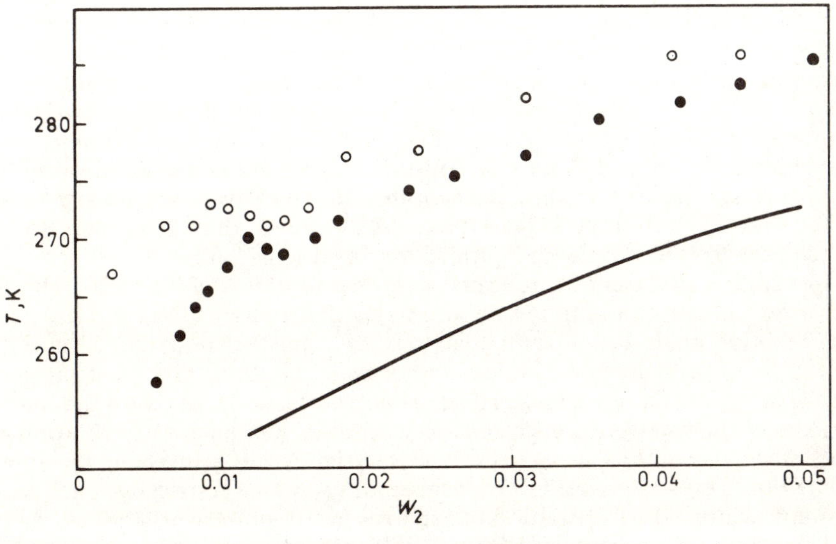

Figure 16. Concentration dependence of the gelation of PBLG (268 000) by visual observation[15] (open circles—onset; filled circles—completion). Solid line is phase boundary determined by polarizing microscope[17] for PBLG (310 000)

In our earliest study of PBLG–DMF we noted the gelation by a simple qualitative, visual observation[15], *Figure 16*. We were unable previously to observe with assurance the appearance of the ordered phase at low concentration. which we know now was due to inadequate temperature control of the microscope stage at low temperature[15] as well as cell design[16]. We can see from *Figure 16* that the visual appearance of a rigid, gel-like solution correlates well with the phase boundary, but anticipates it by several degrees. The viscosity measurements do likewise. With the PBLG–DMF system we now have ample evidence to show that as the transition region in the phase diagram is approached a dramatic change takes place in bulk flow, which precedes the phase boundary as determined by polarizing microscope or nuclear magnetic resonance measurements. In the PCBL–DMF system, where the transition region is shifted 30–40°C, a few qualitative observations indicate a similar phenomenon. The appearance of a low concentration gel-like phase may be a convenient method for deducing at what temperature a given polymer–solvent system approaches the transition region. Be that as it may it is still necessary to explain the existence of a very low concentration gel-like solution in a system of rod molecules already containing maximum intramolecular order. The fact that no meniscus appears upon crossing the transition region phase boundary may be related to this phenomenon. At present we do not understand these observations.

DYNAMICS—INDIVIDUAL ROD MOTION

In the liquid crystal phase there is a correlation of rod motion otherwise the periodicity lines so easily observed[6] would not be present. Each rod has a mean lattice position parallel to its neighbours. Actually the mean positions are not quite parallel as the liquid crystal phase is cholesteric and not nematic in the absence of external fields[31–34]. The twist results from the fact that the polymer is a collection not of molecular rods but of either left or right handed helices. In a given solvent changing the handedness of the helix changes the handedness of the cholesteric screw axis. With screws of a given handedness the cholesteric screw axis may be continuously varied and even reversed by appropriate choice of solvent, a phenomenon which does not appear to be understood. The cholesteric twist is so slight, however, that the mean position of the rods may be considered parallel for our discussion here.

In addition to the cooperative end-over-end tumbling of the rods each rod has a certain amount of motion about its equilibrium position. We saw that local motion made a significant contribution to the thermodynamic stability of the liquid crystal phase. It is of interest to measure this local motion. The axial ratio of a typical experimental polymer is so large that if the centre of mass of the rod were constrained to its equilibrium position, a rotation of only a degree or less would result in physical overlap even in the most dilute liquid crystal phase. The experimental system, of course, does not have this constraint. The distribution of rod axes has been measured indirectly in a 17.5 wt% solution of PBLG in dichloromethane by nuclear magnetic resonance[35]. These studies indicate a substantial spread in rod orientation with 13% lying more than 20° out of parallel alignment with the applied

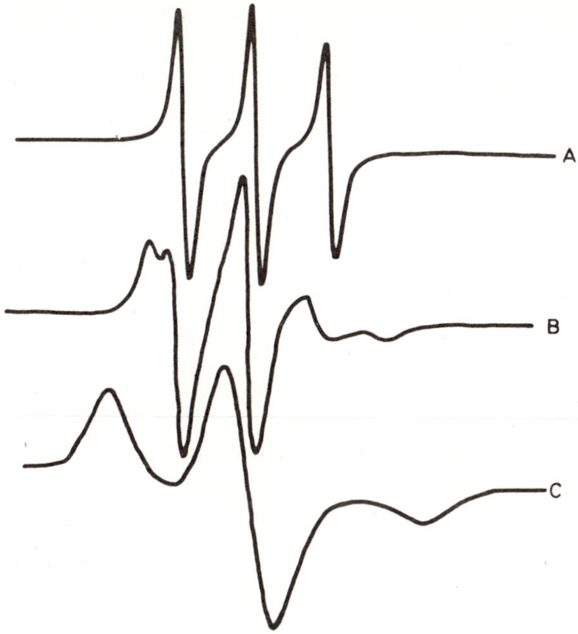

Figure 17. Nitroxide e.s.r. spectra classified as isotropic (A), axially symmetric (B), or rigid glass (C)

magnetic field. In the absence of the magnetic field the amount of disorder should be even larger. It was furthermore found that when a system of aligned rods was placed in a magnetic field their realignment occurred cooperatively and not individually.

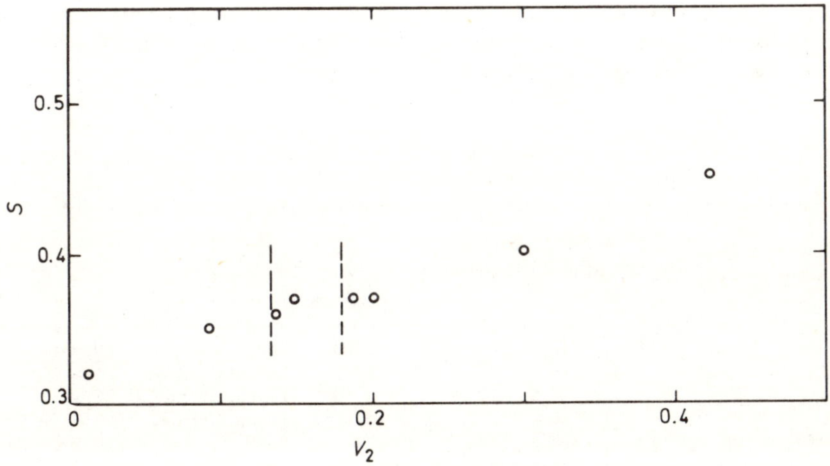

Figure 18. Order parameter (*S*) as a function of PBLG (122 000) concentration for nitroxide labelled PBLG at 25°C[36]. Vertical dashed lines indicate approximate positions of phase boundaries

Table 1. Rotational relaxation of nitroxide in end labelled PBLG (12 000) at 25°C.

Labelled PBLG Conc., vol.%	Nonlabelled PBLG Conc., vol.%	Mol. wt	DMF Conc., vol.%	τ_c^* s
1.0	0	—	99	3.0×10^{-10}
1.0	20	74 000	79	7.0×10^{-10}
1.0	20	310 000	79	7.5×10^{-10}
1.0	20	450 000	79	8.2×10^{-10}

*Calculated according to reference 39.

We have tried to measure the local motion by covalently attaching a nitroxide free radical to the end of each rod. The shape of the e.s.r. spectrum is sensitive to the motion of the nitroxide. The spectra may for our purpose be classified as an isotropic, an axially symmetric, or a rigid glass spectrum (*Figure 17*). For a 122 000 molecular weight PBLG molecule the axial ratio is sufficiently large to yield an axially symmetric spectrum at all compositions except pure or nearly pure polymer[36]. From the spectrum the magnitude of the microorder parameter[37] S may be deduced. The results are shown in *Figure 18*. The e.s.r. magnetic field is too weak to orient the rods[36] so these results should more nearly represent the mean orientation in the absence of an applied field. However, the mean orientation of the rods is considerably larger than the S values deduced from the e.s.r. spectra, as a contribution from internal motion of the nitroxide has not been removed. This internal motion is not concentration dependent and the relative values of the order parameter are meaningful. If the results for the isotropic phase are extrapolated into the liquid crystal phase, the polymer molecules in the ordered phase are seen to have more local freedom than the same composition isotropic phase. This strengthens our belief that molecular asymmetry alone is responsible for stabilizing the ordered over the disordered phase at least in the PBLG–DMF system. Additionally the concentration dependence of the order parameter in the liquid crystal phase is in semiquantitative agreement with the disorientation index y of the rigid, impenetrable rod model[36].

Table 2. Rotational relaxation of nitroxide in side chain labelled PBLG (310 000) in DMF at 25°C.

Polymer concentration Volume %	Polymer concentration Weight %	Molar solvent / Molar (Monomer) polymer	η^* centipoise	Phase	$\tau_c^\dagger$ s
6.4	8.4	32	3700	Isotropic	1.9×10^{-10}
12	15.5	16	2100	Liquid Crystal	2.6×10^{-10}
71	77	0.9	$>10^4$	Liquid Crystal‡	6.8×10^{-10}
86	89	0.4	$>10^4$	Liquid Crystal‡	1.8×10^{-9}
100	100	0	—	Solid	2.0×10^{-8}

*At a shear rate of $19.4\ s^{-1}$.
†Calculated according to reference 39 (first four), or reference 40.
‡Actually a 'wet' solid which is presumably ordered.

Studies on low molecular weight, end labelled PBLG are informative with regard to the hindrance of rod motion due to molecular asymmetry. In very dilute solutions 22 000 molecular weight PBLG rotates sufficiently rapidly to give an isotropic type spectrum (*Figure 17A*). As the concentration is increased to only a few per cent polymer the e.s.r. spectrum undergoes a qualitative change to an axially symmetric spectrum (*Figure 17B*), a result of reduced end-over-end tumbling[38]. That this is not some sort of 'general viscosity' effect can be seen from a series of measurements on a 12 000 molecular weight end labelled sample which has a rigid rod axial ratio of about five. The rotational correlation times, not corrected for any internal nitroxide motion, are shown in *Table 1*. The short spin labelled rod does rotate more freely in dilute solution, but is not affected greatly by the presence of 20 per cent of high molecular weight PBLG. Each of the 20 per cent solutions is liquid crystalline, and also differs in bulk viscosity as a result of the molecular weight difference. As the rod axial ratio approaches unity the rod motion should reflect the local viscosity and not the bulk viscosity which is dominated by the long rods.

Free nitroxide in PBLG–DMF solutions has a rotational correlation time which is only slightly dependent on phase or bulk viscosity[38]. This is consistent with the observations on short rods. These results lead us to believe that the motion of the solvent, DMF in our case, is not significantly phase dependent. This cannot be strictly true, otherwise direct polar coupling observed in the solvent proton magnetic resonance should be observed also in the isotropic phase. However, the rotational relaxation of a nitroxide attached to a glutamate side chain is little affected by bulk properties over a wide range of concentration, as can be seen in *Table 2*. These as well as nuclear magnetic resonance results[26] show that exceedingly little solvent is necessary in order for the helix side chain and the solvent to be in rapid motion.

REFERENCES

1 F. C. Bawden and N. W. Price, *Proc. Roy. Soc* **123B**, 274 (1937).
2 H. Freundlich, *J. Phys. Chem.* **41**, 1151 (1937).
3 J. D. Bernal and I. Fankuchen, *J. Gen. Physiol.* **25**, 111 (1941).
4 G. Oster, *J. Gen. Physiol.* **33**, 445 (1950).
5 A. E. Elliott and E. J. Ambrose, *Discussions Faraday Soc.* **9**, 246 (1950).
6 C. Robinson, *Trans. Faraday Soc.* **52**, 571 (1956).
7 L. Onsager, *Ann. N. Y. Acad Sci.* **51**, 627 (1949).
8 A. Isihara, *J. Chem. Phys* **19**, 1142 (1951).
9 P. J. Flory, *Proc. Roy. Soc. (London)* **A234**, 60 (1956).
10 P. J. Flory, *Proc. Roy. Soc. (London)* **A234**, 73 (1956).
11 P. Doty, J. H. Bradbury and A. M Holtzer, *J. Amer. Chem. Soc.* **78**, 947 (1956).
12 J. Applequist and P. Doty, in M. Stahmann (Ed.), *Polyamino Acids, Polypeptides and Proteins*, p. 161, University of Wisconsin Press: Madison (1962).
13 G. Spach, *C. R. Acad. Sci. (Paris)* **249**, 543 (1959).
14 A. Nakajima, T. Hayashi and M. Ohmori, *Biopolymers* **6**, 973 (1968).
15 K. D. Goebel and W. G. Miller, *Macromolecules* **3**, 64 (1970).
16 E. L. Wee and W. G. Miller, *J. Phys. Chem.* **75**, 1446 (1971).
17 W. G. Miller, J. H. Rai and E. L. Wee, in *Liquid Crystals and Ordered Fluids*, Vol. II, p 243. R. Porter and J. Johnston, eds. Plenum: New York (1974).
18 G. Santee and W. G. Miller, to be submitted to *Macromolecules*.

[19] J. H. Rai and W. G. Miller, *J. Phys. Chem.* **76**, 1081 (1972).
[20] J. H. Rai and W. G. Miller, *Macromolecules* **6**, 257 (1973).
[21] J. H. Rai and W. G. Miller, *Macromolecules* **5**, 45 (1972)
[22] P. J. Flory and W. J. Leonard, *J. Amer. Chem. Soc.* **87**, 2102 (1965).
[23] G. Spach, L. Freund, M. Daune and H. Benoit, *J. Mol. Biol.* **7**, 468 (1963).
[24] H. Fujita, A. Teramoto, K. Okita, T. Yamashita and S. Ikeda, *Biopolymers* **4**, 769 (1966): **4**, 781 (1966).
[25] W. G. Miller and P. J. Flory, *J. Mol. Biol.* **15**, 298 (1966).
[26] J. H. Rai, W. G. Miller and R. G Bryant, *Macromolecules* **6**, 262 (1973).
[27] E. L. Wee and W. G. Miller, in preparation.
[28] J. T. Yang, *J. Amer. Chem. Soc.* **80**, 1783 (1958).
[29] J. T. Yang, *J. Amer. Chem. Soc.* **81**, 3902 (1959).
[30] J. Hermans, *J. Colloid Sci.* **17**, 638 (1962).
[31] C. Robinson and J. C. Ward, *Nature* **180**, 1183 (1957).
[32] C. Robinson, J. C. Ward and R. B. Bevers, *Discussions Faraday Soc.* **25**, 29 (1958).
[33] C. Robinson, *Tetrahedron* **13**, 219 (1961).
[34] C. Robinson, *Molecular Crystals* **1**, 467 (1966).
[35] R. D. Orwoll and R. L. Vold, *J. Amer Chem. Soc.* **93**, 5335 (1971).
[36] E. L. Wee and W. G. Miller, *J. Phys. Chem.* **77**, 182 (1973).
[37] A. Saupe, G. Englert and A. Pova, *Advan. Chem. Ser.* No. 63, 51 (1967).
[38] C. C Wu and W. G. Miller, to be submitted to *Macromolecules.*
[39] J. H. Freed and G. K. Frankel, *J. Chem. Phys.* **39**, 326 (1963); **40**, 1815 (1964).
[40] S. A. Goldman, G. V. Bruno and J. H. Freed, *J. Phys. Chem.* **76**, 1858 (1972).

RECENT STUDIES OF LIQUID CRYSTALS OF BIOLOGICAL IMPORTANCE

D. CHAPMAN

Department of Chemistry, Sheffield University, S3 7HF, UK

ABSTRACT

Phospholipids and glycolipids are important components of the structure of many cell membranes. They exhibit a melting process associated with the lipid chains at temperatures considerably below the final melting point. In the presence of water they also exhibit lyotropic mesomorphism, and various phases exist. The relevance of this behaviour to the structure and function of cell membranes is discussed.

Phospholipids and glycolipids are important molecules in the construction of cell membranes, and include a variety of classes such as the lecithins (phosphatidylcholines), phosphatidylethanolamines, phosphatidylserines and sphingomyelins (Scheme I). Mixtures of these lipids occur in biological

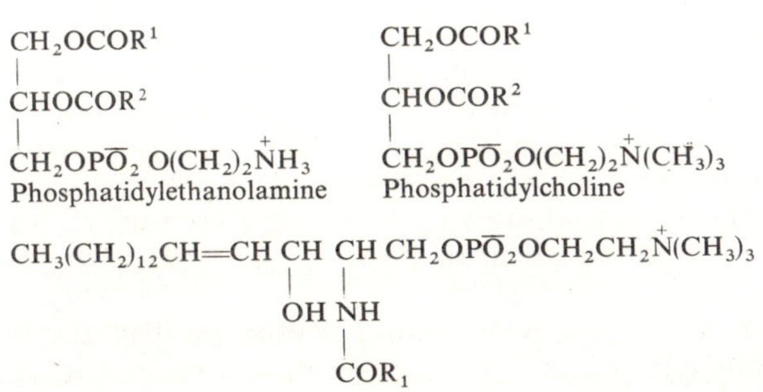

Sphingomyelin

Scheme 1. Molecular structures for some phospholipids

membranes[1] and the proportions of these classes vary from one membrane system to another[2]. A range of fatty acids is usually found associated with each class of lipid. These fatty acids vary in chain length and unsaturation. Stearic, palmitic, myristic are common among the saturated acids, and oleic acid is common amongst the unsaturated acids. Unlike simple soaps which contain one fatty acid, the phospholipids contain two fatty acids. The idea that membranes contain bilayers of lipid had been a view held for about

40 years. Recent evidence is in favour of some regions of membranes containing a bilayer structure[3] but there is still uncertainty concerning the position of protein.

Our own studies of phospholipid molecules began following an earlier study of simple soap systems. Using infra-red spectroscopy we showed a marked change in the spectrum of sodium stearate corresponding to a change from a spectrum characteristic of a crystal to a spectrum typical of a liquid some 200°C below the final melting point of the soap[4]. We also showed that a similar effect[5] occurs with phospholipid molecules, i.e. that phospholipids exhibit thermotropic mesomorphism.

THERMOTROPIC MESOMORPHISM

The capillary melting points of a number of pure phospholipids have been determined and are quite high, e.g. the value for the diacyl phosphatidylethanolamines is about 200°C, while for the phosphatidylcholines these melting points are 230°C.

In addition to the capillary melting point, other phase changes occur with phospholipids at lower temperatures. Thus, when a pure phospholipid, dimyristoylphosphatidylethanolamine, containing two fully saturated chains, is heated from room temperature to the capillary melting point, a number of thermotropic phase changes (i.e. phase changes caused by the effect of heat) occur. This was first shown by infra-red spectroscopic techniques[5], then by thermal analysis and has now been studied by a variety of physical techniques[6, 7].

Above this temperature the spectrum loses all the fine structure and detail which were present at lower temperatures and becomes similar to the spectrum obtained with a phospholipid dissolved in a solvent such as chloroform.

Differential thermal analysis (d.t.a) shows that a marked endothermic transition (absorption of heat) occurs at this transition temperature[8]. An additional heat change occurs at -135°C and only a small heat change is involved near the capillary melting point of the lipid. This behaviour is similar to that which occurs with liquid crystals, such as *p*-azoxyanisole or cholesteryl acetate which form nematic and cholesteric liquid crystalline phases.

The main conclusions from various studies are that (a) even with the fully saturated phospholipid at room temperature, some molecular motion occurs in the solid. This is evident from the p.m.r. spectra and from the i.r. spectra taken at liquid-nitrogen and at room temperatures. (b) When the phospholipid is heated to a higher temperature, it reaches a transition point, a marked endothermic change occurs and the hydrocarbon chains in the lipid 'melt' and exhibit a very high degree of molecular motion. This is evident both in the appearance of the i.r. spectrum and also in the narrow n.m.r. linewidth. The broad diffuse appearance of the i.r. spectrum is consistent with the chains flexing and twisting and with a 'break-up' of the all-planar *trans* configuration of the chains.

Only one main 'melting of the chains' occurs even when there are *two* different types of chain present in the phospholipid molecule. The transition temperatures for different phospholipid classes vary even though they contain exactly the same fatty acid residues.

EFFECTS OF WATER

Small amounts of water can have unusual effects upon the mesomorphic behaviour. Thus, the diacylphosphatidylcholines (lecithins) exhibit additional liquid crystalline forms between the first transition temperature and the capillary melting point. The intermediate liquid crystalline form exhibits x-ray spacings consistent with a cubic phase organization. On the other hand, if all the water is removed from the phospholipid, the lipid will no longer exhibit this cubic phase[9].

When phospholipids are examined in increasing amounts of water, the various physical techniques, such as microscopy, n.m.r. spectroscopy or differential thermal analysis, show that as the amount of water increases, the marked endothermic transition temperature for a given phospholipid falls. It does not fall indefinitely, but reaches a limiting value independent of the water concentration. We can understand this if we regard the effect of water as leading first to a 'loosening' of the ionic structure of the phospholipid crystals. This, in turn, affects the whole crystal structure and a reduction, up to a certain limit, of the dispersion forces between the hydrocarbon chains. Large amounts of energy are still required to counteract the dispersion forces between the chains and quite high temperatures are still required; to cause the chains to melt. These limiting transition temperatures parallel the melting point behaviour of the analogous fatty acids becoming lower with increasing unsaturation.

There are a number of important features associated with the transition temperature for the lipid when it is in the presence of water[10]. The first is that the ability to disperse the lipid in water increases markedly above this transition temperature. Only those phospholipids which have transition temperatures when placed in water below or near to room temperature, spontaneously form myelin figures. Fully-saturated phospholipids which have high transition temperatures do not form myelin figures at room temperature. However, if the temperature is raised to the transition point, myelin figures are formed by these phospholipids. In the presence of an excess of water, phospholipids, such as 1,2-dipalmitoylphosphatidylcholine, spontaneously form myelin figures at 42°C.

This ability to form myelin figures with saturated phospholipids has been confirmed by optical- and electron-microscopic investigations. It has been used to attempt to provide information about the situation of the osmium used after osmium tetroxide fixation procedures[11]. N.M.R. and e.s.r. spectroscopic studies[12–14] have been made of these lamellar lecithin systems and the molecular motion of the $N(CH_3)_3$ group and of the different methylene groups along the chain studied. The molecular motion is greatest at the methyl end of the hydrocarbon chains.

A second feature of lipid + water systems is their monolayer behaviour.

Monolayers obtained with phosphatidylcholines are observed to be much more expanded than are the corresponding phosphatidylethanolamines containing the same acyl chains[15]. These results can be compared with the d.t.a. results discussed earlier[16]. A high transition temperature for liquid crystal formation is, in general, correlated with a condensed-type monolayer and a low transition temperature with an expanded film. The d.t.a. transition temperatures are higher for the phosphatidylethanolamines than for the corresponding phosphatidylcholines. Phospholipids containing *trans* (elaidoyl) unsaturated chains have higher transition temperatures than those containing *cis* (oleoyl) chains. Phospholipids containing one fully saturated chain and one *trans* unsaturated chain give condensed monolayers similar to those observed with completely saturated phospholipids.

Chapman has recently collaborated with Czech scientists to study n.m.r. spectra of these lipid molecules when in the liquid-crystalline form and when arranged to spin at the 'magic angle'. These results show that at spinning frequencies which are quite slow, marked narrowing occurs. The differing signals narrow at different spinning frequencies.

Phospholipids in the presence of water can also show various types of lyotropic mesomorphism, i.e. they can exhibit different types of liquid crystalline organization[18, 19]. Above the transition temperatures when water is present, phospholipids can form different types of phase organization, e.g. some phospholipids give lamellar and hexagonal structures[19, 20]. The lecithins appear to exhibit only a lamellar structure over a large range of water concentration[9]. Other phospholipids, such as phosphatidylethanolamines and samples of brain lipid, appear to be able to exist in both hexagonal and lamellar organization, depending upon the concentration in water.

NATURAL MEMBRANES

Lipids in water may show a marked thermal transition associated with a phase change from gel to liquid crystal but the presence of cholesterol can modify or remove this transition.

Recent studies by Steim *et al.*[21] have been with a membrane which does not contain cholesterol. This is the membrane of *Mycoplasma laidlawii*. With this membrane a strong endothermic transition is indeed observed, as in the simple lipid–water systems. The transition temperature is the same as that of the transition of the extracted lipids dispersed in water.

Studies have also been made with membranes which contain cholesterol, e.g. myelin membranes[22]. With wet myelin, thermal transitions are not detectable. In this case the cholesterol and other lipids appear to be organized in a single phase. To maintain the organization of the lipid in myelin a critical amount of water appears to be required. This water is unfreezable at 0°C and may correspond to 'bound' water. On drying the myelin, the cholesterol and other lipids crystallize and precipitate. Endothermic transitions associated with the cholesterol and lipid can then be observed. The total lipid extract in water does *not* show a detectable endothermic transition but the cholesterol-free lipid does. In the absence of cholesterol, part of the myelin lipid is crystalline at body temperature. These results are understandable if there are regions of lipid bilayer present in the membrane with the cholesterol preventing lipid chain crystallization.

The fluidity of membranes may be related to the transition temperatures of their components. Thus, membranes which contain phospholipids having little unsaturation will have less fluidity than those having phospholipids with much unsaturation. This control of fluidity may then be related to diffusional characteristics of molecules passing into and out of the cell[7]. In line with this concept is a recent study comparing *in vivo* and *in vitro* phase transitions. This study was made with a double mutant of *Escherichia coli* which can incorporate fatty acids with various hydrocarbon structures into its phospholipids[23].

Chapman and Oldfield[24] have pointed out that in some membranes there are lipid chains which are stiff or rigid as well as having fluid regions. Recently membranes containing only rigid lipid chains have also been observed.

REFERENCES

1 D. Chapman. *Biological Membranes* (Academic Press, 1968).
2 L. L. M. van Deenen, *Progress in the Chemistry of Fats and other Lipids* (Pergamon Press, 1965), ed. Holman, 8.
3 M. H. F. Wilkins, A. E. Blaurock and D. M. Engelman, *Nature*, **230**, 72 (1971).
4 D. Chapman, *J. Chem. Soc.* 784 (1958).
5 P. Byrne and D. Chapman, *Nature*, **202**, 987 (1964).
6 R. M. Williams and D. Chapman, *Progress in the Chemistry of Fats and other Lipids*, **11**, 1 (1970).
7 D. Chapman, P. Byrne and G. G. Shipley, *Proc. Roy. Soc. A*, **290**, 115 (1966).
8 D. Chapman and D. T. Collin, *Nature*, **206**, 189 (1965).
9 D. Chapman, R. M. Williams and B. D. Ladbroke, *Chem. Phys. Lipids*, **1**, 445 (1967).
10 M. C. Phillips, R. M. Williams and D. Chapman. *Chem. Phys. Lipids*. **3**. 234 (1970).
11 D. Chapman and D. J. Fluck, *J. Biophys. Biochem. Cytol.* **30**, 1 (1966).
12 Z. Veksli, N. J. Salsbury and D. Chapman, *Biochem. Biophys. Acta*, **183**, 434 (1969).
13 W. L. Hubbel and H. M. McConnell, *Proc. Nat. Acad. Sci., Wash.* **64**, 20 (1969).
14 E. Oldfield, J. Marsden and D. Chapman, *Chem. Phys. Lipids* **7**, 1 (1971).
15 L. L. M. van Deenen, U. M. T. Houtsmuller, G. H. de Haas and E. Mulder, *J. Pharmac. Pharmacol.* **14**, 429 (1962).
16 D. Chapman, N. F. Owens and D. A. Walker, *Biochem. Biophys. Acta*, **120**, 148 (1966).
17 D. Chapman. E. Oldfield, D. Doskocilova and B. Schneider, *FEBS Letters*, **25**, 261 (1972).
18 V. Luzatti, in *Biological Membranes*, ed. D. Chapman, p. 103 (1968).
19 V. Luzatti and F. Husson, *J. Cell. Biol.*, **12**, 207 (1962).
20 T. Gulik-Krzywicki, E. Riv-as and V. Luzatti, *J. Mol. Biol.* **27**, 303 (1967).
21 J. Steim. M. E. Tourtellote, J. C. Reinert, R. D. McElhaney and R. L. Radar, *Proc. Nat. Acad. Sci., Wash.* **63**, 104 (1969).
22 B. D. Ladbrooke, T. J. Jenkinson, V. B. Kamat and D. Chapman, *Biochim. Biophys. Acta.*, **164**, 101 (1968).
23 P. Overath, H. U. Schairer and W. Stoffel, *Proc. Nat. Acad. Sci., Wash.* **67**, 606 (1970).
24 E. Oldfield and D. Chapman *FEBS Letters*, **23**, 255 (1972).
25 K. R. Naqvi *et al.*

STUDIES ON THE ORGANIZED STRUCTURES IN SOLUTIONS AND GELS OF SOME SYNTHETIC MACROMOLECULES

KANG-JEN LIU

Chemical Research Center, Corporate Research and Development, Allied Chemical Corporation, Morristown, New Jersey, USA

ABSTRACT

This presentation reviews our work on the organized structures of some synthetic macromolecules in solutions and gels. Various techniques, such as spectroscopic, viscometric, x-ray analytic and molecular relaxation measurements were used. Emphasis is on the *concept of solvation* of macromolecules. The following subjects are included: (*i*) The intramolecular conformational structures of some *simple polymers*, such as polyethylene and polyethylene oxide, are presented. It is clearly revealed that even an isolated chain molecule in an infinitely dilute solution is subjected to both intra-polymer and polymer–solvent interactions that determine *its final structure*—organized or disorganized. (*ii*) The nature of macromolecular association is discussed in terms of polymer–polymer and detailed polymer–solvent interactions. In particular, we discuss the effect of molecular association based on stereospecific complementarity of the interacting species on the local segmental environments of chain molecules. This leads to *a deep insight* into the intricate nature of macromolecular associations in gel or in solutions. (*iii*) Interactions of solvent with macromolecules in organized and random structures are compared and significant differences are observed. It therefore suggests a *new route* to the study of macromolecular structure by examining the behaviour of the interacting solvent molecules.

INTRODUCTION

In recent years there has been an increasing interest in studying biological principles in synthetic polymer systems. This is particularly true in the area of molecular structures and interactions. Organized structures long recognized in solutions of biopolymers may no longer be characteristic of biopolymers only. There are counterparts in synthetic polymers. In this presentation we review our work on the organized structures of some synthetic macromolecules in solutions and gels.

STRUCTURAL STUDIES OF CHAIN MOLECULES

It is well known that the conformational structure of many biopolymers in solution is determined by the solvent in which they are dissolved. Relatively little work has been done on synthetic polymers. Let us start with the 'simplest' polymer, polyethylene (PE). It does not contain chemical groups that interact in a specific pattern, such as the hydrogen bonding in polypeptides. Its structure in solution was, therefore, taken for granted as a typical representation of random coil[1, 2]. This concept was questioned by Liu. A high-resolution proton-magnetic resonance (p.m.r.) study of the intramolecular structure of PE in solution was then carried out[3–5].

P.M.R. spectra of various *n*-alkanes in α-chloronaphthalene are shown in *Figure 1.* These compounds are, of course, oligomers of polyethylene. The first peak to the left (I) is due to the external reference; peaks II and III are due to the methylene and methyl protons, respectively. The complicated spectral pattern arising from the methyl protons III is caused by spin–

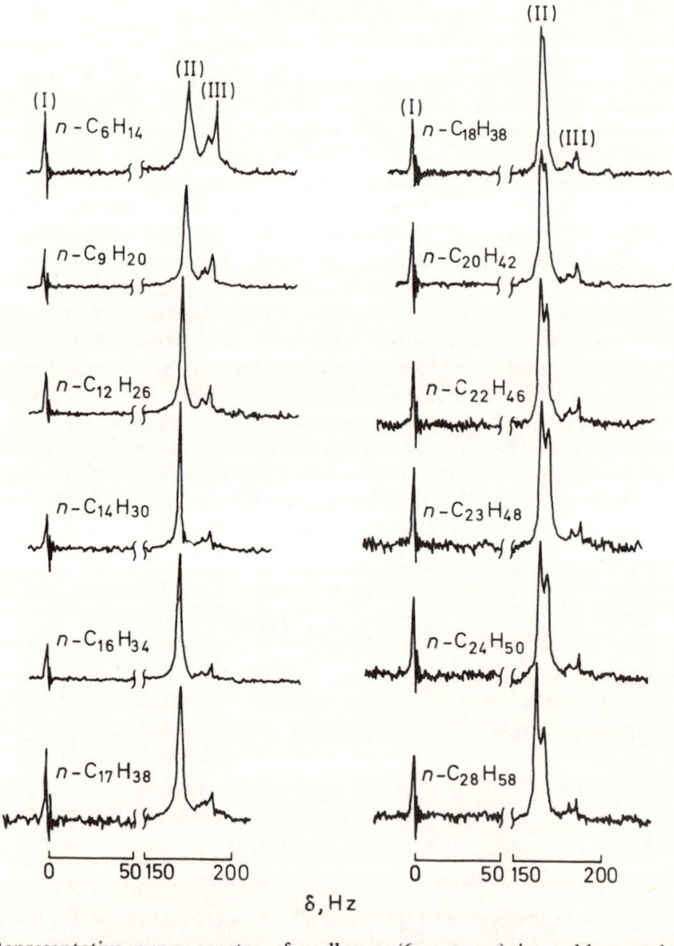

Figure 1. Representative p.m.r. spectra of *n*-alkanes (6 per cent) in α-chloronaphthalene at 35°C. Peaks I, II, and III are external reference (8 per cent dioxane in carbon tetrachloride) and methylene and methyl absorption lines, respectively.

spin interactions with the adjacent methylene protons. Similar splittings should appear for the methylene protons adjacent to the methyl groups. These splittings are presumably masked in the present spectra by the internal methylene protons (IMPs) contributing to peak II.

It is noteworthy that all IMPs give rise to a single p.m.r. peak for alkanes with carbon number $n \leqslant 16$, and two partially resolved peaks (on an expanded scale) when $n \geqslant 17$. The presence of two distinct IMP resonances for the higher *n*-alkanes is a definite indication of some new structural arrangement in these compounds that is not present in the lower members of the series. Detailed analysis shows that the new peak grows in relative intensity as the number of chain units is further increased, and it can be identified with the p.m.r. peak of high-molecular-weight polyethylene.

The formation of this new structure in the higher *n*-alkanes is strongly solvent-dependent. Aromatic solvents, such as α-chloronaphthalene and 9-chloroanthracene, were found to produce the split IMP resonance for $n \geqslant 17$. No splitting was evident on the same compounds when they were dissolved in carbon tetrachloride or in deuterated *n*-alkanes. These results do not seem to be related to 'ring current' effects in the aromatic solvents: addition of a small amount of carbon tetrachloride to a solution of $C_{24}H_{50}$ in α-chloronaphthalene drastically modifies the new IMP resonance.

The structure of the internal methylene units responsible for the additional IMP resonance cannot be derived from p.m.r., although the explanation of the experimental results requires some ordered structural feature. A chain-folding mechanism has been proposed that is consistent with thermodynamic consideration[6]. It should be noted that whatever the structure may be, it cannot be static: the narrow p.m.r. linewidths preclude this. Rather, it must be a dynamic arrangement with fluctuating sequences of more-or-less ordered regions. As a result of this, little hydrodynamic rigidity is given to the molecules, which behave in many respects as random coils.

In addition to the observed conformational change of polyethylene in certain bulky aromatic solvents at C-17, we have also found evidence of the existence of a particular conformational change for these chain molecules in both their pure state and benzene solutions at C-9[5]. This is in agreement with the Raman results reported by Schaufele[7].

P.M.R. has been used to study the intramolecular structure of poly-(ethylene glycol), PEG[8,9]. The spectra were measured as a function of chain length at different concentrations in various solvents. Representative spectra, obtained in benzene, are shown in *Figure 2*. In 50 per cent solutions, the PEG chains are in various states of association, and all their ether-type protons contribute to a structureless peak. In contrast, a great deal of structure is apparent in the p.m.r. spectra of the 5 per cent solutions. The internal ether-type protons contribute to a single peak for the trimer, tetramer, and pentamer but form a doublet for the hexamer. The p.m.r. spectrum of the PEG heptamer contains a new peak that is absent in lower members of the series. As the chain length is increased beyond the heptamer, this peak grows in relative intensity. Thus, it appears that a local polymeric environment is revealed in an oligomer that is seven units long. P.m.r. studies of PEG in carbon tetrachloride yield similar results, although the splitting of the various peaks is not as great in this solvent.

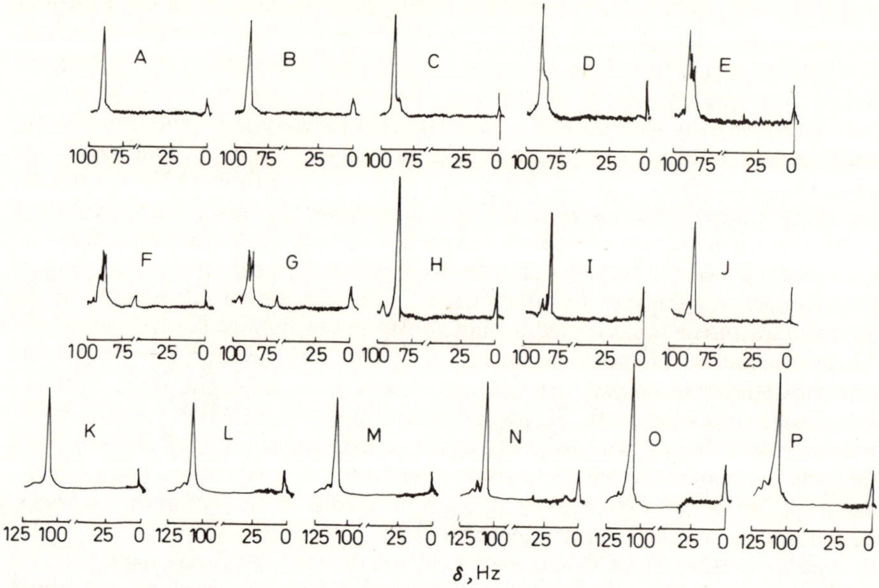

Figure 2. P.M.R. spectra of PEGs in benzene (A–J, 5 per cent; K–P, 50 per cent) at 35°C. (A) PEO–20 000, (B), PEO–4 000, (C) PEO–1 000, (D) PEO–600, (E) PEO–400, (F) heptamer, (G) hexamer, (H) pentamer, (I) tetramer, (J) trimer, (K) octamer, (L) heptamer, (M) hexamer, (N) pentamer, (P) trimer.

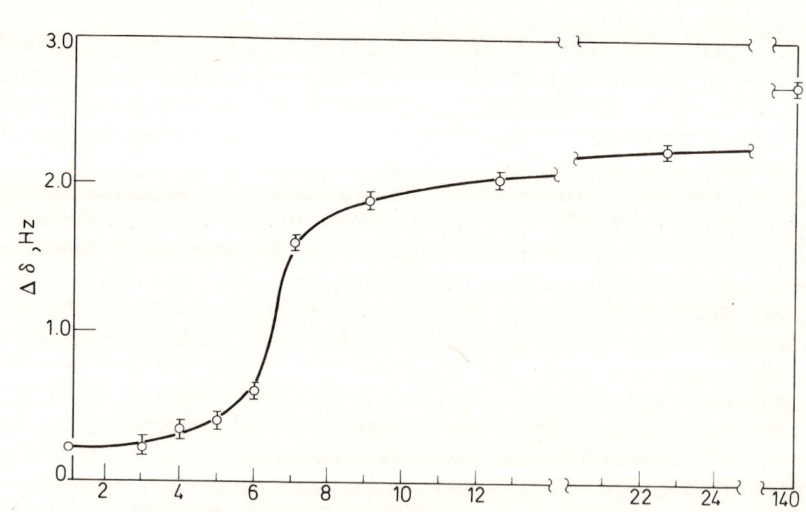

Figure 3. Plots of the difference of chemical shifts (with bulk magnetic susceptibility correction) of PEO (1 per cent) in methanol with and without potassium iodide (0.05 M) as a function of the chain length (n) at 35°C.

Additional evidence of ordered conformations of PEG is provided by chemical-shift studies of the PEG oligomers in methanol[10]. The difference in PEG chemical shifts in methanol with and without added potassium iodide is plotted as a function of chain length in *Figure 3*. It is evident that ion–dipole interactions with the lower PEG oligomers ($n \leqslant 6$) are substantially weaker than interactions with oligomers of greater chain length[11]. The distinct transition in these chemical-shift data with chain length is consistent with a new PEG conformation in chains containing seven or more units.

We have further carried out conformational studies of PEG by p.m.r. and i.r. measurements[12]. The p.m.r. results indicate that the average segmental environment of PEG is very sensitive to water concentration up to about 50 vol% and then remains almost constant for any further addition of water. Three molecules of water are required for the hydration of each ethylene oxide unit in a PEG chain. The temperature dependence of the hydrate formation in aqueous PEG solutions is small. Without specific interaction between the polymer and the solvent, the segmental environment changes linearly with the solvent content in chloroform or benzene over the entire composition range. In concentrated chloroform or benzene solution, PEG shows strong molecular association through intermolecular hydrogen bonds. The i.r. results contribute further information about the nature of the solvent effects on the conformational structures of PEG. Significant differences were observed in the i.r. spectrum of PEG in an aqueous medium as compared with that in benzene. The conformation of PEG is believed to be in a more ordered form in an aqueous medium than in either benzene solution or in the melt, and favours the TGT conformation for the COCCOC sequence. This specific conformation change of PEG in an aqueous medium was also reflected by a number of our proton magnetic relaxation measurements[13, 14].

ORGANIZED STRUCTURE INDUCED BY POLYMER–POLYMER INTERACTIONS

The association of stereo-regulated poly(methyl methacrylate) (PMMA) isomers is an interesting subject since the interaction is a consequence of the steric complementarity of the two interacting species, utilizing a principle similar to that which accounts for such striking specific interactions as those of an enzyme with its substrate or an antigen with its antibody[15]. Studies of this association phenomenon have been carried out by a number of investigators[16–23]. According to Liquori[19], the stereo-association between the tactic PMMAs is promoted in polar solvents through the affinity of the side methyl groups attached to the polymer backbones. Our studies on detailed molecular interactions indicate that the interaction between a nonpolar solvent benzene and the backbone of the PMMA is very weak because of steric hindrance of the polymer chain[24]. The association may therefore be expected to persist even in certain nonpolar media. Here, we report the results of our studies on these polymer–polymer interactions.

When relatively dilute benzene solutions of syndio- and iso- ($M_v = 9.32 \times 10^5$, iso-PMMA-1) PMMA were mixed, changes in solution viscosity occurred, which varied as a function of constituent polymer composition at constant total concentration. Results, obtained at equilibrium, are

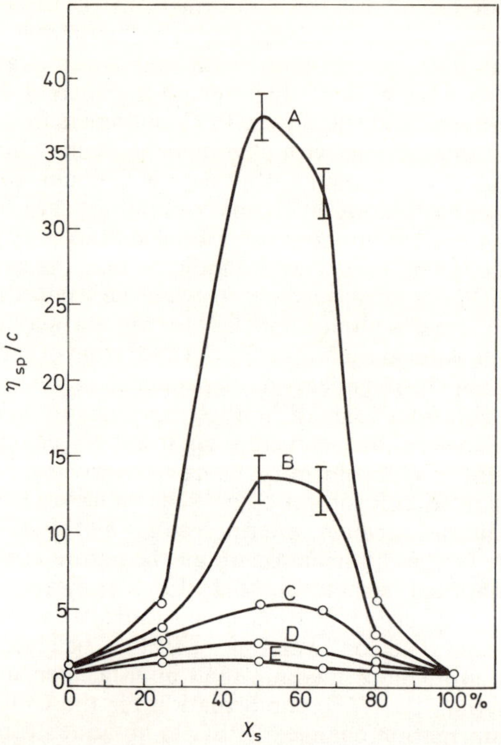

Figure 4. Plots of equilibrium reduced viscosity (η_{sp}/c) as a function of syndiotactic PMMA content (x_s) in the syndio- ($M_v = 1.1 \times 10^5$) and iso- ($M_v = 9.3 \times 10^5$) PMMA mixtures in benzene at 24°C. The total polymer concentrations are (A) 0·5 per cent, (B) 0.4 per cent, (C) 0.3 per cent, (D) 0.2 per cent and (E) 0.1 per cent.

presented in *Figure 4*. Since the association rate is slow, the reduced viscosity (η_{sp}/c) was linear in the composition of the mixed stereoregular PMMA immediately after mixing.

When the polymer concentration was less than 0.4 per cent, the maximum viscosity was found in samples containing a 1 : 1 syndio : iso ratio (*Figure 4*). More concentrated samples (0.4 and 0.5 per cent of PMMA in benzene) yielded values of reduced viscosity greater than 10 in some cases. The measurements were less accurate, as indicated in *Figure 4*, and it is difficult to identify precisely whether samples of PMMA of 2:1 or 1:1 syndio:iso ratio had the highest viscosity. Nevertheless, the enhancement of the reduced viscosity in the more concentrated samples is so overwhelming that the association phenomenon is easily apparent.

When a lower molecular weight iso-PMMA–2 ($M_v = 1.68 \times 10^5$) was used in the polymer mixtures, the maximum viscosity is definitely obtained with solutions with a 2 : 1 syndio: iso ratio at concentrations greater than 0.5 per cent; the maximum occurs again at 1 : 1 syndio : iso ratio for more dilute solutions.

An interesting aspect of the molecular weight dependence is that the 2 : 1 syndio : iso mixture (0.5 per cent polymer) with iso-PMMA–1 had a viscosity on initial preparation about double that obtained with the corresponding mixture containing iso-PMMA–2; on equilibration the viscosity ratio was greater than 20.

When a still lower molecular weight iso component was used ($M_v = 6.88 \times 10^4$, iso-PMMA–3), stereo-association between this iso-PMMA–3 and the syndio counterpart could hardly be detected, even in relatively concentrated samples (0.5 and 1.0 per cent of PMMA). These results clearly indicated that the effect of the stereo-association between the two tactic PMMAs on the viscosity behaviour of the polymer solution depends profoundly on molecular weight.

Plots of the reduced viscosity of the polymer mixture against concentration were characterized by a strong upward curvature. Such plots at a constant 2 : 1 syndio : iso PMMA ratio are shown in *Figure 5*. In sample mixtures containing iso-PMMA–1, the reduced viscosity increased sharply when the PMMA concentration was above 0.3 per cent (curve A, *Figure 5*). When the

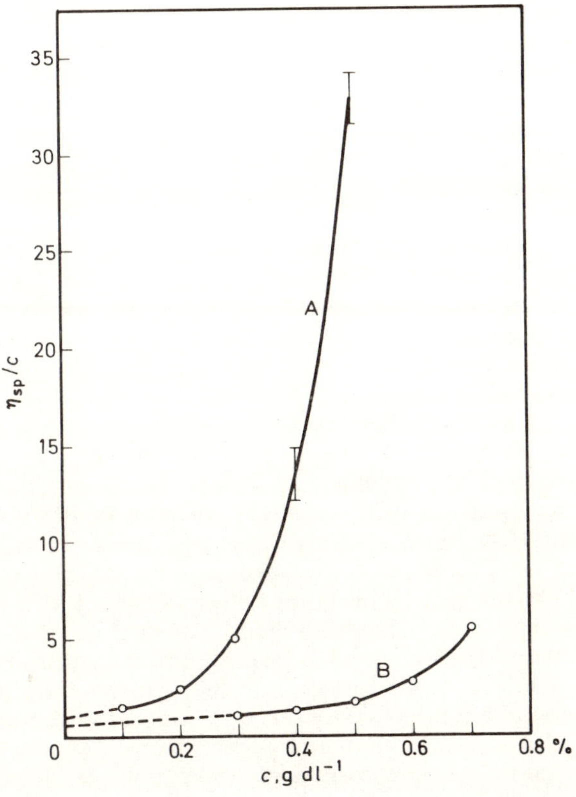

Figure 5. Plots of equilibrium reduced viscosity (η_{sp}/c) as a function of polymer concentration for samples containing a 2:1 syndio:iso PMMA ratio in benzene at 24°C; (A) syndio–iso-PMMA–1 system; (B) syndio–iso-PMMA–2 system.

lower molecular weight iso-PMMA–2 was used in the mixtures, the reduced viscosity did not rise appreciably until the PMMA concentration was above 0.6 per cent (curve B, *Figure 5*). This indicates that the effect of the polymer complexation on solution viscosity is highly dependent on polymer concentration as well as on molecular weight.

Stereochemical association between syndio- and iso-PMMAs was also observed in toluene solution, another nonpolar medium. In relatively more polar solvents, such as chlorobenzene and chloroform, molecular association between these tactic PMMAs was not detected by viscometric measurement. Plots of the reduced viscosity (at a constant polymer concentration of 0.5 per cent) as a function of the composition of these two tactic PMMAs were linear.

In solvents of higher polarity than those mentioned above, i.e. acetone and acetonitrile, phase separation of the iso–syndio-PMMA complex took place. However, viscometric measurements could be obtained in PMMA–acetone series before any sign of phase separation took place. The highest viscosity is found in the sample of a 2:1 syndio:iso PMMA ratio. It was interesting to note that this same sample was also the one which showed the earliest phase separation in this series.

Although the stereocomplex of PMMA precipitated from acetonitrile, the complex remained in solution in a mixture of 20 per cent chloroform and 80 per cent acetonitrile. A maximum viscosity is again found in the sample of a 2:1 syndio: iso PMMA ratio. By constrast, the addition of 20 per cent chloroform to benzene seems to eliminate the formation of the stereocomplex completely. Particularly striking are the results in the benzene–acetonitrile systems. The solvent system, made of 80 per cent acetonitrile and 20 per cent benzene, led, as expected, to complex formation of the two tactic PMMAs. However, it is surprising to note that a solvent system made of 20 per cent acetonitrile and 80 per cent benzene was unexpectedly found to destroy completely the stereocomplex. Further viscometric results indicate that, although both pure solvents support complex formation, no complex exists in mixtures of the two solvents over a certain concentration range of 50 per cent to 80 per cent of benzene.

Studies on the stereo-association of these tactic PMMAs using a high-resolution p.m.r. technique yielded further information on the nature of this PMMA stereocomplex. For simplicity, only two representative sets of high-resolution p.m.r. peaks of the ester-methyl protons of PMMA (5 per cent) in chloroform and in benzene with various syndio : iso ratios are shown in *Figure 6*. The polymeric peak was significantly reduced in the PMMA complex, and was affected by the composition of tactic PMMAs in benzene. This type of p.m.r. peak reduction reflects a restriction of local segmental motion when the PMMA complex is formed in benzene. On the other hand, the two tactic PMMAs independently coexist in chloroform; little reduction of the polymeric p.m.r. peak was observed in chloroform solutions. Similar results were also obtained in many other solvent systems. These p.m.r. results are in complete agreement with those obtained from viscometric determinations, regarding the presence or absence of the PMMA complex in various solutions.

It may be beneficial to mention here briefly the significance of employing high-resolution p.m.r. spectroscopy and viscometric measurements for the detection of association between polymers. Even very slight segmental association may well be detected by viscometric determinations, while this is difficult to reveal by a high-resolution p.m.r. technique. It is known that in systems of lightly crosslinked PMMA networks, the crosslinking strongly affects the viscosity of the swollen gel, but hardly influences the segmental motion[25]. Under the present experimental conditions, the p.m.r. technique appears to be more effective than viscometric determinations, only when the association involves a large number of polymer segments. Therefore, results obtained by these two techniques complement each other in a useful way to help to elucidate the intricate nature of the stereo-association between the two tactic polymers.

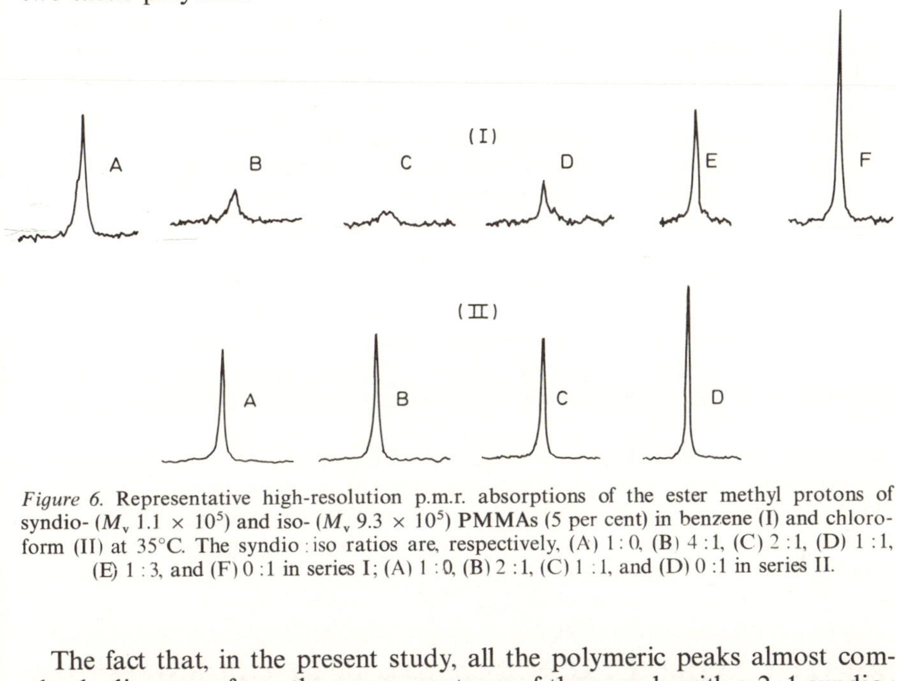

Figure 6. Representative high-resolution p.m.r. absorptions of the ester methyl protons of syndio- (M_v 1.1 × 10^5) and iso- (M_v 9.3 × 10^5) PMMAs (5 per cent) in benzene (I) and chloroform (II) at 35°C. The syndio : iso ratios are, respectively, (A) 1 : 0, (B) 4 :1, (C) 2 :1, (D) 1 :1, (E) 1 : 3, and (F) 0 :1 in series I; (A) 1 : 0, (B) 2 :1, (C) 1 :1, and (D) 0 :1 in series II.

The fact that, in the present study, all the polymeric peaks almost completely disappear from the p.m.r. spectrum of the sample with a 2 :1 syndio : iso PMMA ratio in complex-supporting solvents implies that the association between these two tactic PMMAs must be so strong as to restrict highly polymeric motion at the segmental level. There, according to the p.m.r. results, the formation of this PMMA stereocomplex probably involves the association of blocks of the polymer segments rather than certain isolated entanglements at a few points along the polymer chains.

It is important to note that viscometric data are by themselves insufficient to characterize the degree of polymer association. According to viscometric determinations, relatively dilute samples (0.5 and 1.0 per cent) of the syndio- and iso-(iso-PMMA–3) PMMAs in benzene gave little sign of stereocomplexation. However, this does not conclusively exclude the possible presence of the stereocomplex. P.M.R. data may be used to supplement the information obtained by viscometry.

It has been shown that when the ambient temperature is raised, the PMMA complex begins to dissociate; consequently, the segmental motion of the constituent tactic PMMAs should be expected to increase. This is true as illustrated in the p.m.r. spectra obtained at various high temperatures. The p.m.r. peaks gradually increase as the temperature is raised from 35 to 60°C. A sharp increase of the p.m.r. peak was seen as the temperature increased from 60 to 70°C. Melting of the PMMA complex is clearly demonstrated in these p.m.r. spectra. This agrees with what has been found from viscometric determinations.

We have further carried out x-ray diffraction measurements on samples of individual tactic PMMAs and 2:1 syndio:iso PMMA mixtures prepared under various conditions. The results can be summarized as follows: (*i*) All the mixed 2:1 syndio:iso PMMA solid samples prepared under various conditions (from acetone, acetonitrile and benzene solutions) give the same type III[26] x-ray diffraction pattern. (*ii*) A poor diffraction pattern of the PMMA mixture was obtained for the sample prepared by freeze drying the solution immediately after the mixing of the two tactic benzene solutions, while a sharp diffraction pattern was observed for the same sample obtained by freeze drying the mixed solution after a long period of standing time. (*iii*) The concentrated (50 per cent) 2:1 syndio:iso PMMA mixture in benzene yields the same diffraction pattern as those solid samples.

According to these x-ray results, it seems clear that the stereocomplexation of the two tactic PMMAs in benzene involves a crystallization of segments of PMMA chains, whose structure is identical with that of the solid crystal. It seems also clear that if a period of standing time is not allowed, the crystallization of the PMMA complex in dilute benzene solution is very poor, even when promoted by lowering the temperature to a dry-ice–acetone temperature. Although the growth of the organized structure cannot quantitatively be evaluated from the present x-ray measurements, it is clear that the structural organization of the PMMA complex in benzene is time-dependent. This is consistent with our previous viscosity and n.m.r. results.

SOLVENT INTERACTIONS IN SYSTEMS OF ORGANIZED AND RANDOM POLYMERS

Synthetic polypeptides are frequently used as models for proteins since they adopt similar conformational structures in solution. Transitions among these conformations have been extensively studied by techniques such as o.r.d. and dielectric relaxation. Recently, p.m.r. has thrown further light on this interesting subject. However, the emphasis has almost always been on the p.m.r. peaks of the polymer[27–33]. Very little attention has been paid to examining the chemical shifts of helix-breaking solvents, such as trifluoroacetic acid (TFA), in the system, although the important role of this solvent component for controlling the helix–coil transition of PBLG has been well recognized and discussed[32–38]. We therefore suggested studies of this nature[39]. This is particularly useful for polypeptides of high molecular weight and/or high polymer concentration, because no clear high-resolution p.m r. peaks of PBLG may be observed in the helical (or even partially helical) form under these conditions.

We measured chemical shifts of the solvent peak (TFA) in various poly (γ-benzyl-L-glutamate) (PBLG) solutions. These results were compared with the corresponding values of TFA in the solvent mixtures without the presence of the polypeptide. Obviously, the difference (Δ in Hz) is due to the presence of the polypeptide. A representative plot of Δ versus TFA content in solvent mixtures (TFA–benzene) is shown in *Figure 7*. The corresponding optical activity data are also included in the same figure for comparison. Clearly, Δ changes abruptly in the range when the polypeptide converts from an organized structure to a random coil.

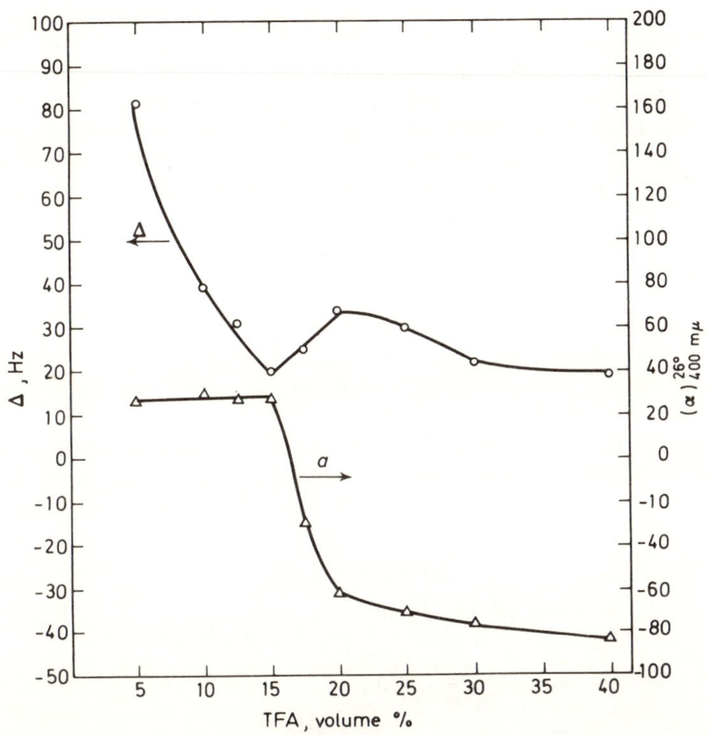

Figure 7. Plots of Δ and $[d]_{400\ m\mu}$ versus TFA content in benzene–TFA at 26°C in the presence of 10 per cent PBLG.

It is understandable that Δ decreases as the TFA content increases at constant polymer concentration, because the relative polymeric effect on the chemical shift of TFA decreases. This is true when the polypeptide is either in a helical form or in a random coil structure. Δ varies very differently when helical PBLG starts to transform into a random coil form; it actually increases with increasing TFA content. When the transformation has been completed, it varies normally again; it decreases with increasing TFA content.

This peculiar variation of Δ during a helix $\rightarrow$ random coil transition may be understood in terms of the different solvation behaviours of PBLG in the helical form and in the random coil structure. More TFA molecules are likely to be used for the solvation of PBLG in the random coil form than in the helical structure where intramolecular association prevails. Therefore, when the polypeptide starts to transform from a helix into a random coil, a certain excess amount of TFA is used, thus leading to a larger proportion of solvated TFA than is found at slightly lower TFA concentration. As a consequence, the observed relative chemical shift Δ increases rather than decreases with TFA, the latter being the normal course in the absence of the helix → random coil transition. Of course, when the transformation of PBLG from helix to random coil has been completed, the polypeptide is solvated as a random coil molecule, and further addition of TFA contributes only a normal concentration effect. This is exactly what has been experimentally observed.

Solvent interactions with organized and random macromolecules in systems other than synthetic polypeptides have also shown very interesting results. The spin–lattice relaxation time of the solvent benzene was measured as a function of tactic composition at constant total PMMA concentration. Our preliminary results indicate that the solvent relaxation time is affected more by the organized PMMA complex than the individual PMMA at the same polymer concentration.

CONCLUSION

In this presentation, we reviewed some of our work which was stimulated by the principles existing in biological macromolecules. It is not my intention to convince all my polymer colleagues to convert to biologists, although the temptation is high. I am only trying to show how we may learn more principles from nature to guide us for the future development of synthetic polymers.

REFERENCES

[1] R. Chiang, *J. Phys. Chem.* **69**, 1645 (1965).
[2] A. Abe, R. L. Jernigan and P. J. Flory, *J. Amer. Chem. Soc.* **88**, 631 (1966).
[3] K. J. Liu, *J. Polymer Sci.* **A2**, **5**, 1209 (1967).
[4] K. J. Liu, *J. Polymer Sci.* **A2**, **6**, 947 (1968).
[5] K. J. Liu, *Polymer (London)*, **10(12)**, 951 (1969).
[6] K. J. Liu and R. Ullman, *J. Polymer Sci.* **A2**, **6**, 451 (1968).
[7] R. F. Schaufele, *J. Chem. Phys.* **49**, 4168 (1968).
[8] K. J. Liu, *Macromolecules*, **1**, 213 (1968).
[9] K. J. Liu, *Makromol Chemie*, **116**, 146 (1968).
[10] K. J. Liu, *Macromolecules*, **1**, 308 (1968).
[11] R. D. Lundberg, F. E. Bailey and R. W. Callard, *J. Polymer Sci.* **A1**, **4**, 1563 (1966).
[12] K. J. Liu and J. L. Parson, *Macromolecules*, **2**, 529 (1969).
[13] K. J. Liu and J. E. Anderson, *Macromolecules*, **2**, 235 (1969).
[14] K. J. Liu and J. E. Anderson, *Macromolecules*, **3**, 163 (1970).
[15] H. Morawetz, *Macromolecules in Solution*, Interscience Publishers, New York (1966).
[16] W. H. Watanabe, C. F. Ryan, P. C. Fleisher and B. S. Garrett, *J. Phys. Chem.*, **65**, 896 (1961).
[17] D. L. Glusker, E. A. Galluccio, and R. A. Evans, *J. Amer. Chem. Soc.* **86**, 187 (1964).
[18] C. F. Ryan and P. C. Fleisher, *J. Phys. Chem.* **69**, 3384 (1965).
[19] A. M. Liquori. G. Anzuino, V. M. Coiro, M. D'Alagne, P. DeSantis, and M. Savino, *Nature*, **206**, 358 (1965).

[20] A. M. Liquori, M. DeSantis, M. Savino and M. D'Alagni, *Polymer Letters*, **4**, 945 (1966).
[21] R. Chiang, J. J. Burke, J. O. Threlkeld and T. A. Orofino, *J. Phys. Chem.* **70**, 3591 (1966).
[22] W. B. Van den Berg, B. Hijmans, P. Piet and D. Heikens, *Nature*, **217**, 948 (1968).
[23] H. Z. Liu and K. J. Liu, *Macromolecules*, **1**, 157 (1968).
[24] K. J. Liu, *J. Polymer Sci.* **A2**, **5**, 1199 (1967).
[25] K. J. Liu and W. J. Burlant, *J. Polymer Sci.* **A1**, **5**, 1407 (1967).
[26] T. G. Fox, B. S. Garrett, W. E. Goode, S. Gratch, J. F. Kincaid, A. Spell and J. D. Stroupe, *J. Amer. Chem. Soc.* **80**, 1768 (1958).
[27] F. A. Bovey, G. V. D. Tiers, and G. Filipovich, *J. Polym. Sci.* **38**, 73 (1959).
[28] D. I. Marlborough, K. G. Orrell, and H. N. Rydon, *Chem. Commun.* 518 (1965).
[29] J. L. Markley, D. H. Meadows, and O. Jardetzky, *J. Mol. Biol.* **27**, 25 (1967).
[30] K. J. Liu, J. S. Lignowski, and R. Ullman, *Biopolymers*, **5**, 375 (1967).
[31] E. M. Bradbury, C. Crane-Robinson, H. Goldman, and H. W. E. Rattle, *Nature*, **217**, 812 (1968).
[32] F. A. Bovey, *Pure Appl. Chem.* **16**, 417 (1968).
[33] J. H. Bradbury and M. D. Fenn, *Austral. J. Chem.* **22**, 357 (1969).
[34] W. E. Stewart, L. Mandelkern, and R. E. Glick, *Biochemistry*, **6**, 143, 150 (1967).
[35] S. Hanlon and L. M. Klotz, *Biochemistry*, **4**, 37 (1965).
[36] J. A. Ferretti and L. Paolillo, *Biopolymers*, **7**,155 (1969).
[37] J. Steigman, E. Peggion, and A. Casani, *J. Amer. Chem. Soc.* **91**, 1822 (1969).
[38] J. Steigman, A. S. Verdini, C. Montagner, and L. Strasorier, *J. Amer. Chem. Soc.*, **91**, 1829 (1969).
[39] K. J. Liu and S. J. Lignowski, *Biopolymers*, **9**, 739 (1970).

THEORY OF NUCLEATION AND CRYSTAL GROWTH OF POLYMERS IN CONCENTRATED SOLUTIONS

H. G. ZACHMANN

Institut für Physikalische Chemie der Universität Mainz, Mainz, BRD

ABSTRACT

The process of crystallization in concentrated solutions depends strongly on the rate of crystallization and the rate of long range diffusion of the polymer chains. If the crystallization proceeds slowly compared to diffusion, this type of crystal nucleus will be formed for which the free energy of nucleation is smallest. By taking into account entropy effects which are characteristic for chain molecules one can show that the crystal with smallest free energy of nucleation is a crystal with almost regular chain folds on the surfaces. The influence of supercooling, concentration and molecular weight on the thickness and growth rate of such crystals is discussed.

If the crystallization proceeds rapidly compared to the large range diffusion each part of the chain will crystallize as far as possible at the same place where it is lying in the solution. The nucleation process is not governed by thermodynamics. The type of crystal which is formed depends in this case on the concentration and the order of the chains in solution. A single parameter has to be introduced in the theory: the probability p that a new molecule is incorporated into the crystal. If p is large, crystals with loose loops with comparatively large end-to-end distances will be formed. With decreasing p the case of short regular folds with adjacent reentry is approached. In dilute solutions small values of p can be expected. With increasing concentration p increases, provided that the polymer molecules penetrate each other.

A. INTRODUCTION

Experiments have shown that polymers crystallize from dilute solutions in the form of lamellae. The lamellae have a uniform thickness of about 100 Å. The molecules are oriented perpendicular to the basal plane of each lamella and fold back at the upper and lower surface of the lamella. Possible models for the arrangement of the chains in the crystals are shown in *Figure 1*. Some investigators[1] assume that the folds are regular and as short as possible as indicated in *Figure 1a* (adjacent reentry model), others[2,3] think of loose loops as shown in *Figure 1b* ('switch-board' model). In addition to the folds or loose loops chains with one free end ('cilia') have also to be assumed[4].

Lamellar crystals are also formed in concentrated solutions. Here however the crystals are often bound together forming bigger units which, in some cases, show a structure more or less similar to that of the spherulites obtained from the melt[5]. As a consequence, in addition to loops, folds and chains with free

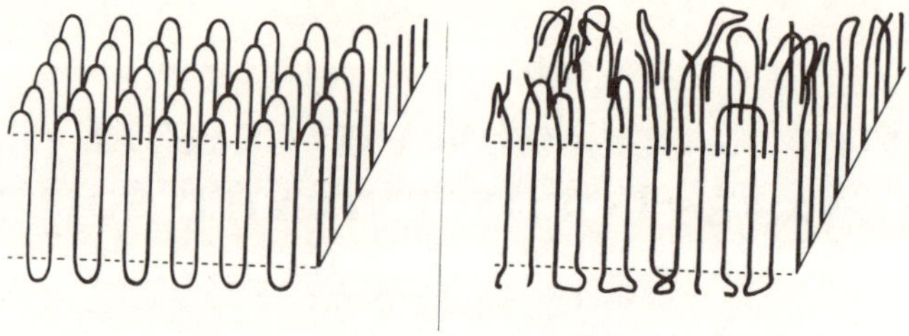

Figure 1. Possible models for the arrangement of the molecules in single crystals grown from solutions[15].

ends, tie molecules connecting two different crystals may also emerge from the crystal surfaces.

How can these results be explained? The theory of the crystallization of polymers has been developed in a number of papers. The first investigations were published by Hoffman and Lauritzen[6]. These authors *assumed* that crystals with regular folds are formed. With the help of this assumption they showed that the crystals must have uniform thickness, namely the thickness of the critical nucleus. A more refined calculation of crystal thickness, taking into account variation in the thickness as well, was performed later by Frank and Tosi[7], Price[8], and Lauritzen, Di Marzio and Passaglia[9]. Quite recently, Sanchez and Di Marzio[10–12] extended the theory to polymers of finite molecular weight and calculated the influence of molecular weight, as well as of the concentration, on the crystal growth rate. But none of these theories explains chain folding: they are theories on crystal growth rates, distribution of crystal thickness, etc. under the assumption that folding of the chains occurs.

How can chain folding itself be explained? In dilute solutions, chain folding results simply from the fact that the molecules crystallize one by one. If a single molecule forms a crystal it *has* to fold more or less regularly. This is not true, however, for concentrated solutions. Here, allowance must be made for two or more molecules crystallizing at the same location. As a consequence, the formation of bundle-like crystals is possible and seems quite probable. Are there also thermodynamic or kinetic reasons which favour chain folding in a system of concentrated chain molecules? If so, what is the most probable distribution of the lengths and the end-to-end distances of the loops?

These questions have been investigated extensively in our laboratory in recent years. It has been shown that chain folding can occur under certain conditions both in concentrated solutions and in the melt. It is a consequence of entropy effects which are characteristic for chain molecules. One has to distinguish between two cases:

(1) Crystallization proceeds slowly compared to long range diffusion. In this case crystals with regular folds are formed because such crystals have a minimum of free enthalpy of nucleation[13,14].

(2) Crystallization proceeds rapidly compared to long range diffusion. In this case each part of a molecule is crystallizing near the place where it is lying in the solution or melt respectively. In dilute solutions, again, crystals with regular folds are formed. In concentrated solutions and in the melt, however,

regular folds are formed only if the molecules do not penetrate each other before crystallization[15], otherwise mainly loose loops and tie molecules are found.

In the following we will review more extensively all the theories mentioned. We shall not discuss them in historical order but shall combine them in such a way that we obtain a systematic theory for the crystallization of concentrated solutions.

B. THEORY OF SLOW CRYSTALLIZATION

1. Which type of crystal is formed?

In a concentrated solution, in principle different types of crystals can be formed: bundle-like crystals (*Figure 2a*), crystals with irregular loose loops (*Figure 2b*), crystals with regular folds (*Figure 2c*), and crystals which are connected to other crystals by tie molecules (*Figure 2d*). Of course, a mixture

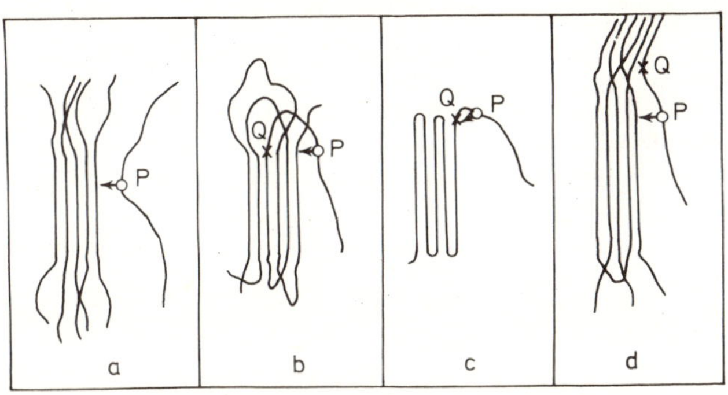

Figure 2. Different types of crystals which may be formed in a concentrated solution.

of all these types can also occur. Which type will actually be formed?

In order to answer this question we have to apply the theory of nucleation. The number of crystal nuclei formed in unit time is given[16] by

$$I = \text{const.} \exp(-\Delta G_A^*/kT) \tag{1}$$

ΔG_A^* is the free enthalpy of nucleation, k is the Boltzmann constant and T is the temperature. We will first calculate ΔG_A^* for a cylindrical crystal of small molecules. The free enthalpy necessary to form such a crystal is given by[6]

$$\Delta G = \pi\rho^2 l\rho_k \, \Delta G^\circ + 2\pi\rho^2\sigma_e + 2\pi\rho l\sigma_t \tag{2}$$

ρ is the radius of the base, l the height, ρ_k the density, σ_t and σ_e the free surface energy at the lateral surface and the two other surfaces respectively. ΔG° is the difference in bulk free enthalpy between the crystal and the melt per unit of mass. ΔG as a function of l and ρ is represented by a plane with a saddle point (see *Figure 3*). For small crystals ΔG increases with increasing l and ρ due to the formation of additional surface, crystal growth is not favoured thermodynamically. After having reached a critical size, ΔG decreases with further growth, and further crystallization is thermodynamically favoured. The amount of free enthalpy ΔG_A, which is necessary to reach the critical size so

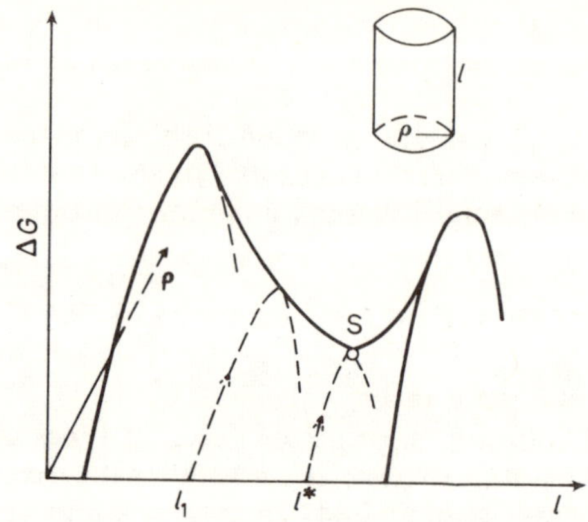

Figure 3. Amount of free enthalpy ΔG for formation of a cylindrical crystal as a function of the crystal dimensions l and ρ.

that further growth is favoured, depends on l. For a certain value of l, denoted by l^*, ΔG_A assumes a minimum ΔG_A^*, called free enthalpy of nucleation. ΔG_A^*, l^* and ρ^* are the values corresponding to the saddle point S. One obtains[6, 17]

$$\Delta G_A^* = \frac{8\pi\sigma_t^2\sigma_e}{\rho_k(\Delta G^\circ)^2} \tag{3}$$

$$l^* = -\frac{4\sigma_e}{\rho_k \Delta G^\circ} \tag{4}$$

and

$$\rho^* = -\frac{2\sigma_t}{\rho_k \Delta G^\circ} \tag{5}$$

In the case of long chain molecules the situation is more complicated. The essential difference between the crystallization of small molecules and of long chain molecules is illustrated in *Figure 4.* With small molecules, in an elementary step of crystallization, it is always a complete molecule which passes from the solution into the crystal (left side of *Figure 4*). The entropy change caused by crystallization is the same for each molecule. With polymers, in an elementary step of crystallization only a small part of the molecule, say a monomer unit, crystallizes (see *Figure 4*, right side). The change of entropy connected with this process is different from unit to unit. This is due to the fact that not only the entropy of the crystallizing unit is changed but usually also the entropy of the rest of the chain which is not crystallized. This can be seen very clearly in the process indicated in *Figure 2b.* When the unit P crystallizes the chain between Q and P becomes a chain with two fixed ends. Before this process it had one end, namely P, free. Therefore the number of conformations and the entropy of the chain QP decrease markedly when P crystallizes.

The entropy change mentioned, ΔS_k, of the part of the chain which is not crystallized but is hanging out of the crystal, varies for different types of

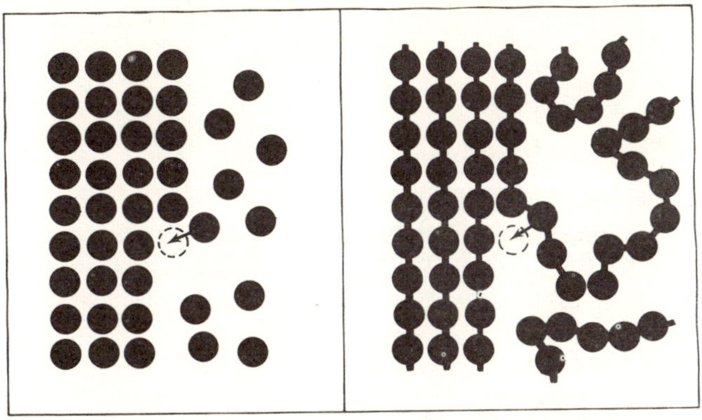

Figure 4. Elementary step of crystallization in a low molecular weight substance and in a polymer.

crystals*. In order to take this effect into account, equations 2–5 have to be modified[13,14]. If ΔS_k is the same for each chain the total additional change of entropy during formation of a crystal is proportional to the number of chains coming out of the surface of the cylinder. Therefore this entropy acts in the same way as an additional contribution to the surface free energy σ_e. If we denote this additional contribution by σ_e' we can write

$$\sigma_e' = \nu_0 k T \Delta S_k \tag{6}$$

where ν_0 is the number of chains emerging from the unit area. Therefore, in order to take the entropy effect mentioned into account we have simply to replace σ_e by $\sigma_e + \sigma_e'$ in equations 2–5. In particular, for the free enthalpy of nucleation we obtain[13]

$$\Delta G_A^* = \frac{8\pi\sigma_t^2(\sigma_e + \sigma_e')}{\rho_k (\Delta G^\circ)^2}. \tag{7}$$

We will now show that for various types of nuclei σ_e' is different[14]. Therefore according to equations 7 and 1 ΔG_A^* and the nucleation rate I are different. This type of crystal, which has the lowest value of σ_e' and therewith the largest value of I, will be formed most quickly and therefore will dominate.

We consider at first the growing of a bundle-like crystal composed of chains hanging out of another crystal. Such a process is indicated in *Figure 2d*. Every time a chain is added to the growing crystal, a chain with two fixed ends (tie molecule) is created, for example the chain QP in *Figure 2d*. The fixing of the chain end is connected with a big loss in entropy ΔS_k. *Figure 5* gives this entropy change as a function of chain length measured in number of units of the chain. The parameter h is the end-to-end distance. When one assumes for example an end-to-end distance of 100 nm and a chain length of 600 nm one obtains $\Delta S_k = 25\,k$. This corresponds in the case of polyethylene to an additional surface free energy $\sigma_e' = 700\ \text{erg cm}^{-2}$.

We consider next a crystal with loose loops and nonadjacent reentry (*Figure 2b*). Here also, during the growth process, chains with fixed ends are

* The considered change of entropy of the part of the molecule which is not crystallized is always negative. Therefore the introduced decrease of entropy, ΔS_k, is always positive.

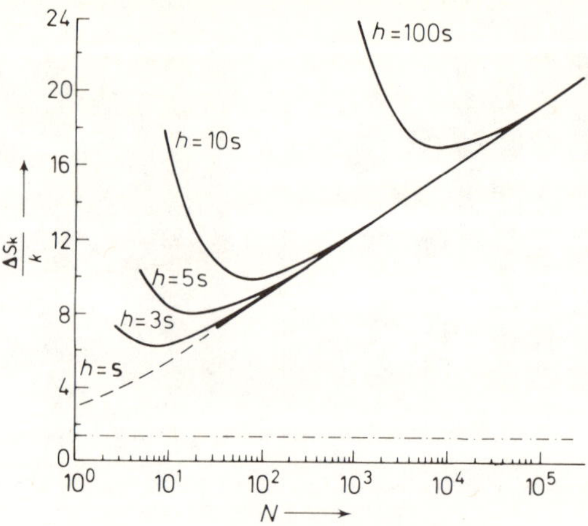

Figure 5. Decrease of entropy occurring with fixing the ends of a chain in a distance h as a function of the number N of monomer units of the chain[13].

created at the crystal surfaces, namely the loops. However, the end-to-end distance of the chain-loops can be considerably smaller than in the first case. Therefore σ'_e also is not so large. When $h = 10$ and $N = 50$ one obtains $\Delta S_k = 9k$ and $\sigma'_e \approx 400$ erg cm^{-2}. σ'_e decreases further with decreasing values of h.

If the loop length N and end-to-end distance h assume the smallest possible values, the loose loop model goes over into the adjacent reentry model shown in *Figure 2c*. The shortest loop with adjacent reentry consists in the case of polyethylene of 4 CH_2-units which are outside the crystal lattice. The fixing of the chain end, in this case, gives rise[18] to a surface free energy of only 30 erg cm^{-2}. We have also to consider that there is an increased number of gauche-conformations in such a short loop, which corresponds to about 70 erg cm^{-2}. Therefore the total additional surface free energy σ'_e is about 100 erg cm^{-2}. A slight increase of loop length and end-to-end distance might decrease σ'_e a little. This cannot be calculated exactly because the chain statistics applied in our calculations are not valid for very short chains. In any event, however, comparatively large loops and end-to-end distances lead to much higher values of σ'_e, as shown above. Therefore one can say that in the case of chain folding the smallest value of σ'_e is obtained for comparatively short loops with adjacent or almost adjacent reentry.

Lastly we consider a bundle-like crystal formed of molecules which were completely amorphous before they were incorporated into the crystal (*Figure 2a*). In this case, on both surfaces chains with one free end hang out of the crystal. Each of these chains loses entropy when the molecule to which it belongs is attached to the crystal. There are two reasons for the entropy loss:
(*i*) The number of conformations of the chain considered is reduced because the volume available to this chain is limited by the crystal. Owing to this effect conformations like that shown in *Figure 6* are no longer possible. *Figure 7* shows the loss of entropy ΔS_k as a function of chain length. If, for example, the

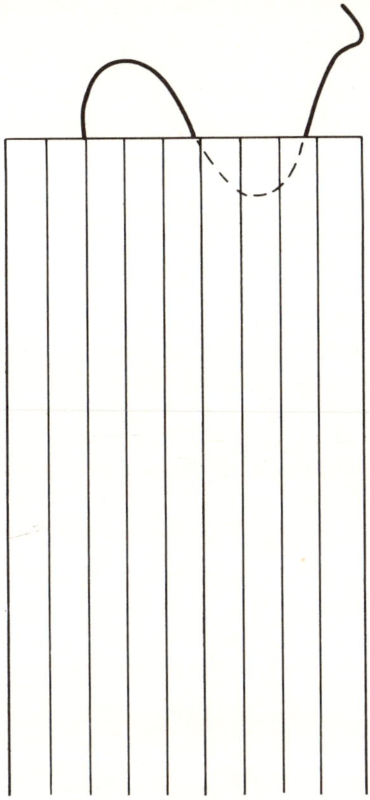

Figure 6. Example of an impossible conformation of a chain hanging out of a crystal[19].

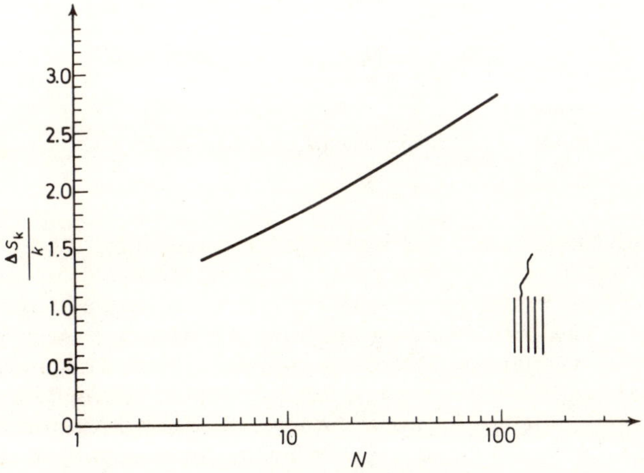

Figure 7. Decrease of the entropy of a chain with one free end caused by the limitation of the volume as a function of the number of units of the chain[20].

chain hanging out of the crystal consists of 1000 units, the loss of entropy $\Delta S_k = 4k$ corresponding to an additional surface free energy of 120 erg cm^{-2}. (*ii*) The chain as a whole can no longer perform a translational diffusion after part of it is incorporated into the crystal. The entropy caused by diffusion is small compared to the conformational entropy of the whole chain. It is not small however compared to the conformational entropy of a few units which are incorporated into the crystal. Therefore it cannot be neglected in nucleation processes. The considered entropy change is given according to Flory–Huggins theory by[21]

$$\Delta S_t = k \ln N_0/v_2 \tag{8}$$

where N_0 is the number of units of the chain and v_2 the mole fraction of polymer. For $N_0 = 2000$ and concentrated solution the factor v_2 can be neglected and one obtains $\Delta S_t/2 = 3.80\,k$. This value corresponds to an additional surface free energy of $\sigma_e' \approx 120$ erg cm^{-2}. Both effects together give a value of $\sigma_e = 240$ erg cm^{-2}. According to Hoffman and Lauritzen[22] one has to consider also a third contribution of about 100 erg cm^{-2} from the special packing conditions.

The results obtained are summarized in *Figure 8*. Here again one can see

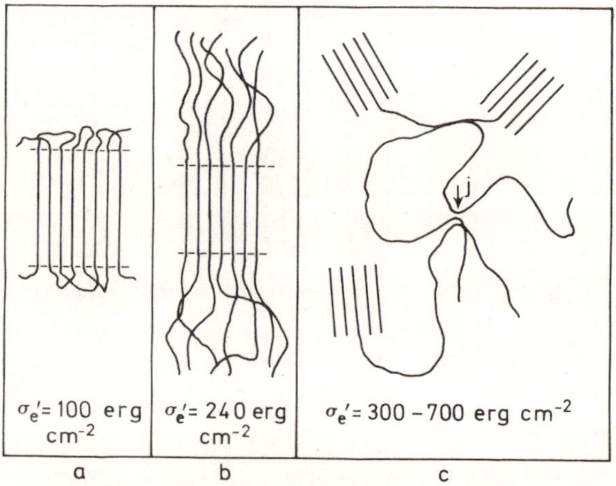

Figure 8. Additional surface free energies for different types of crystals[14].

the crystal with short loops and almost adjacent reentry, the bundle-like crystal formed of chains which were completely amorphous before incorporation into the crystal, and a crystal which is formed of chains hanging out of another crystal. The corresponding additional surface free energies σ_e' are given below each crystal type. One sees that the crystal with folds has the smallest value of σ_e'. We have to stress here, that in our estimation of the σ_e' values we have already got an upper limit for the folded chain crystal and lower limits for the other types of crystals. From this result we conclude that the crystal with short folds with almost adjacent reentry is thermodynamically favoured.

2. Crystal thickness

After having seen that crystals with almost regular folds are also formed in concentrated solution, we can apply the theories developed for folded chain crystallization. We will first calculate the thickness of the crystals according to the theory of Lauritzen and Hoffman[6]. The theory starts with the statement that crystals with chain folds can grow only in a lateral direction; an increase in thickness is combined with difficult diffusion and is very slow. Therefore during crystal growth l is constant and ρ is increased. The change in free enthalpy during crystal growth with a constant value of l is indicated for some values of l by the dashed lines in *Figure 3*. One sees that the amount ΔG_A of free enthalpy which is necessary to create a stable crystal varies with l. ΔG_A assumes the minimum value ΔG_A^* if l equals the thickness l^* of the critical nucleus defined by the saddle point. In principle crystals with all values of l will be formed. But according to equation 1 the nucleation rate for crystals of different thickness l is given by

$$I = \text{const.} \exp(-\Delta G_A/kT) \qquad (9)$$

Therefore the overwhelming number of crystals will have the thickness l^* for which ΔG_A is at a minimum and I at a maximum.

As mentioned in the Introduction, this theory has been refined by considering also variations in thickness during growth. Without going into details, we show only some results. The curves in *Figure 9* show the calculated thickness

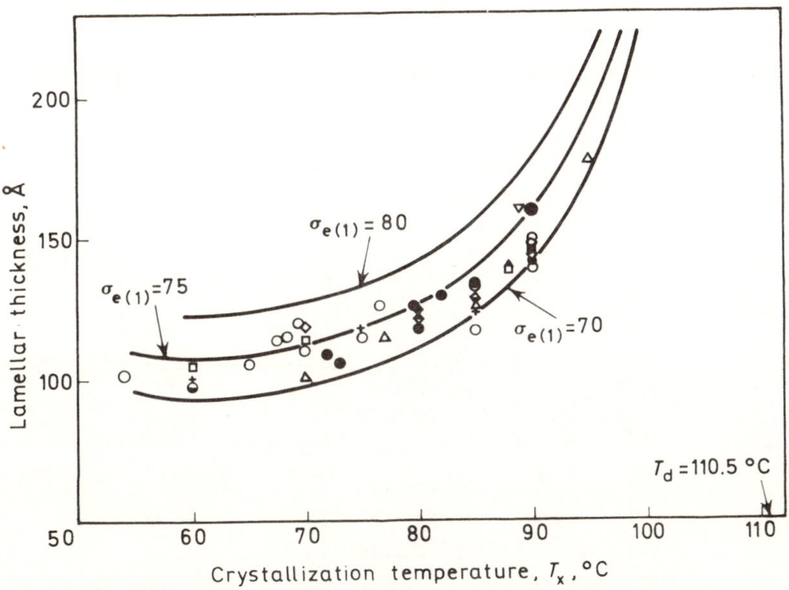

Figure 9. Lamellar thickness of chain-folded polyethylene single crystals as a function of crystallization temperature. The curves represent calculated values for different values of surface free energy σ_e. The points represent the measured values[23].

of the lamellae as a function of crystallization temperature for different values of the surface free energy. The points have been obtained experimentally for crystallizations from dilute solutions. No experimental results are known for concentrated solutions.

3. Growth rate

Assuming chain folding, the theory of Sanchez and DiMarzio[10–12] for crystal growth rates can also be applied. Growth rate is governed mainly by the rate of secondary nucleation. Sanchez and DiMarzio distinguish between two different nucleation mechanisms: nucleation by cilia, that is by chain ends hanging out of the crystal, and nucleation by chains which lie completely in the melt. If S^c and S^s are the two nucleation rates respectively, and w^c and w^s the probabilities for the two processes, the total nucleation rate is given by

$$S = w^c S^c + w^s S^s \tag{10}$$

S^c and S^s have been calculated separately as a function of molecular weight and concentration for the case of polyethylene in *p*-xylene. The results are shown in *Figure 10*. One sees that both S^c and S^s increase with increasing

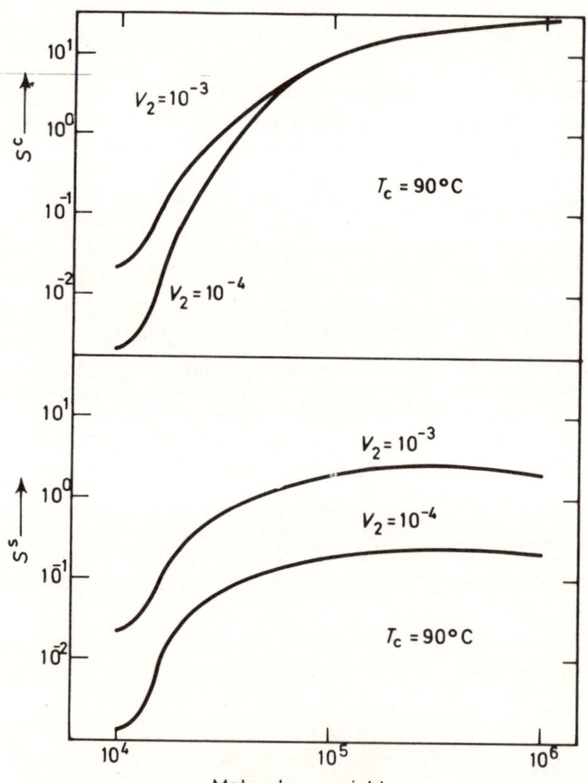

Figure 10. The relative effect of concentration on nucleation rate of solution molecules S^s and cilia S^c as a function of molecular weight[11].

molecular weight. The influence of concentration on S^c is not the same as on S^s. S^s increases with concentration, S^c however is almost independent of concentration. Unfortunately only very low concentrations have been considered in the calculation.

The values w^s and w^f have not been calculated by Sanchez and DiMarzio. According to our considerations in part 1, w^s has to be much smaller than w^f.

C. THEORY OF RAPID CRYSTALLIZATION

We will next assume that crystallization proceeds rapidly compared to long range diffusion[15]. In this case a part of the molecule which is reached by the crystal surface will be incorporated by the crystal, independently of whether the thermodynamically most favoured configuration is reached by this process. In *Figure 11* we see two molecules penetrating each other, before

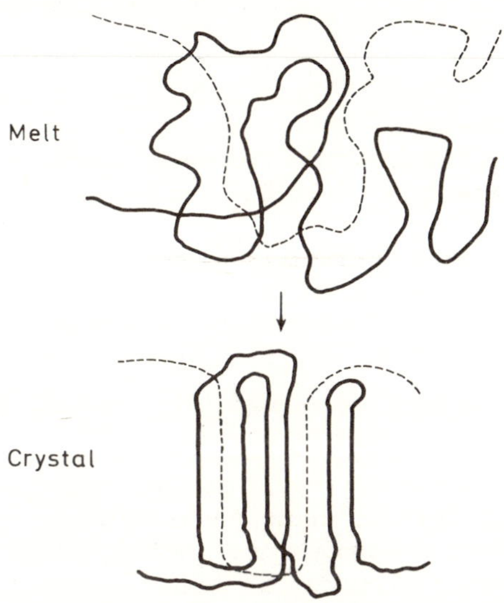

Figure 11. Schematic representation of the crystallization process if crystallization proceeds rapidly compared to long range diffusion[15].

crystallization (above) and after crystallization (below). Generally, those parts of the molecules which are neighbours in the concentrated solution remain also neighbours in the crystal.

Figure 12 illustrates the different crystallization processes we have to consider. Crystal growth has proceeded so far that at location P a chain has to be added. In any stage of crystallization one will find on the crystal surface sharp folds, loops with nonadjacent reentry and chains with one free end. The amounts of the different types of chains may vary but in principle all these types have to be considered. We have to distinguish between four possible processes:

(a) A part of a new chain, which before lay completely in the melt, is incorporated in the crystal. In this case a chain with one free end is created both in the upper and in the lower surface.

(b) A part of a chain hanging out of the upper surface can be incorporated in

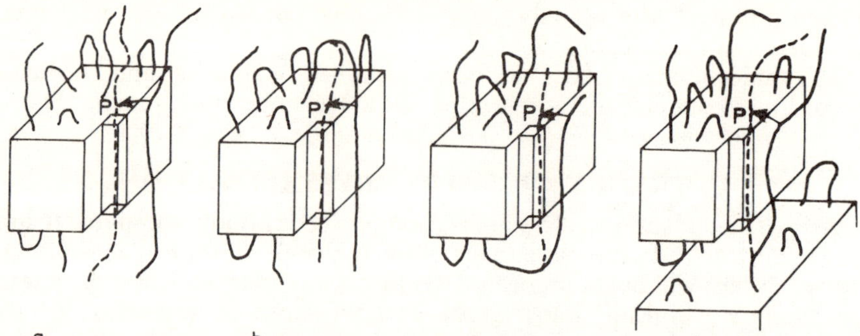

Figure 12. The different processes during crystal growth[15].

the crystal. In this case a loop is formed at the upper surface and a chain with one end free at the lower surface.
(c) A part of a chain hanging out of the lower surface can be incorporated, thus forming a loop in the lower surface and a chain with a free end in the upper surface.
(d) A chain hanging out of another crystal can be crystallized, thus creating a tie molecule and a chain with one free end.
For the following calculation we need only one parameter for the characterization of the system, which we will call p; p is given by

$$p = p_1 + \tfrac{1}{2}p_2 \tag{11}$$

where p_1 is the probability that a new molecule, previously lying completely in the solution, is added to the crystal, and p_2 is the probability that a tie molecule is formed. In the case of dilute solutions p will be almost zero because no other molecule is in the vicinity of the molecule considered. In the case of concentrated solutions p increases with increasing concentration, provided that the molecules penetrate each other. If the molecules do not penetrate each other p is also small for concentrated solutions. Using this parameter we will derive the end-to-end distance distribution function $\omega(h)$ of the loops, that is, the number of loops which have an end-to-end distance between h and $h + 1$. The meaning of h can be seen in *Figure 13*.

For the calculations the following assumption is made: the probability that a loop of end-to-end distance h is formed at a certain point P is given by the number of possibilities of forming such a loop divided by the number of possibilities of forming a loop of any end-to-end distance. If $\tilde{\mu}(h)$ denotes the density of chains with one end free in distance h from point P one can write

$$\omega(h) = \frac{\tilde{\mu}(h)\, h \sum_N Z^*(N, h)}{\int_a^{h_{max}} \tilde{\mu}(h) \sum Z^*(N, h)\, \mathrm{d}h} \tag{12}$$

$Z^*(N, h)$ is the number of conformations of a chain of length N and end-to-end distance h. h_{max} is the average length of chains with one free end and a the

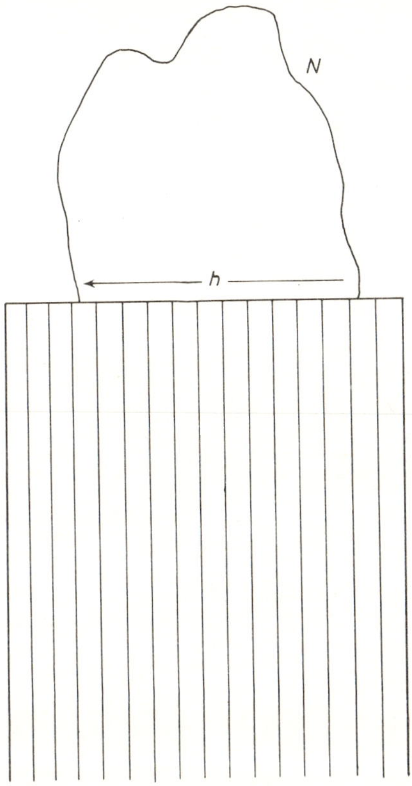

Figure 13. Definition of h and N for a loop[15].

distance of neighbouring chains in the crystal. The function $\tilde{\mu}(h)$ is not known. It can be derived however from the condition that the numbers of chain ends which disappear must be equal to the number of loops formed. This condition yields a differential equation

$$\tilde{\mu}'(h) + \beta(1 - p)\sum_N Z^*(N, h)\,\tilde{\mu}(h) = 0 \tag{13}$$

with

$$\beta = \nu\left[Z\tilde{a}b\pi\int_a^{h_{\max}}\tilde{\mu}(h)h\sum_N Z^*(N, h)\,\mathrm{d}h\right] \tag{14}$$

By solving this equation $\tilde{\mu}(h)$ is determined and then with help of equation 12 the function $\omega(h)$ is obtained.

The influence of the probability p on the end-to-end distribution function $\omega(h)$ for molecules consisting of $N_0 = 2000$ units is shown in *Figure 14*. If p has a very small value, for example 10^{-10}, which is the case in a dilute solution or in a concentrated solution in which the molecules do not penetrate each other, almost all molecules have an end-to-end distance which is smaller than

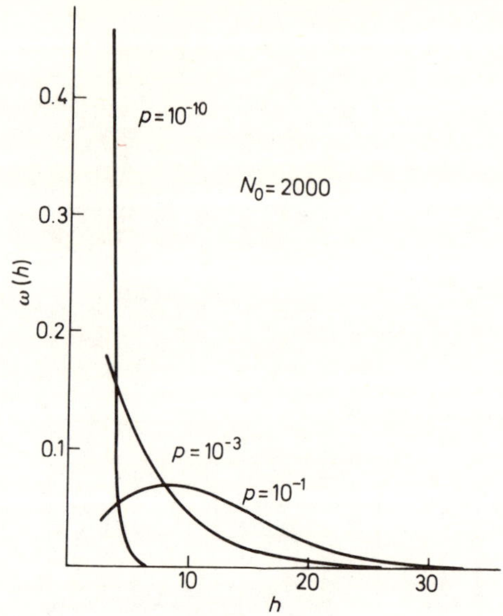

Figure 14. End-to-end distance distribution function ω(*h*) for different values of the probability *p*. N_0 = number of monomer units of a molecule.

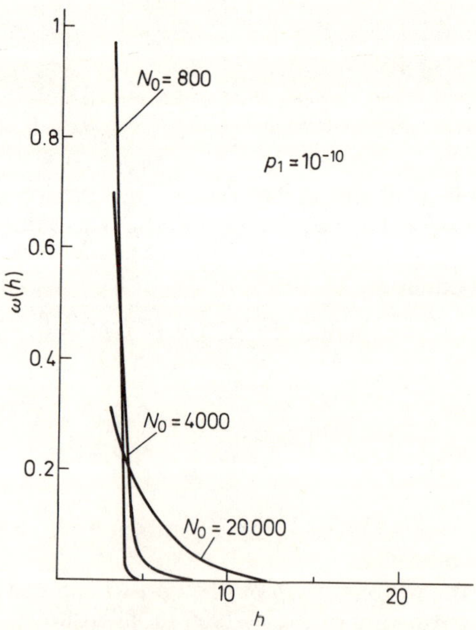

Figure 15. End-to-end distance distribution function ω(*h*) for molecules of different length N_0 when $p = 10^{-10}$. The length N_0 is measured in monomer units[15].

5. This means that one has almost adjacent reentry. With larger values of p the distribution becomes broader and loops with greater and greater end-to-end distances occur. p increases when the concentration increases and the molecules penetrate each other. As the molecules actually penetrate each other under usual conditions, one can conclude that, with increasing concentration of the solution, the end-to-end distances of the loops become larger.

The end-to-end distance distribution is also influenced by the length of the molecules, that is by the molecular weight. *Figure 15* shows $\omega(h)$ for molecules of different lengths N_0 when p equals 10^{-10}. With increasing lengths the end-to-end distances become larger. The same is true also for other values of p as shown in *Figure 16*.

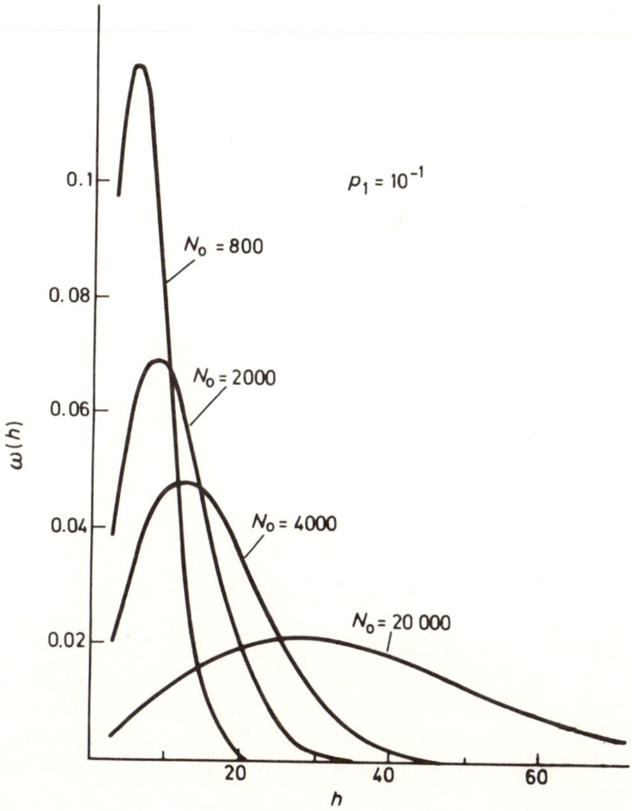

Figure 16. End-to-end distance distribution function $\omega(h)$ for molecules of different length N_0 when $p = 10^{-1}$. The length N_0 is measured in monomer units[15].

Up to now, we have discussed only the end-to-end distances. The loop lengths have not been calculated. This is because the kinetic factors during crystallization do not influence the final loop lengths. The end-to-end distance distribution established during crystallization cannot be changed later without melting the whole crystal. The loop lengths however can always be changed by comparatively simple diffusion processes and the melting of small surface layers of the crystal. Therefore the final loop lengths are not determined by kinetic but by thermodynamic factors. The kinetic parameter p influences the loop length only as far as it determines the end-to-end distances of the loops.

Calculations of the average equilibrium loop length assuming fixed ends of the loops* have been performed by methods described elsewhere[15,18,24]. Some results are shown in *Figures 17* and *18*. The average loop length $<N>$ increases with increasing temperature because of partial melting. Compared at the same temperature, $<N>$ increases with increasing p and N_0. From these results we can conclude that the loop lengths and the melting range of the crystals will increase with increasing concentration of the solution in which the crystals were grown.

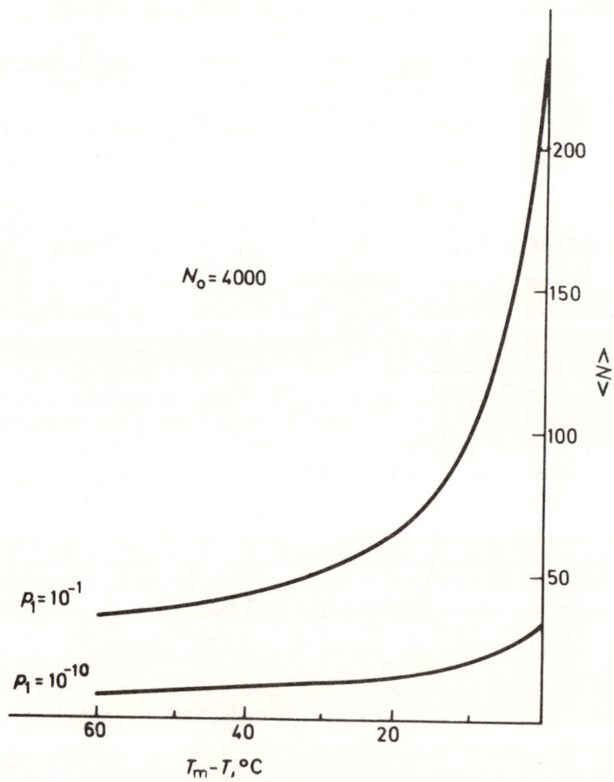

Figure 17. Average loop length $<N>$ as a function of temperature for molecules consisting of N_0 = 4000 monomer units[15].

* Calculations of the end-to-end distance distribution functions and the lengths of the loops when the end-to-end distance of each loop may vary, have been performed by Roe[25] as well as by Zachmann and Peterlin[24]. We do not think, however, that the assumption of various end-to-end distances is fulfilled under usual experimental conditions (see discussion in Ref. 25).

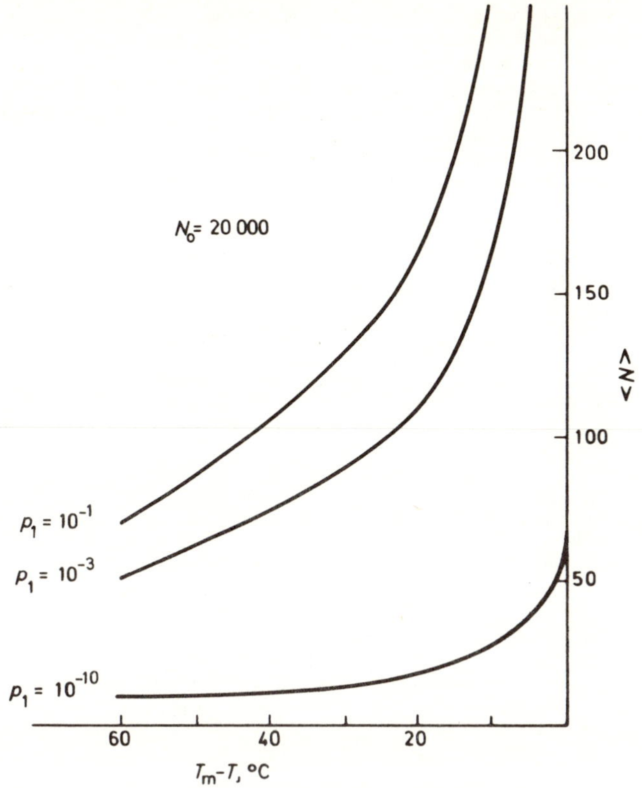

Figure 18. Average loop length $<N>$ as a function of temperature for molecules consisting of $N_0 = 2000$ monomer units[15].

REFERENCES

1 A. Keller, *Phil. Mag.* **2**, 1171 (1957).
2 P. I. Flory, *J. Amer. Chem. Soc.* **84**, 2857 (1962).
3 E. W. Fischer and R. Lorenz, *Kolloid-Z. Z. Polymere* **189**, 97 (1963).
4 P. H. Lindenmayer, *Kolloid-Z. Z. Polymere* **231**, 593 (1969).
5 F. Khoury, *J. Res. Nat. Bur. Standards* **70A**, 29 (1966); I. Heber, *Kolloid-Z. Z. Polymere* **189**, 112–114 (1963).
6 J. I. Lauritzen Jr and J. D. Hoffman, *J. Res. Nat. Bur Standards* **64A**, 73 (1960).
7 F. C. Frank and M. Tosi, *Proc. Roy. Soc.(London)* **A263**, 323 (1961).
8 F. P. Price, *J. Chem. Phys.* **35**, 1884 (1961).
9 J. I. Lauritzen Jr, E. A. Di Marzio and E. Passaglia, *J. Chem. Phys.* **45**, 4444 (1969); J. I. Lauritzen Jr and E. Passaglia, *J. Res. Nat. Bur. Standards* **71A**, 261 (1967).
10 I. C. Sanchez and E. A. Di Marzio, *J. Chem. Phys.* **55**, 893 (1971).
11 I. C. Sanchez and E. A. Di Marzio, *Macromolecules* **4**, 677 (1971).
12 I. C. Sanchez and E. A. Di Marzio, *J. Res. Nat. Bur. Standards* **76A**, 231 (1972).
13 H. G. Zachmann, *Z. Naturforschung*, **20a**, 719 (1965).
14 H. G. Zachmann, *Kolloid-Z. Z. Polymere*, **216–217**, 180 (1967).
15 W. M. Ewers, H. G. Zachmann and A. Peterlin, *Kolloid-Z. Z. Polymere*, **250**, 1187 (1972).
16 D. Turnbull and J. C. Fisher, *J. Chem. Phys.* **17**, 71 (1949).
17 H. G. Zachmann, *Fortschr. Hochpolym. Forsch.* **3**, 581 (1964).
18 H. G. Zachmann and A. Peterlin, *J. Macromol. Sci.* **B3**(3), 495 (1969).

[19] H. G. Zachmann, *Kolloid-Z. Z. Polymere*, **231**, 504 (1969).
[20] H. G. Zachmann and P. Spellucci, *Kolloid. Z. Z. Polymere*, **213**, 39 (1966).
[21] H. G. Zachmann, to be published.
[22] J. D. Hoffman and J. I. Lauritzen Jr, *J. Res. Nat. Bur. Standards*, **65A**, 297 (1961).
[23] J. D. Hoffman, J. I. Lauritzen Jr, E. Passaglia, G. S. Ross, L. J. Frolen and J. J. Weeks, *Kolloid-Z. Z. Polymere*, **231**, 564 (1969).
[24] W. M. Ewers, H. G. Zachmann and A. Peterlin, *J. Macromol. Sci.* **B6**, 695 (1972).
[25] R. J. Roe, *J. Chem. Phys.* **53**, 3026 (1970).

PULSE-INDUCED CRITICAL SCATTERING (PICS) FROM POLYMER SOLUTIONS

KENNETH W. DERHAM, JOHN GOLDSBROUGH and MANFRED GORDON

Institute of Polymer Science and Department of Chemistry
University of Essex, Wivenhoe Park, Colchester, CO4 3SQ, UK

ABSTRACT

Pulse-induced critical scattering (PICS) is a new, fast technique for determining thermodynamic and kinetic parameters of polymer solutions. Relative intensities of scattered laser light are measured over periods of seconds after a fast temperature step into the region of Debye critical opalescence. In the mode of operation using thermal pulses to progressively lower temperatures, the intensity pulse shape gives sensitive information on the fluctuation equilibrium and on the kinetics of phase separation with formation of emulsions of sub-wavelength particles. Spinodal and critical loci for the system polystyrene–cyclohexane (PS–CH) are measured about 10 times faster and more accurately than in recent work based respectively on conventional light scattering equipment and on turbidimetry. The classical Debye theory of scattering near a critical point, or Scholte's extension to spinodal points, gives excellent semi-empirical extrapolations to spinodal temperatures T_s. Though there are signs that the classical theory needs amendment in the sense of the generalized non-linear theory of Fisher, Chu, and co-workers, this would generally give results on T_s in close agreement with the simpler classical theory in the experimental range accessible to PICS. The effects of heterodispersity on the phase diagram are studied, comparing data by Scholte and new data by PICS on PS–CH. The evidence suggests that spinodal loci are governed mostly by M_w (as suggested by recent theories). Marked secondary effects seem, however, to arise, not (as recently suggested) from M_n, but from M_z. Results obtained by the technique of approaching phase equilibrium from the heterogeneous side, using an emulsion carefully 'seeded' by a thermal step, suggest that supercooling is more serious in cloud-point determinations of *dilute* polymer solutions than hitherto thought. The formation of sub-wavelength droplets follows kinetics bearing more than formal resemblance to that for the nucleation and growth of spherulitic crystals in bulk.

1 INTRODUCTION

Useful structural and thermodynamic information on polymers can be derived from semi-empirical fits of measurements on phase equilibria of solutions, a subject much advanced by Koningsveld[1] and his co-workers. Recent progress in technology, especially the advent of the laser, has wrought much improvement in the relevant measurement techniques. Pulse-Induced

PAC—38—2—E

Critical Scattering (PICS) was developed as a result of these stimuli. Evidence is presented that such an instrument measures spinodal and cloud points of polymer solutions faster than, but at least as accurately as, the best equipment has achieved hitherto. The reliability of the quantitive interpretation of scattering intensities is increased by correlating thermodynamic and kinetic quantities. In Section 2, we recall the typical phenomenology and underlying molecular causes of polymer solutions near a Θ-point. The rationale of applying rapid thermal steps, and examining the effect on light scattering over a time-scale of seconds, is firmly rooted in the basic molecular events, according to the theory of critical scattering by Debye and co-workers[2–4], recently improved by Vrij and van den Esker[5]. Scholte[6] extended this classical theory to the determination of spinodals of polymer solutions, and Chu, Schoenes and Fisher[7]—and later Chu and co-workers[8]—made more accurate measurements and used a generalized non-linear theory to fit them. For the purposes of this work, the classical linear theory is found to be adequate.

2 THE TYPICAL PHASE DIAGRAM

A schematic phase diagram of temperature T vs. concentration (volume fraction ϕ) of polymer is shown in *Figure 1* for a polymer solution just below its Flory Θ-point, e.g. polystyrene (PS)–cyclohexane (CH). As is familiar, such a diagram is merely rotated through 180° about a line parallel to the

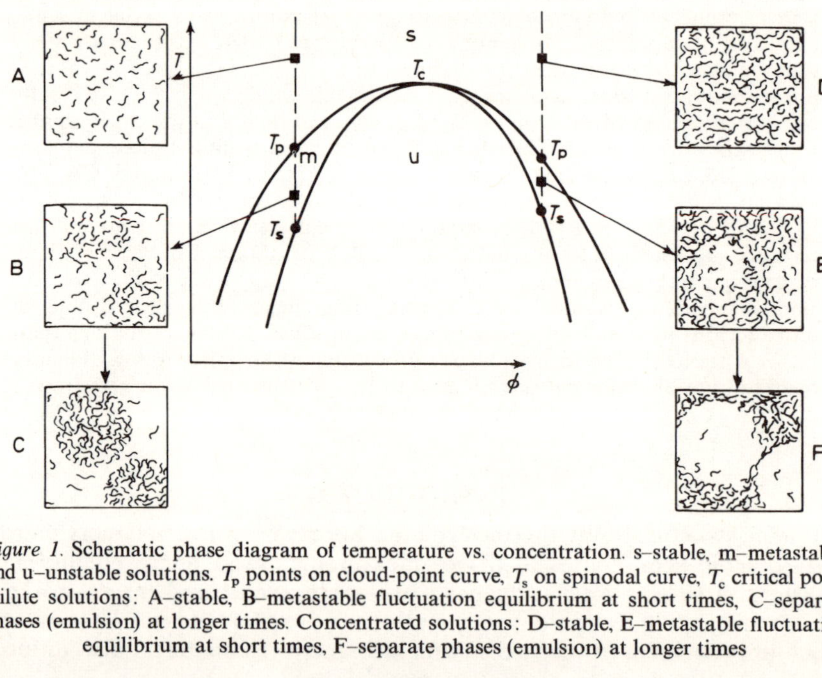

Figure 1. Schematic phase diagram of temperature vs. concentration. s–stable, m–metastable, and u–unstable solutions. T_p points on cloud-point curve, T_s on spinodal curve, T_c critical point. Dilute solutions: A–stable, B–metastable fluctuation equilibrium at short times, C–separate phases (emulsion) at longer times. Concentrated solutions: D–stable, E–metastable fluctuation equilibrium at short times, F–separate phases (emulsion) at longer times

concentration axis, when applied to the region of a Rowlinson Θ-point. In such cases, e.g. polyethylene–octane, upper critical solution temperatures are replaced by lower ones, and PICS requires heating pulses instead of cooling ones (see below), but nothing is changed in principle as far as the present work is concerned.

Figure 1 shows the three temperature–concentration regions which generally occur in a quasi-binary phase diagram. Stable solutions at high temperatures are separated from metastable solutions at lower temperatures by the binodal or cloud-point curve; the metastable region is in turn separated from that of unstable solutions at even lower temperatures by the so-called spinodal. The spinodal is immensely important in technology, e.g. of metals and ceramics, as well as in polymer science, because of the 'spinodal decomposition' which can occur there—for a recent review see Goldsbrough[9]. Thermodynamics requires that the cloud point curve and the spinodal share a common tangent at the critical point (ϕ_c, T_c). For a truly binary (homodisperse) solution, T_c occurs at the maximum temperature of the binodal; for heterodisperse polymers it occurs to the right of the maximum (Chermin *et al.*[10]).

The molecular events are schematized in the figure, both for solutions more dilute (A, B, C), and more concentrated (D, E, F) than ϕ_c. At short times of residence in the metastable region between cloud point and spinodal, fluctuation regions are seen, which normally reach a metastable fluctuation equilibrium (cf. Section 5). The closer the temperature to the spinodal, the less time is available to study this metastable state. Soon a local fluctuation of high (B) or low (E) concentration ϕ will chance to exceed the critical size for nucleation of a droplet, and such droplets rapidly grow to macroscopic dimensions and develop sharp phase boundaries (C, F). Thus phase separation has occurred. The concentration of the droplet phase separating at $\phi < \phi_c$ is high (C); at $\phi > \phi_c$ it is low (F).

Light scattering permits the study of inhomogeneities in refractive index, whether occasioned by diffuse fluctuation regions (B, E) or by droplets circumscribed by sharp phase boundaries (C, F). To observe scattering from metastable fluctuation regions, we must work rapidly, because metastable solutions below the cloud point curve have lifetimes which progressively shorten as the spinodal is approached. The requisite apparatus is described in Section 3.1.

3 INSTRUMENTATION AND MATERIALS

3.1. The PICS instrument

The instrument for creating thermal pulses and collecting scattering intensity and temperature data has been developed over a number of years. *Figure 2* shows a simple version with a single scattering cell. A more sophisticated multi-cell instrument has recently been built. The thin-walled capillary cell of diameter 1 mm holds a few microlitres of solution. The light source is a low-power He–Ne laser beam passing axially along the cell. Light scattered from the cell is collected by two light guides, one (I_{30}) sampling over scattering angles centred at $\theta = 30°$, the other (I_{90}) at 90°. The light-guides pass the light to a solid-state detector, while temperatures are measured

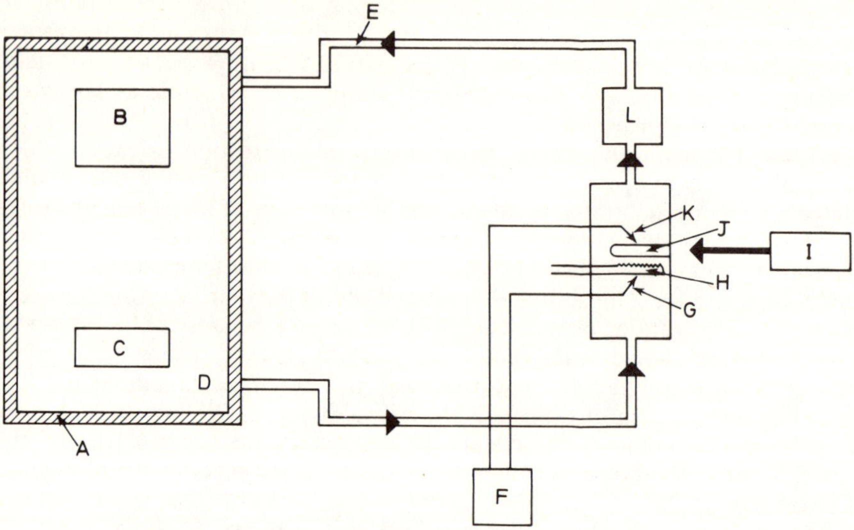

Figure 2. Simplified diagram of PICS apparatus. A–bath, B–immersion heater, C–refrigeration unit, D–heat transfer fluid, E–flow system, F–data handling system, G–light guide, H–heater, I–helium–neon laser, J–scattering cell, K–thermistor, L–pump

by a thermistor. A purpose-built data collection system, based on a Solartron Data Transfer Unit, copes with the high speed of accumulation of measurements. A purpose-built sequence-control unit arranges for the collection of intensity data at the base-line, and at the end of a pulse, later processed by computer. The intensities I_{30} and I_{90} are also displayed in real time on a recorder chart. A typical pattern for measurements in the stable region just above the cloud point is shown in *Figure 3*.

The cell is placed in a water flow system with a small heater situated upstream from the cell to maintain it at a temperature above the cloud point. The temperature step is produced by switching off the heater. The cell contents then cool down rapidly to the temperature of the flowing water. With the most up-to-date method of producing cooling pulses, the thermal response time is about 1 s, as checked by placing a bead thermistor inside the cell. This agrees with the typical delay of about 5 s (*Figure 3*) before maximum fluctuation scatter is attained at the beginning of a pulse ($e^{-5} \approx 0.007$). Fluctuation scattering equilibrium, either stable above the cloud point or metastable below the cloud point, is always attained in less than 5 s. It is not, therefore, at present possible to study relaxation from state A to state B, say, in *Figure 1* after a thermal step. The kinetics of phase change, from state B to C, however, is readily followed by the pulse technique (see Section 6.2). Temperature calibration of the thermistor is against a platinum resistance thermometer calibrated by the National Physical Laboratory. Maximum drift over a 14 months period was 0.02°C over the range 20 to 40°C, and the small corrections resulting from drift were applied.

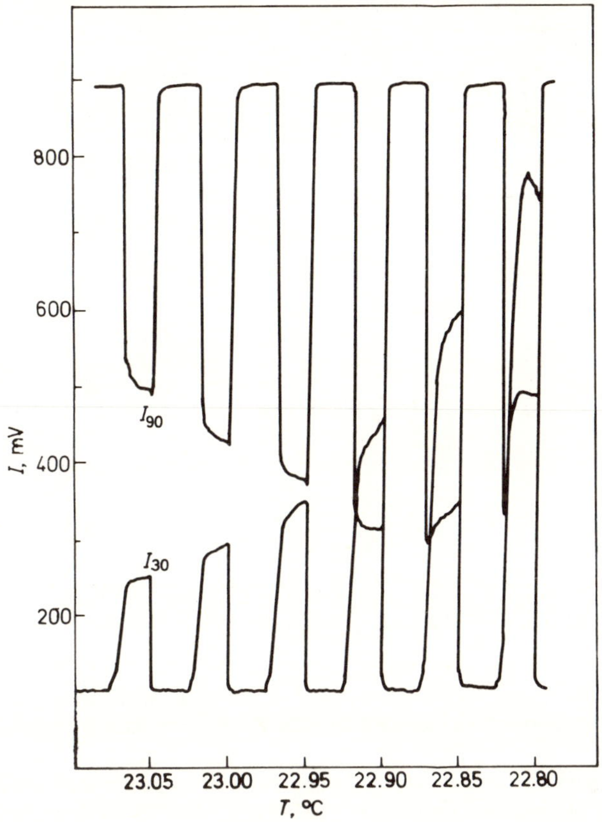

Figure 3. Typical PICS light scattering intensity vs. temperature diagram during cooling run at high concentration (0.15 w/w PS200). Top: intensity at 90° (scale increasing downward). Bottom: intensity at 30° (scale increasing upward). The millivolt output scales for I_{30} and I_{90} are not directly comparable, as I_{90} is amplified more strongly

3.2 Materials

Cyclohexane (CH) was the British Drug Houses spectroscopic grade, dried over molecular sieve. The polystyrene samples were sharp fractions purchased from Pressure Chemicals, with the maker's specifications given in *Table 1*. The fractions were coded PS527 for the sample of $\overline{M}_w = 5.27 \times 10^5$ etc., see *Table 1*. Mixtures of these fractions were made up and their molecular weight averages calculated as shown in *Table 1*. All solutions were kept in sealed ampoules for more than 24 hours at *ca.* 50°C with occasional shaking before filling and sealing the capillary cells. These cells themselves were stored at 50°C when not in use.

Table 1. Samples of polystyrene studied in this work and by Scholte[11]

Sample	Composition	Mol. weights × 10^{-3} M_n	M_w	M_z	M_w/M_n	Ratios M_z/M_w
Narrow fractions						
PS 33	—	36	33	—	~1	—
PS 51	—	49	51	55	1.04	1.08
PS 111	—	111	111	—	~1	—
PS 166	—	154	166	181	1.08	1.09
PS 200	—	193	200	—	1.04	—
PS 527	—	490	527	593	1.08	1.12
Bimodal mixtures						
1 (Scholte[11])	—	70	100	131	1.43	1.31
2 (Scholte[11])	—	38	100	150	2.63	1.50
3 (Scholte[11])	—	22	100	157	4.62	1.57
PSM 6	0.534 g PS33 + 0.466 g PS200	54	110	173	2.05	1.56
PSM 5	0.716 g PS33 + 0.284 g PS527	45	165	432	3.68	2.62

4 LOCATION OF CRITICAL POINTS (DEBYE ET AL.)[2–4] AND SPINODAL POINTS (SCHOLTE)[6]

The precise nature of the mathematical singularity at critical points is not settled. However, the sensitive way in which the scattering intensity varies beyond the immediate neighbourhood of the singularity may be taken from the classical theory of Debye *et al.*[2–4] who derived a formula of the following form, which is practically linear in $1/(T - T_c)$ over the small ranges of T concerned:

$$I = TP(\theta)/[a(T - T_c) + b\sin^2(\theta/2)] \quad (1)$$

Here $P(\theta)$ is the particle scattering factor and a and b are constants. According to this equation, the intensity would appear to diverge at $T = T_c$ in the direction of the beam ($\theta = 0$), while at finite angle θ this catastrophe would be averted by the term $b\sin^2(\theta/2)$ in the denominator. A plot of $1/I$ vs. T would remain almost linear until $b\sin^2(\theta/2)$ became comparable with $a(T - T_c)$. For $\theta = 30°$, this is calculated to occur at $T - T_c < 0.03$°C within the range of molecular weights of PS in CH here studied. The calculation is based on the revised characteristic lengths l given by Vrij and Van den Esker[5], e.g. *ca.* 3 nm at $M = 5 \times 10^5$. Linear plots, unaffected by the angle-dependent term $b\sin^2(\theta/2)$, are therefore predicted up to very close to the critical point. Scholte propounded the corresponding theory for spinodal points, and he found experimentally plots of ΔI^{-1} vs T^{-1} to be linear for PS–CH. Here ΔI is the difference in scattering intensity between solution and solvent. Deducting the small solvent scattering effect is equivalent to eliminating the small amount of scatter attributable to density fluctuations, which are superposed on the concentration fluctuations of interest. Scholte's results, obtained with standard light-scattering equipment, are here broadly confirmed and extended by PICS to a wider range of PS samples, with an increase in accuracy from a few tenths to a few hundredths of a degree. We have eliminated the extrapolation to $\theta = 0°$ scattering angle because the angle effect on the apparent spinodal temperature is certainly small at $\theta = 30°$, and would at worst produce a very small parallel shift of the spinodal locus in a phase diagram. We comment on the work of Chu *et al.*[8] in the Discussion (Section 7).

5 MEASUREMENTS OF SPINODAL POINTS BY PICS

The rationale of locating a spinodal point by PICS rests on extrapolation of the Scholte plot of ΔI^{-1} to T_s^{-1} on the T-axis. As we are concerned with plots ranging over <2 degrees at ~ 300 K, we simplify throughout by plotting $(\Delta I)^{-1}$ against T rather than T^{-1}. A large number of experimental points at temperature intervals of 0.01 degree or less are collected by a

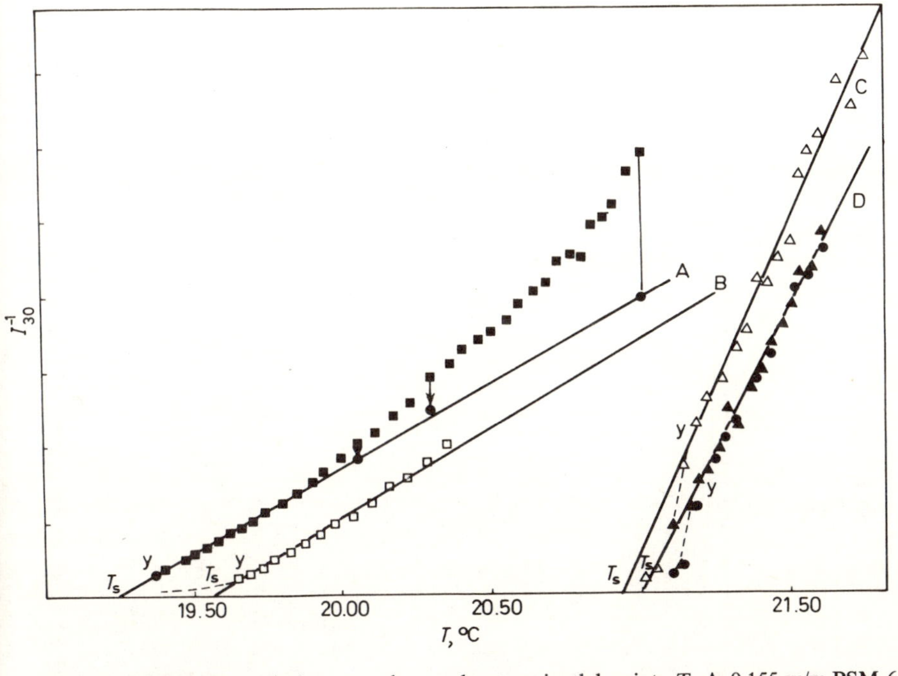

Figure 4. Typical extrapolation procedure to locate spinodal points T_s. A–0.155 w/w PSM 6; B–0.140 w/w PSM 6; C–0.154 w/w PS 111; D–0.144 w/w PS 111. Plot A illustrates the empirical linearization by the parameter ${}^\prime c$ for four selected points ■→●; the point nearest T_s is not visibly shifted by this correction. In plot B, points at high temperature omitted for clarity. In plot D, ● and ▲ refer to replicate runs. y marks the last acceptable pulse in each run before phase change invalidates the pulse profile (see *Figure 3*). According to the second-approximation theory (Section 7.3), plots A and C should have the same T_s, and the same would apply to B and D, since each pair share almost exactly the same ϕ and M_w

controlled cooling rate of the circulating water, e.g. at 0.1 deg min^{-1}. An average of four intensity measurements at the end of a 20 s cooling pulse gives a measure of I_{30} of the solution. The value of I_{30}^0 for the solvent can be obtained approximately from the same solution at sufficiently high temperature to suppress fluctuation scattering, or by placing a similar cell containing cyclohexane in the beam. In practice, we prefer to make an empirical correction to such an estimate of I_{30}^0, typically of order 10–30% of the lowest intensity pulse included in an extrapolation plot, and <4% of the most intense pulse. The magnitude of the correction c is determined by linearizing the plot of $(I_{30} - c)^{-1}$ vs T. It has little effect (*Figure 4*) on the extrapolated value of T_s. Interference by incipient phase change with the recorded pulse-height intensities at the high-intensity end of the plot would have the most serious effect on the extrapolation to T_s. Such interference generally leads to artificially raised I_{30} at low concentration ϕ, and artificially low I_{30} at high ϕ. Fortunately, in both cases, the onset of the trouble is sharp, and well signalized by changes in the profile of the intensity pulses. In *dilute solutions*, the separation of even a small amount of droplets of a concentrated solution phase always causes a sudden marked rise in I_{30}, above the low level due to fluctuation scattering. This is shown in *Figure 5*. For a time the fluctuation scattering is still seen as a shoulder on the rising peaks due to scattering from the emulsion, but further cooling rapidly

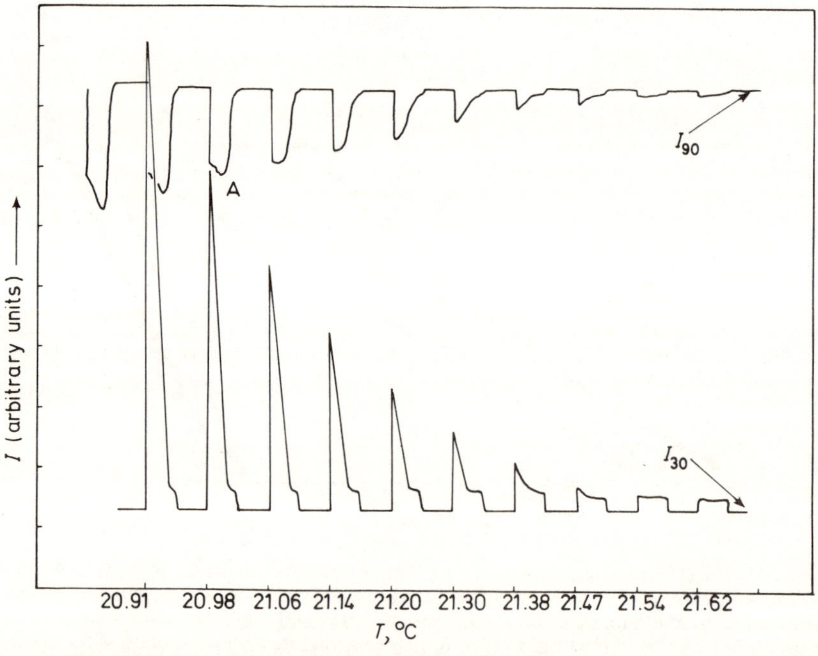

Figure 5. Typical PICS cooling run on dilute solution (0.0100 w/w PS 166). Note low intensity fluctuation-scattering pulses at high temperature, progressively superseded by emulsion scattering spike on I_{30} at lower temperatures. Decline in pulse intensity profile on I_{90} after A is due to the droplet diameter of the emulsion approaching the wavelength (~600 nm)

accelerates the phase separation process, so that the shoulder becomes obliterated. In *concentrated solutions*, on the other hand, where critical (fluctuation) scattering is generally stronger, and the separating droplet phase much more dilute, the scattering intensity I_{90} will drop sharply as a result of phase separation. This is illustrated in *Figure 3*. The drop in intensity is attributed to the decrease in critical scattering of the continuous phase, as its spinodal temperature will decrease as a result of the change in composition. The clear and intelligible change in pulse shape allows the limit of approach to T_s to be determined accurately, to which the critical scattering pulses are valid.

Figure 4 is important in providing evidence for effects due to polydispersity. This is taken up in the Discussion (Section 7.3). The extrapolations may be claimed to be reliable to about 0.02 degree K, which is about the standard deviation of replicate runs on the same cell. As noted by Scholte, the determination of T_s becomes progressively less accurate in more dilute solutions.

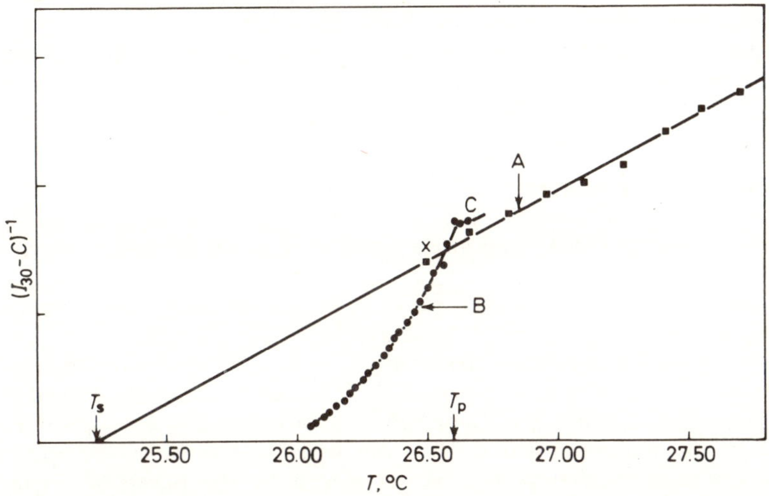

Figure 6. Combined plot of $(I_{30} - c)^{-1}$ vs. T for a relatively dilute solution 0.029 w/w PS527. A-pulsed cooling run, B-continuous heating of emulsion. A and B meet at cloud pount T_p. At C, slight displacement of points belonging to stable solution from line A due to experimental error (base line shift)

The accessible critical scattering range is then separated from T_s by a widening gap in which fast phase separation masks the fluctuation scatter. It is known that the cloud point curve rises further above the spinodal curve (cf. *Figure 1*) in dilute polymer solutions, the greater the polydispersity of the polymer. Accordingly, the accuracy of T_s determinations for dilute polydisperse samples is even less than for sharp fractions. We have not succeeded in making measurements of T_s for the mixtures of sharp fractions listed in *Table 1* at $\phi < 0.06$ w/w with our present light detection system.

Figure 6 illustrates the case of a longer extrapolation in a relatively more dilute solution, when phase separation is seen to interfere with the validity of the critical scattering pulses approximately 1.3 degrees above the spinodal. Here the accuracy of T_s is estimated to be ± 0.2 degree C.

6 CLOUD POINT DETERMINATION BY THE PICS INSTRUMENT

Thermodynamic binodal or cloud points can be determined rapidly, and practically without trouble from hysteresis, by approaching the equilibrium from the two-phase side, i.e. by heating PS–CH emulsions. The PICS equipment affords the necessary control over the thermal programming, and the location of the cloud point is deduced from scattering measurements. It is useful to 'seed' the emulsion by a sudden step from above to just below the cloud point, e.g. 0.1°C below the cloud point T_p in concentrated solutions, or 1°C below T_p in dilute solutions. Circumstantial evidence from the scattering behaviour suggests that such emulsions are stable during the short period required for determining the cloud point. We shall illustrate the best procedure for relatively concentrated solutions using as example 0.0843 w/w of PS 111 in CH, and the best procedure for relatively dilute solutions using 0.0144 w/w of PS 111 in CH.

6.1 Concentrated solutions

In concentrated solutions, the problem of cleanly separating the emulsion scattering from the substantial intensity of residual fluctuation scattering, is best achieved using intensity measurements at $\theta = 90°$. As shown in *Figure 3*, I_{90} declines suddenly during a pulse in which the formation of emulsion intervenes. The temperature T_e, say, at which this decline occurs for the time during the PICS run for measuring T_s, is generally suitable for forming the emulsion with which to measure T_p. The cell is suddenly cooled from well above T_p to T_e, and held there until the intensity has decayed to about $\frac{1}{4}$ or $\frac{1}{3}$ of its maximum value, typically in 20 s. The linear heating rate ($\sim$0.2°C/min) is then switched on. As shown by the points in *Figure 7*, the intensity then rapidly increases with temperature (and time) until the original critical scattering intensity is recovered, which happens at or very near to the sharp maximum in the scattering intensity when plotted against temperature. The position of the maximum is reproducible to 0.02°C or better and is taken as the cloud-point. As shown by *Figure 7*, a 3-fold increase in the heating rate shifted the maximum only slightly (0.025°C), and the intensity of the scattering lagged behind the positions of the critical scattering pulses, possibly a manifestation of a little hysteresis. The slight increase of the intensities during the slow heating run (■) above the critical scattering curve is not explained, but probably represents experimental error. Drifts due to variations in laser intensity are mostly, though perhaps not totally, compensated by the base line shift between the beginning and end of the heating run. The extent of this shift is shown at A. The smoothness of the heating runs (■ and □) shows that short-term intensity stability is excellent, and temperature stability 0.001°C or a little better.

The total time taken for the cloud-point measurements, comprising the cooling run to obtain pulses, and the formation and re-heating of the emulsion, is about 30 minutes.

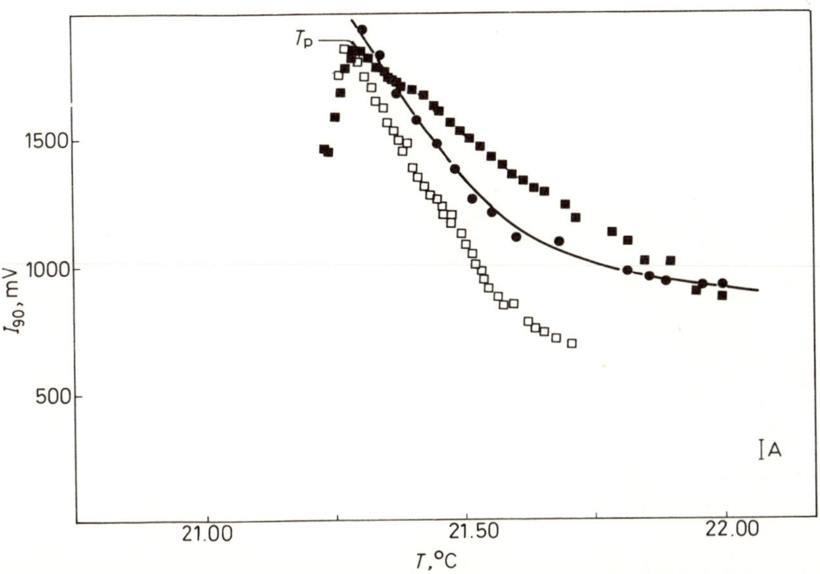

Figure 7. Locating cloud point T_p by the maximum method at high concentration, using a plot of I_{90} vs. T. Solution of 0.084 w/w PS 111, ■ slow heating rate; □ heating rate 3 times greater. ● Fluctuation scattering intensity I_{90} by pulsed cooling run

6.2 Dilute solutions

In dilute solutions, especially at low molecular weights, I_{90} is weak but I_{30} is very strong (*Figure 4*), and the procedure just outlined for concentrated solutions can be carried out, *mutatis mutandis*, with I_{30}. The first pulse to show evidence of emulsion formation—now by a sudden increase in I_{30} rather than a decline in I_{90}—indicates a suitable temperature T_e for forming the emulsion. The solution is rapidly stepped to T_e and held for about a minute to form the emulsion and to obtain a strong enough I_{30}. Switching on the heating rate of $\sim 0.2°C\ min^{-1}$ (*Figure 8*) now leads to a rapid decline in I_{30}. The cloud-point T_p is determined to somewhat better than 0.1°C as the point where I_{30} returns to the base line. This base line does not measurably decrease on further heating, i.e. the diminution of fluctuation scattering with temperature is not detected at this great distance from T_s. *Figure 8* also shows that T_p so found is independent of the degree of supercooling over the range of 0.7 to 1.1°C employed. The four plots superpose within experimental error (cf. section 7.2). They fit a curve of type

$$I_{30} = a(T_p - T) + b(T_p - T)^2 \qquad (2)$$

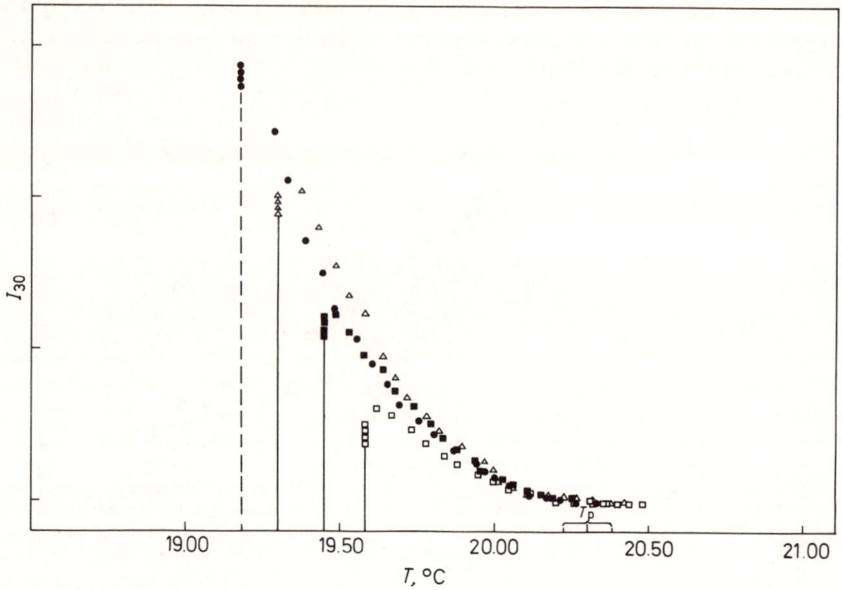

Figure 8. Location of cloud point T_p in dilute solution (0.014 w/w PS 111). Note superposability of four curves especially near T_p, obtained from different degrees of supercooling. ● 19.18; △ 19.30; ■ 19.45; □ 19.58°C

derived from the simplest theory. The theory makes I_{30} proportional to the volume of a droplet, which varies $\propto (T_p - T)$, and additionally proportional to the polymer concentration in a droplet, which varies linearly with temperature over the small range below T_p concerned.

6.3 Kinetics of phase separation

In dilute solutions of PS–CH, the speed of phase separation can be measured over a range of supercooling, and the results rationalized on simple assumptions. The appropriate model corresponds to the first term in an expansion of the exponential in the well-known Avrami equation for the growth of spherulites. Assuming heterogeneous nucleation, and the usual linear growth rate, the amount of spherulitic material separated initially varies as the third power of time, though later the rate is reduced by impingement of the spheres. In the separation of droplets, the same laws can be applied initially—again the rate will later decline, here because of the change in composition of the continuous phase. Impingements leading to coagulation seems to be negligible over the ranges of supercooling and observation times here employed. The intensity I_{30} should, for sufficiently small drops, vary as the volume of the droplets, so that the initial rate law would be

$$I_{30} = kt^3 \tag{3}$$

The four experimental curves plotted in *Figure 9* do approximately conform to equation 3, shown as a solid line. All four experimental lines go through the origin and show a prolonged 'induction time'. They have been superposed by scaling the time axis with a deduction of 10 s at the beginning of

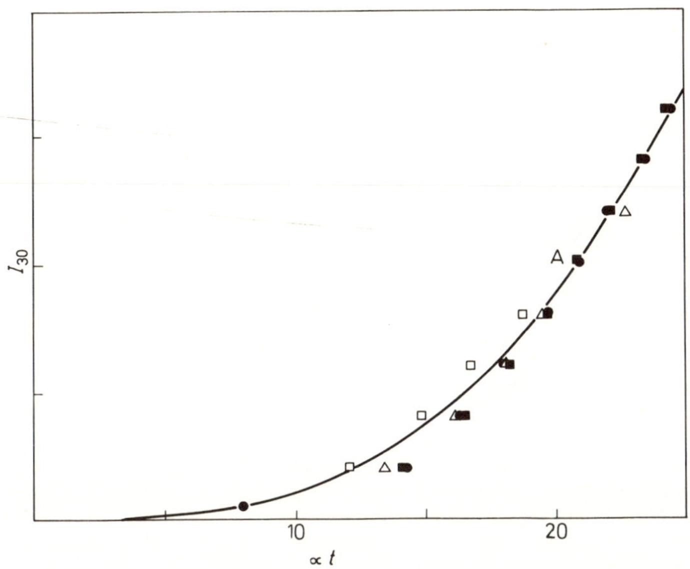

Figure 9. Initial part of plot of intensity I_{30} vs. time in dilute solution (0.014 w/w PS 111). Four runs obtained after sudden stepdown to four different temperatures: ■18.99; ●19.23; ▲19.51; □19.77°C. The curve drawn is the cubic growth law (equation 3). Time axis scaled for each run to bring point A on the curve; time scale in seconds correct for ●

each run for temperature equilibration. The long thermal equilibration time—about 10 times the response time—arises from the great sensitivity to temperature of the rate of phase separation. The appropriate values of k (equation 3) cover a more than 100-fold range. Two typical pulses (*Figure 10A* and *B*) show the 'induction period', the subsequent initial rise in I_{30} which conforms roughly to equation 3, and the levelling off as the phase change approaches completion. The maximum in I_{90} during the rise in I_{30} confirms the interpretation in terms of growth of spherical objects, since such a maximum is expected as the diameter of the sphere approaches the wavelength of the light ($\sim$ 600 nm).

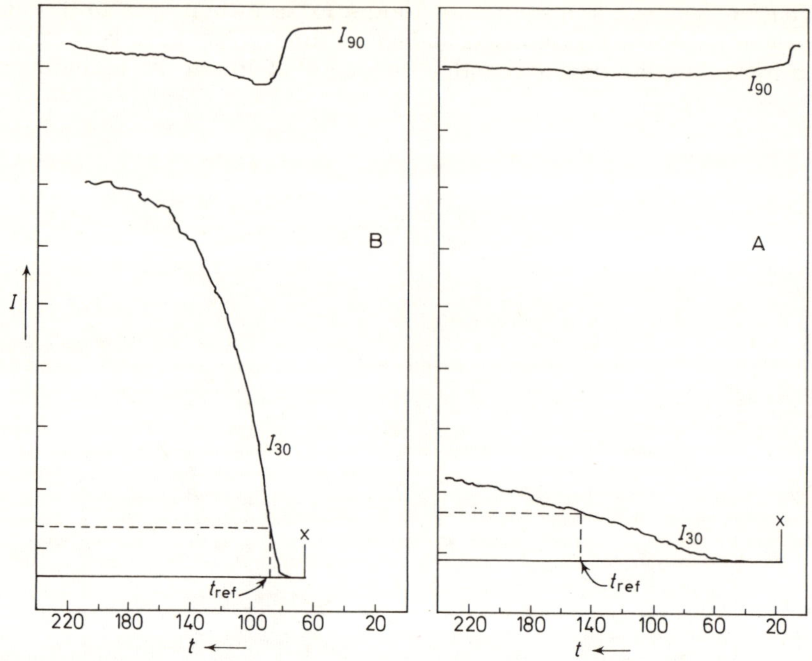

Figure 10. Extended plot of intensity I_{30} (lower curves) and I_{90} (upper curves, inverted) at two different supercoolings on 0.014 w/w PS 111 (cf. *Figure 9*). A 19.58; B 19.18°C. Note the large temperature effect on rate of formation of emulsion. The reference time t_{ref}, taken at an arbitrary low intensity I_{30}, is proportional to the scaling factor in *Figure 9*; its logarithm is plotted as a measure of rate in *Figure 12*

7 DISCUSSION

7.1 Spinodal measurements

Before comparing our results with those of Scholte[6], their relation to the fundamental work of Fisher, Chu[7,8] and their co-workers requires comment. These workers argue that, on the basis of the nature of the mathematical singularities involved at T_c and T_s and on experimental grounds, the classical theory of equation 1 of Debye and co-workers needs to be generalized, essentially to

$$I^{-1} \propto (T - T_s)^{\gamma} \tag{4}$$

This equation is still empirical; γ is no longer unity, but a value of 1.25 may well have universal significance. Their equipment, a purpose-built laser self-beating spectrometer, with excellent temperature control, is suitable for investigating such fundamental points of theory. However, for merely locating T_s, at a concentration not too far from ϕ_c, by extrapolation, it is readily shown that no practical gain is derived from the more elaborate treatment of experimental data. The scattering attenuation correction described by Chu *et al.*[8] has no effect on the apparent T_s, and the volume

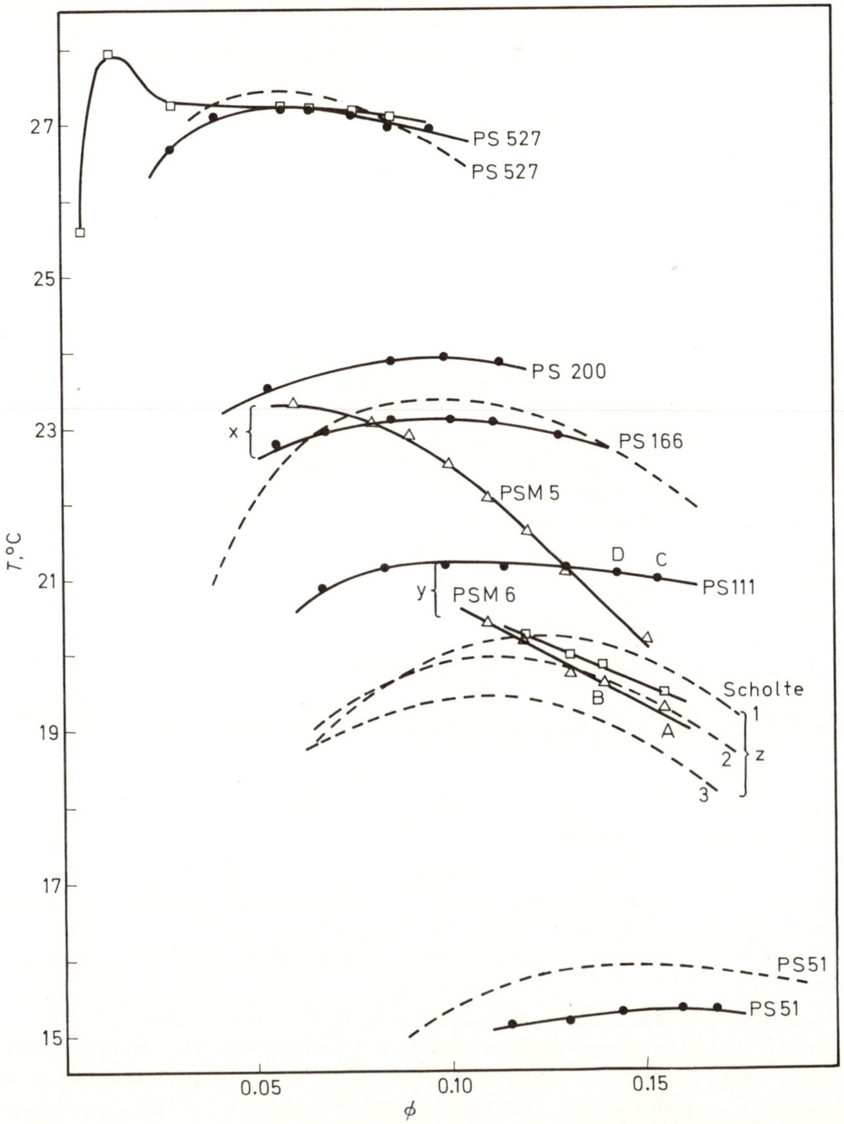

Figure 11. Phase diagram for polystyrene in cyclohexane. ---- Smoothed data by Scholte[6, 11], using conventional light scattering equipment. ——— present data by PICS. Circles: spinodal points; Squares: cloud points. Points A, B, C and D correspond to the points so labelled in *Figure 4*. The braces X, Y, Z link runs at fixed M_w but different M_z. Comparison of the five mixtures PSM 5, PSM 6, and Scholte 1, 2, 3 with their molecular weight averages (*Table 1*) suggests that the skew displacement downward of progressively more heterodisperse mixtures correlates with M_z rather than with M_n. Scholte's results on PS 51, PS 166 and PS 527 lie above the PICS results by a few tenths of a degree almost everywhere. Note that a tenfold range in M_w covers a range of about 12 degrees. The Θ-point, corresponding to $M_w = \infty$, lies some 4 to 6°C above the upper limit of the diagram

correction for measurements at different angles has at most a minute effect. Finally, a constant value is subtracted by these authors from each intensity measured on a given solution, in respect of scattering from density fluctuations. Such a correction is included in our term c (Section 5). We adjust this empirically (*Figure 4*) to linearize the classical plot according to equation 1. Chu *et al.* estimate the correction by an arbitrary extrapolation of the scattering intensity to $T_p + 25$. The curved plot in their *Figure 4*, on a solution of PS–CH with $T_p = 26.98$, can be linearized within experimental error by arbitrarily adjusting c downwards ($c < 0$). This procedure merely changes the intercept T_s from 26.4_8 to about 26.5_6°C. As such a small change is hardly significant for our purpose, our base-line correction c is quite justified over the range of concentrations studied. Its effect near ϕ_c is, of course, quite negligible, because the extrapolation is then very short.

The high quality of measurements possible with the equipment of Chu *et al.* is based on a scattering cell of 8 mm diameter, compared with 1 mm for PICS. Accordingly, the thermal response time must be larger by a factor of $\sim 8^2$, i.e. a minute instead of a second. As judged by the smoothness of the plots, and the number and ranges of experimental points included, the data of Chu *et al.* are indeed excellent. It is therefore worrying that the results in their *Figure 1* on PS of $M_w = 4.15 \times 10^5$ in CH, are not in agreement with the approximately concordant results of Scholte and of the present work. Thus their $T_c = 26.95_7$ for PS of $M_w = 4.98 \times 10^5$ is approximately located upon our spinodal for PS of $M_w = 5{\cdot}27 \times 10^5$. This discrepancy is quite unconnected with differences in extrapolating data, and will need further investigation.

Turning to the comparison between Scholte's results and ours, they generally agree to a few tenths of a degree C (*Figure 11*), the precision claimed by Scholte. He used standard light-scattering equipment with a modified temperature control system. He generally collected six intensity measurements on a given solution, covering a 5- to 10-fold range of intensities over a 30 degree C range in temperature. In a later[11] paper, 8 points were spaced at 1°C intervals. In contrast, PICS collects a large number of points closer to T_s, covering (in a few minutes) about a 1 degree C range in temperature, but a similar or larger range of relative intensities (*Figure 5*). Accordingly, PICS gives about 10 times greater precision to the extrapolated value to T_s, and pushes the lower limit of experimentally attainable temperatures as close to T_s as possible, normally a little below the cloud point. With a stiffer polymer than PS, much closer approach to T_s would be possible, as illustrated by the work of Van Aartsen, Smolders and co-workers[12] on the spinodal of solutions of polyoxyphenylene.

7.2 Cloud-point measurements

Nucleation would be expected to control the rate of droplet formation in dilute polymer solutions. Comparison of *Figures 10A* and *B* confirms this, since no other cause is likely for such a large temperature effect implying a negative activation energy. As a result the standard practice of obtaining cloud-points of dilute solutions by cooling clear solutions leads to substantial errors, not eliminated merely by reducing the speed of cooling ever further. The situation is similar to the nucleation of spherulites in bulk polymers just below T_m. No method exists for studying that equilibrium on any reasonable time-scale, by approaching it from either side. Therefore, a rather long extrapolation, based on nucleation and growth rate theory, is one method for locating T_m. The present work confirms experimentally that phase equilibrium in dilute polymer solution is readily approached from the two-phase (emulsion) side. A simple calculation shows why this would be so. The diffusion of a substrate out of a sphere of radius r into a medium containing a lower constant concentration of that substrate, will be more than 96% complete at the centre of the sphere in r^2/D seconds[13]. The diffusion constant of PS 111 in a dilute solution in CH should be $<5 \times 10^{-7}$ $cm^2\ s^{-1}$. Therefore, equilibrium is practically attained in <1 s for droplets of diameter 10^{-3} cm. Judging by the position of the maximum in I_{90} in *Figure 10A*, our droplets are of diameter approximately equal to the wavelength, say 5×10^{-5} cm, about the time t_{ref} which we use as a measure of their growth rate constant (*Figures 10A* and *B*). They would not grow to a diameter of 10^{-3} cm within the range of the experiments reported. The achievement of equilibrium is strongly supported by the superposability of $I_{30}(T)$ for experiments at different degrees of supercooling (*Figure 8*).

The comparison with nucleation-controlled bulk crystallization may be sketched as follows. Mandelkern[14] found that the rates of crystallization of three polymers, polydecamethylene adipate, polyethylene oxide, and poly-*N,N′*-sebacoylpiperazine, respectively passed into the measurable range 5.5, 6, and 12°C below their T_m, and increased by a factor of 10^{12} by further supercooling through respectively 5, 7 and 13°C. In comparison, our rate of droplet growth was measurable *ca.* 0.7°C below T_p and increased by a factor 10^2 over about a further 0.4°C. The absolute rate of droplet formation at fixed $T_p - T$ is much faster than that of spherulite formation at the same $T_m - T$. This reflects the faster transport of chains through the interface in solution, when compared to the bulk. The temperature dependence of the process is much higher for droplets than for spherulites, which reflects the smaller latent heat, and probably also the larger specific surface free energy, accompanying the formation of solution droplets.

The plot in *Figure 12* of the logarithm of the overall rate constant of droplet growth against T has the same kind and degree of curvature as Mandelkern's[14] corresponding plots for the three crystallization systems. He discusses the linearization of the plots by the traditional adjustment upwards (here about 3°C) of the assigned T_m. Our plot in *Figure 12* would require a similar increase in T_p to achieve linearity. Since we contend that the equilibrium point T_p is measurable to about 0.08°C (*Figure 8*), this linearization is not open to the interpretation of phase separation below T_p in solution. However, the curvature may be explained, e.g. because the specific surface free energy of the droplets would increase on cooling, as a result of increasing difference in concentration between the two phases.

Koningsveld and Staverman[15] noted a hysteresis of 0.6°C in apparent T_p on cooling a 1 per cent solution of polyethylene in diphenyl ether, compared to the result on heating the cloudy emulsion. As their turbidimetric measurements were aimed at an accuracy of a few tenths of a degree, this was not considered serious. However, Chu and co-workers aimed at accuracy about 100 times greater, using their purpose-built light scattering photometer for

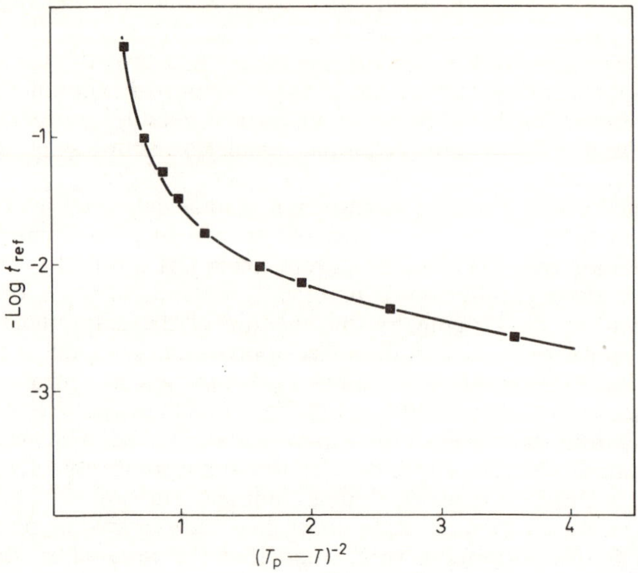

Figure 12. Plot of $-\log t_{ref}$ vs. $1/(T_p - T)^2$ for formation of emulsion from 0.014 w/w PS 111 (cf. *Figure 10*)

measuring T_p by the cooling method only. They reduced the cooling rate to 0.001°C min^{-1}, over ten times slower than that employed by Koningsveld and Staverman. Even so, we believe that their measurement at lowest concentration (~ 0.02 w/w) is likely to be significantly too low because of the expected hysteresis due to nucleation control. Much useful information on transport properties should be obtained by extending our measurements on nucleation and growth kinetics to samples of polymers of different molecular weight, branching, etc.

7.3 The phase diagram

It was realized quite early (see, e.g. Tompa[16]) that the Flory–Huggins (first-approximation) theory—despite its enormous usefulness and its role as a paradigm in polymer science—cannot predict with any accuracy properties which depend on higher derivatives of the free energy of mixing. The spinodal depends on the first Gibbs determinant (cf. Koningsveld[1]) involving second derivatives, and the critical point on the second determinant, which involves third derivatives, with respect to concentration. Even though phase diagrams in general are thus poorly predicted by the crude theory, it can be used to explore, as a heuristic model, for instance, the effects of polydispersity on the cloud point curve (Šolc[17], Koningsveld and Staverman[15]). A great step forward was taken by Koningsveld and his co-workers, when they showed that the interaction parameter χ of the Flory–Huggins model can be turned into an arbitrary function of overall polymer concentration, without unduly complicating the mathematics. From this much more general second-approximation theory, it was deduced that the spinodal of solutions of linear polymers depends only on M_w, and the critical point only on M_w and M_z. As a third approximation (Ref. 18), even effects of end-groups, branch units and copolymerization remain tractable within the framework of this theory, but some dependence on M_n is thereby introduced. At least one further factor is certain to require attention before a complete theory of phase diagrams can emerge: the effects introduced when one of the phases in an equilibrium is too dilute to conform to models of the Flory–Huggins type, which require substantial intermingling of polymer coils in solution. This situation is commonly encountered in practice. Fortunately, a promising theory—not costly in terms of adjustable parameters—has just been developed to meet it (Koningsveld, Stockmayer, Kennedy and Kleintjens[19]).

The most obvious application of a phase diagram, such as *Figure 11*, is to the characterization of polymer distributions, especially by reference to M_w and M_z. Despite their relatively high speed, PICS measurements are seen to give results of high accuracy as shown by the relative smoothness of the spinodal loci. The phase diagram suggests that the M_w of a sharp fraction could, after suitable calibration, be measured to 3 per cent at $M_w = 5 \times 10^5$ and to 1 per cent at 5×10^4. As this is somewhat better than the primary calibration of the M_w of the fractions, PICS measurements would be suitable for smoothing primary calibrations.

The two curves PSM5 and PSM6 show that the spinodal of highly heterodisperse samples deviate strongly from that of a sharp fraction of equal M_w (respectively PS 111 and PS 166), contrary to the result of the second-approximation theory. The same effect was observed by Scholte[11], who thought that this might be taken into account by a term depending on M_n in the free enthalpy of mixing (cf. the end-group effect[18]). However, comparing Scholte's results with ours (*Figure 11* and *Table 1*), it is clear that the deviation of the spinodal of a heterodisperse mixture from that for the sharp fraction of the same M_w, is governed much more exactly by M_z/M_w than by M_w/M_n. The deviations from the second-approximation theory exposed by *Figure 11*, do not invalidate the principle of characterizing polymer distributions by T_s and T_c. Clearly, the measurements are even *more* sensitive to the molecular weight averages than previously thought, and at worst this calls for more extensive calibration procedures, which are not time-consuming with PICS.

The spike at low concentration on the T_p curve of PS527 is interesting. It would easily be missed by approaching the equilibrium from the homogeneous side. The technique of heating a carefully 'seeded' emulsion, establishes the existence of the spike beyond doubt (though its detailed course has not been accurately charted). Koningsveld and Staverman found by theoretical calculation that such spikes would result from, and be sensitive to, heterodispersity. PS527 is a reputable sharp but by no means homodisperse (*Table 1*) fraction. The experimental finding of the spike is a further confirmation, not only of the sensitivity of phase equilibrium measurements, but of their successful modelling by theoretical calculations. The combined measurements by Scholte, using conventional light scattering, and ours using PICS, in *Figure 11*, should help in improving the theory of the free enthalpy function beyond the second approximation.

ACKNOWLEDGEMENT

We gratefully acknowledge financial assistance of this work by Applied Research Laboratories Limited, and a studentship for K.W.D. from ICI Plastics Division. We thank Mr B. Ready for designing and building electronic circuits and Mr Gary Stoneham for technical assistance.

REFERENCES

[1] R. Koningsveld, *Disc. Faraday Soc.* **49** (1970); R. Koningsveld and L. A. Kleintjens, *Macromolecular Chemistry—8* (K. Saarela, ed.), p. 197, Butterworths, London (1973).
[2] P. Debye, *J. Chem. Phys.* **31**, 680 (1959).
[3] P. Debye, H. Coll and D. Woermann, *J. Chem. Phys.* **33**, 1746 (1960).
[4] P. Debye, B. Chu and D. Woermann, *J. Chem. Phys.* **36**, 1803 (1962).
[5] A. Vrij and M. W. J. Van den Esker, *J. Chem. Soc., Faraday II* **68**, 513 (1972).
[6] Th. G. Scholte, *J. Polymer Sci., A-2*, **9**, 1553 (1971).
[7] B. Chu, F. J. Schoenes and M. E. Fisher, *Phys. Rev.* **185**, 219 (1969).
[8] N. Kuwahara, D. F. Fenby, M. Tamsky and B. Chu, *J. Chem. Phys.* **55**, 1140 (1971).
[9] J. Goldsbrough, *Sci. Prog., Oxf.* **60**, 281 (1972).
[10] H. A. G. Chermin, M. Gordon and R. Koningsveld, *Macromolecules* **2**, 207 (1969).
[11] Th. G. Scholte, *J. Polymer Sci., C* **39**, 281 (1972).
[12] C. A. Smolders, J. J. van Aartsen and A. Steenbergen, *Kolloid-Z. Z. Polymere* **243**, 14 (1971); P. T. Van Emmerik and C. A. Smolders, *J. Polymer Sci., C* **38**, 73 (1972).
[13] J. Crank, *The Mathematics of Diffusion*, p 86, Clarendon Press, Oxford (1964).
[14] L. Mandelkern, *Chem. Rev.* **56**, 903 (1956).
[15] R. Koningsveld and A. J. Staverman, *J. Polymer Sci., A-2*, **6**, 349 (1968).
[16] H. Tompa, *Polymer Solutions*, p. 208 Butterworths, London (1956).
[17] K. Šolc, *Macromolecules*, **3**, 665 (1971).
[18] J. W. Kennedy, M. Gordon and R. Koningsveld, *J. Polymer Sci. C* **39**, 43 (1972).
[19] R. Koningsveld, W. H. Stockmayer, J. W. Kennedy and L. A. Kleintjens, in press.

THERMOKINETICS OF FORMATION OF ORDERED STRUCTURES IN POLYMER SOLUTIONS AND GELS

S. Frenkel

Institute of Macromolecular Compounds,
Academy of Science of the USSR, Leningrad, USSR

ABSTRACT

A general approach is proposed allowing the reduction to a 'common denominator' of different forms of formation of ordered structures in polymer–solvent systems. This 'common denominator' results from the combination of the equivalence principle of thermokinetics and the concept of diphility of macromolecules. In the first sections the discussion deals with a precise definition of order and with the foundation of the equivalence principle and related physical formalisms.

It is shown that the consideration of any mode of structure formation can be reduced to the analysis of phase diagrams and free energy–temperature diagrams, the possibility of such reductions being connected with two main assumptions (supported by all known experimental data):

(*i*) Due to inherent diphility the Flory–Huggins parameter χ_1 can change continuously or discontinuously with the concentration of components leading to the occurrence of preferential chain conformations or abrupt intramolecular conformational transitions. In cases of pronounced explicit diphility (in random or block copolymers) the preference of interactions also leads to intramolecular segregation which on increasing polymer concentration develops as a supermolecular order.

(*ii*) One can connect these conformational transformations with the dependence of Flory's flexibility parameter f on χ_1. Taking into account this dependence makes it possible to interpret identically 'antiplastification' phenomena, the formation of liquid crystalline mesophases with different morphologies and the formation of binary crystals.

Hence, various manifestations of molecular and supermolecular forms of ordering in solutions and gels can be treated by analogy with the usual polymorphism in polymers, in the same way as in the Flory–Mandelkern crystallization theory.

The equivalence principle allows us to consider from quite similar positions the nucleation and growth of crystals or liquid crystals in the presence of external fields. In these cases the flexibility parameter f turns out to be dependent on the external field and all further considerations can be reduced to the already analysed situation with variable and interdependent χ_1 and f parameters. A more profound analysis shows that this equivalence is not formal but reflects the real state of events. The uncoiling of macromolecules in a shear or longitudinal field and the increase of their interactions means in reality the increase of the interaction parameter χ_1 and ultimately the loss of solubility, or desolvation.

The analysis from those positions of a large amount of experimental data shows that in some cases organized structures may exist only in the presence of a solvent which stabilizes them thermodynamically rather than only stimulates kinetically their formation.

When organized structures are formed due to external fields they can persist on the elimination of the field and removal of the solvent, on condition that the energy 'pumped over' in the system from the external field is compensated for as the latent crystallization heat (i.e. is simply used for crystallization in the strict sense of the word). In the opposite case (in an atactic polymer) the oriented state is unstable due to developing internal stresses. In special cases these stresses lead to an 'orientational catastrophe'—the self-annihilation of the sample.

---

1 INTRODUCTORY NOTES

1.1 On the concept of order in polymers

Before summarizing the different data presented at this Symposium one must try to make some general statements about what we understand as 'order' in polymer–solvent systems. This is necessary not only to avoid possible confusion and misunderstanding but also to limit oneself to the discussion of the most realistic and important (or, maybe, typical) cases of ordered structures formation.

Probably the first limitation must be connected with the *range* of order. Although in discussing molecular and supermolecular order in polymers many authors include in the definition of order *short range* order, for instance in amorphous polymers, this does not seem to be correct or in accord with the general principles of molecular physics. Of course the appearance of short range order in the liquid state is a necessary step in the transition from the total disorder presented by the gaseous state to a total order which is attainable only in cases of covalent crystals of the diamond type (which immediately returns us to polymers, the non-covalent simple molecular crystals being always imperfect!). On the other hand, one still makes a distinction between ordered and unordered simple liquids, for instance liquid crystals and isotropic liquids, some anisotropic liquids being intermediates within the liquid state, which is an intermediate state itself.

Thus both everyday logic and a need for correct formulations lead us to the *long range order* as a 'representative of Order' in polymer–solvent systems. These didactic problems that I have mentioned very superficially are of the same kind as the problem of comparison of liquids with gases and/or solids[1]. In the case of polymers an additional peculiarity, both physical and didactic, immediately arises, the polymer chains *per se* (even atactic) presenting a sort of unidimensional order the range of which is determined by the degree of polymerization. However, if we adopt Huggins's struction concept[2] to describe not only the molecular but also the supermolecular mode of distribution of chain units in space, then most difficulties can be avoided, and we come back to the definition of order as some kind of long range packing, the structons being formed by repeating units, chain segments or, in some cases, by the whole macromolecules. Here, by 'packing' we denote only the type of spatial distribution of characteristic units, without further limitations concerning immobilization, or mode of thermal motion, or closeness (of packing) etc.

The given definition of order includes three-dimensional crystalline lattices as well as systems with only lateral or only longitudinal (or both) long range order. Thus, it includes automatically every kind of liquid crystalline mesophase. In addition—and this is important—one can take into account *different levels* of order, the role of 'elementary units' being played in those cases by crystallites, fibrils or other morphoses having a sufficient amount of *internal* long range order. Such an approach will lead us to the concept of structons of 'higher ranges' which coincides in sense with Hosemann's paracrystalline concept of matter[3].

Two more words should be written about the magnitude of structural units that we still are allowed to call ordered. Just as in the case of heterophase fluctuations[1] existing in equilibrium with the main phase, one should start with stable crystallization nuclei which for some reason or other cannot grow, polyvinyl alcohol in water giving a most striking example of such behaviour. One can treat these nuclei as 'frozen' heterophase fluctuations, and later we shall discuss their nature in some detail.

As to the linear dimensions of these minimal ordered units, the criterion of order will be connected only with the possibility or impossibility of checking the presence of order by means of x-ray or electron diffraction.

1.2 The thermokinetic approach

By thermokinetics I mean the physical kinetics of phase transitions and changes in order of polymer systems. However, the term by itself hints at some kind of hybridization of thermodynamics and physical (and in some cases also chemical) kinetics. One can alternatively define thermokinetics as a form of statistical mechanics where the partition function and all the related values are treated from the *structural* point of view. Correspondingly, one must consider the time derivatives of the partition function or other time-dependent external or internal parameters, for in reality there exist no physical structures that can change instantaneously.

Thus the mentioned time derivatives of the partition function (and consequently of free energy, entropy, etc.) and other time dependent factors must figure explicitly in the equations describing phase transitions, changes in phase equilibria and in general any changes in any systems under the influence of any kind of external fields. In this sense thermokinetics is not 'tied' specially to polymer systems. However, if for ordinary systems (as for instance in the kinetic theory of gases) these time-dependent factors play a part of only small corrections that can be usually neglected, in polymer systems these 'corrections' may turn out to be the dominating factors defining both the character of the transition and the morphology (and related properties) of the resulting state.

Thus, I try to advance a concept rather than a new term. The source of this concept can be found in the *Kinetic theory of liquids*[1], the kinetic theory of phase transitions being a specific part of the general theory.

The thermokinetic approach to polymer systems as an extension of the above mentioned theory to any kind of structural changes was first proved by the author and G. K. Elyashevich in two papers[4, 5] where the necessary details can be found. In what follows the same approach will be used to discuss some typical cases of formation of order in polymer solutions and gels.

It should be added that some of the confusion concerning the phase transitions in polymers is largely due to the fact, usually neglected, that polymerization itself is a very specific and fundamental phase transition involving at least three groups of symmetry, namely those of translations, rotations and reflections[6]. All physical changes in a polymeric system proceed so to say on a background of this already completed 'main' transition. Thus, in some instances they are not sufficiently pronounced and as a result are mistreated, and in some other cases critical effects appear where they should not be present if we try to apply for polymeric systems the laws derived for simple bodies[7–9]. Though this is still a point of discussion, in my opinion most contradictions connected with the nature and/or order of phase (and other, for instance glass) transitions in polymers are due to the same neglect of the phase nature of polymerization.

To make the following description more illustrative I shall use diagrammatic representations rather than formulae. The axiomatics of thermokinetics allows different formalisms, the most suitable of them (for the problem under discussion) being the 'deformation of phase diagrams', the latter presented by the coexistence or free energy–temperature curves. In further sections we shall show the application of this formalism to concrete examples.

2 THERMODYNAMIC AND KINETIC STABILITY OF AQUEOUS POLYVINYL ALCOHOL SOLUTIONS AND NON-GROWING STABLE CRYSTALLIZATION NUCLEI

The coexistence curve for PVA–water systems has been a subject of controversy for a long time. The main experimental difficulty was the practical impossibility of obtaining even at high dilutions, a purely stable solution at appropriate temperatures. Always after some time colloid 'gel-particles' were generated, the process being quite similar to that of amylose retrogradation. Filtration or centrifugation does not allow us to eliminate these particles: after some time new ones are generated. On the other hand, at some sufficiently high concentrations, for instance in spinning solutions, and consequently at sufficiently high viscosities one still can obtain a fairly stable state without gelation or formation of microgels for an appropriately long time.

The problem was discussed in detail by the Saratov group (Klenin *et al.*) in a series of studies where turbidity spectra, light scattering and electron microscopy were used as main methods of observation[10, 11]. The main results of these studies can be summarized as follows.

The second virial coefficient A_2 of cleared and gel-particle-free solutions of PVA decreases with increasing temperature, the extrapolation pointing to a 'reversed' (Rowlinson) Θ-temperature near 160°C. However, direct measurements (for experimental details see Refs. 10–11) show that at temperatures close to 130°C, A_2 starts to increase again. Thus one must conclude that the PVA–water systems are represented by a superposition of binodals leading to the so-called 'flower vessel like' phase diagram (*Figure 1*). This does not explain, however, the generation of gel-particles in the would-be stable region on the left side of the diagram at room or even elevated temperatures. The only possible explanation is that in the whole attainable concentration range the equilibrium melting temperature is higher than 100°C.

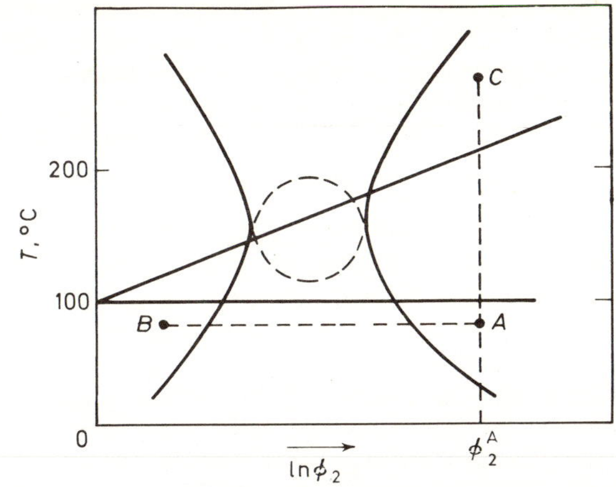

Figure 1. The coexistence diagram PVA–water. The ordinary and 'inverted' binodals overlap. The thick line represents the dependence of the melting temperature on concentration. On the abscissa the logarithm of concentration is plotted in order to compensate for the usual strong skewness of phase diagrams of polymers

Then a continuous nucleation is unavoidable; however, two new questions arise: (1) why this nucleation does not lead to formation of lamellar single crystals or spherulites? and (2) why at higher concentrations no crystallization occurs?.

To answer the first question one must introduce the notion of inherent diphility which applies not only to PVA. (It should be kept in mind that we discuss here the behaviour of a sufficiently pure PVA only, containing less than 1 per cent of acetate groups). The concept of inherent diphility has the same physical meaning as the concept of diphility, or amphiphility in the case of copolymers. For instance, in the case under consideration there is a kind of 'conflict' between the non-polar backbone of the PVA chain and the polar side OH-groups. This 'conflict' is amplified in the presence of water, the backbone trying to attain a supercoiled conformation. Of course such a conformation does not favour the evolution of crystalline growth, the crystallization being terminated henceforth at the nucleation step, but the nuclei remaining stable in the same way as micelles. The size of these nuclei is sufficiently large to make them observable by x-ray diffraction. As was shown in our work with Baranov and Volkov[12], as well as in some earlier Japanese work (see Ref. 12), one can eliminate the effects of inherent diphility by the use of suitable solvents modelling to some extent the structure of short parts of the PVA chain. For example, it is readily possible to obtain fairly developed spherulites from heated glycerol or ethyleneglycol solutions of PVA.

I would like to draw attention to the fact that the appropriate choice of a solvent may be essential not only for the character but for the possibility itself of the formation and development of ordered structures. This aspect of the problem is discussed in greater detail in the main lecture of Kang-Jen Liu[13]; however, we shall return to it when considering the so-called specific ordered states resulting from specific polymer–solvent interactions. Here as two other examples of inherent diphility as a factor that can hinder crystallization, I could quote the behaviour of polymethacrylic acid in water and in methyl alcohol[14] or of tactic polymethylmethacrylate in the presence of precipitants having a preferential affinity either to the backbone or to the side ester groups, crystallization proceeding normally in the first case only[15].

Having thus answered the first question we still have not explained why on the right side of the coexistence diagram (*Figure 1*) one can obtain a practically stable solution. Take for instance, the state described by point A. The morphology of this state depends strongly on the way by which the point was attained. One possible way will be to start on the left side of the diagram (point B) and increase the concentration gradually by evaporation. Even in this simplest case the final condition will depend on the rate of evaporation. Obviously at least two events may happen: (a) The rate is sufficiently low: then nucleation may start, and in the unstable (towards the 'amorphous' phase equilibrium) region within binodals a gel will be formed with stable nuclei playing the part of the nodes of the gel framework. (b) When the rate is very high one can 'jump over' the amorphous instability range and attain the state A before both nucleation and spinodal phase decomposition have had time to start. This is a very primitive though illustrative example of a substantially 'thermokinetic behaviour' of the PVA solution: depending on the rate of increase of polymer concentration two distinctly different states can be attained, the second one being still thermodynamically unstable towards crystalline–amorphous phase separation, but *kinetically* stable toward this separation, and again thermodynamically stable toward amorphous phase separation.

In terms of the Flory–Mandelkern theory of the crystallization of polymers[7] we simply attain a state of supercooling where the nucleation rate becomes zero. However, this state exists, and it has its own partition function and free energy with corresponding configurational terms which differ drastically from those resulting in a low rate process.

In other words, if we ascribe to the initial state a free energy value F_B, the change of free energy, ΔF, *will not be* an unequivocal function of polymer concentration φ_2 depending strongly on the derivative $\partial\varphi_2/\partial t$, t being time. Thus,

$$\Delta F = F_A - F_B = \int_0^t \left(\frac{\partial F}{\partial \varphi_2}\right) \frac{\partial \varphi_2}{\partial t} \, dt \tag{1}$$

where $(\varphi_2)_{t=0}$ corresponds to the initial free energy, and $\varphi_2(t)$ though being the same in all cases, corresponds to *different* free energies and different stable or metastable morphologies. Hence the essential derivative in equation (1) is not $\partial F/\partial\varphi_2$ but $\partial\varphi_2/\partial t$, all this being a primitive though correct demonstration of the thermokinetic approach. It should be pointed out that we deal with the usual thermodynamic quantities without applying the formalism of irreversible thermodynamics, but introducing explicitly some time derivatives. Equation 1 can also be presented in the form

$$\Delta F = \Delta F_0 \pm \delta(\Delta F) \tag{2}$$

ΔF_0 being the 'equilibrium' value of free energy difference between states A and B corresponding to $\partial\varphi_2/\partial t \to 0$, and $\delta(\Delta F)$ being the above-mentioned 'correction term'. $\delta(\Delta F)$ can be of the same order of magnitude or even larger than ΔF_0.

The same state can be attained if one keeps φ_2 constant and lowers the temperature, T. Again, the form 2 of the change in free energy will hold, equation 1 being transformed into

$$\Delta F = \int_0^t \left(\frac{\partial F}{\partial T}\right) \frac{\partial T}{\partial t}\, dt \qquad (3)$$

which can be interpreted in an analogous manner. The essential point of all this discussion consists, however, not in the 'tricks' we do to avoid the usual description of relaxation phenomena accompanying supercooling, but in the reality of very different final morphologies of the system allowing different degrees of order (or disorder) that can be expressed in quantitative terms, for instance number of nuclei and degree of crystallinity.

3 'DEFORMATION OF THE BINODAL' AND RELATED CRITICAL PHENOMENA

In the lecture by Professor M. Gordon[16] very fine experiments are presented concerned with an abrupt interception of the binodal due to very rapid cooling. Though in Gordon's discussion the 'order' in the strict sense as noted in Section 1 is not formed, the process by itself is of great interest being closely connected with the possibility of different modes of nucleation.

The thermokinetic approach has an additional advantage allowing us to investigate the almost inverse situation, when due to an external field a point

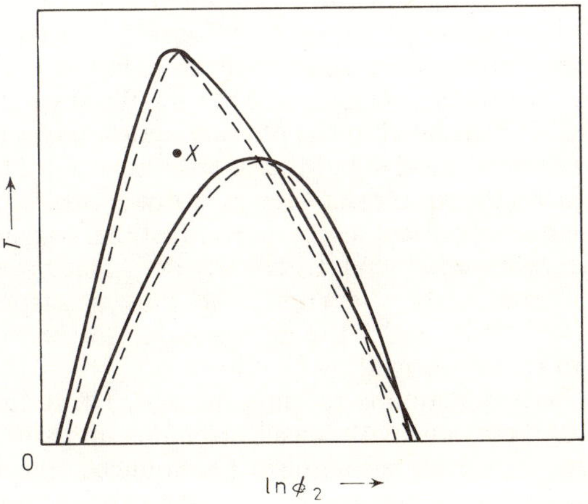

Figure 2. The deformation of the binodal. For details see text. The 'new' binodal is on the left

(*Figure 2*) primarily situated in the one-phase region suddenly finds itself under the spinodal, φ_2 and the temperature being kept constant. In this case not only a pulse-induced critical scattering may occur but a spinodal phase separation can start in the same manner as described by van Aartsen and co-workers[17–19]. (The idea of the 'binodal deformation' was inspired by the excellent study by Pennings[20, 21] which led me to a reconsideration and new interpretation of some of my earlier work[22, 23].)

Imagine a case of a steady Couette flow with the concentration of polymer in the solution being sufficiently low and the viscosity of the solvent sufficiently high to produce some uncoiling due to the velocity gradient $\dot{\gamma}$. (It should be kept in mind that $\dot{\gamma}$ has the dimension s^{-1}, i.e. time^{-1}; hence again time dependent variables are introduced explicitly in the following considerations.) We suppose further that at a given value of $\dot{\gamma} = \dot{\gamma}_1$ the molecular coils do not overlap, the solution thus being 'dilute'. At a larger value of $\dot{\gamma} = \dot{\gamma}_2$ due to uncoiling the co-ordinational spheres of the macromolecules begin to overlap, and following the classical definitions of Debye and Flory[24] the solution can no longer be considered as dilute. The uncoiling and overlapping unavoidably bring relatively large portions of different chains into contact; hence due to the external field the configurational entropy of the solution as a whole is changed and consequently its free energy is also changed. One can postulate that the behaviour of the solution at a velocity gradient $\dot{\gamma}_2$ is equivalent to that of a solution at $\dot{\gamma} = \dot{\gamma}_1$ with an increased Flory–Huggins interaction parameter χ_1 (i.e. with a decreased solubility) and with increased rigidity of chains to account for the increase of coil dimensions. Later we shall see that this equivalence is not a formalistic mode of description but a fundamental physical fact. At any rate, and this is already an experimental fact, the gradient $\dot{\gamma}$ can attain a critical value at which the chains start to precipitate due to overlapping and contacting; in the case of crystallizable polymers a crystalline highly ordered precipitate forms. Just this is the main result obtained by Pennings for the case of stirring polyethylene solutions which were quite stable prior to stirring.

We return now to the phase diagram (*Figure 2*). If at $\dot{\gamma} = \dot{\gamma}_1$ the solution of concentration φ_2 is stable at a temperature T, and if $\dot{\gamma}_2$ is the critical value of $\dot{\gamma}$, then at the same concentration and temperature but at $\dot{\gamma} = \dot{\gamma}_2$ the solution becomes *unstable*. However, this means that now the state described by the same point $X(\varphi_2, T)$ is situated in the phase diagram *under* the binodal and, possibly, also under the spinodal. In other words, to account for the gradient-induced change in phase equilibria we must draw a new binodal shifted to the left and with a higher critical temperature. Just this transformation of the coexistence curve represents what I call 'the deformation of the binodal'. If one changes $\dot{\gamma}$ from $\dot{\gamma}_1$ to $\dot{\gamma}_2$ abruptly the state described by the point (φ_2, T) also abruptly 'falls' under the new binodal, so the overall process is truly equivalent to that described by M. Gordon.

Proceeding in this way Pennings recently succeeded in attaining crystalline-network films of polyethylene with tensile strengths up to 100 kilograms per square millimetre. However, even more pronounced effects of the same 'binodal deformation' kind are observed in the case of longitudinal flow: we shall discuss them in more detail in the section concerned with the stream–thread transition.

4 'ANTIPLASTIFICATION' AND SPECIFIC ORDERED STATES DUE TO SELECTIVE POLYMER–SOLVENT INTERACTIONS

4.1 The primitive kinetic approach; crystallization induced by plastification

The main aim of this sub-section is to remind one that in many instances the diluent plays a role of a 'kinetic stimulant' for crystallization. For example, it is very difficult to start nucleation approaching the crystallization temperature from the side of the glassy state, polyethyleneterephthalate being perhaps the only positively established case when such crystallization is possible. This possibility is due to the abnormally large difference between the melting and glass transition temperatures, T_m and T_g. If in the case of any other polymer by addition of a diluent one can decrease T_g more than T_m, a situation is frequently attained which is favourable towards crystallization. There exists a well known practical rule: to induce crystallization in a polymer that does not crystallize due to kinetic hindrances, one must add a diluent which is a poor solvent for this polymer[7]. Being a poor solvent (the interaction parameter χ_1 is large) such a diluent practically does not affect T_m but lowers sufficiently T_g thus providing the amount of mobility of the chains necessary for nucleation.

From the point of view of molecular physics the situation with such polymers is just the same as in the case of kinetically stable solutions of PVA: thus we can call otherwise crystallizable polymers that fail to crystallize solely for kinetic reasons, 'kinetically stable' in the amorphous state. One such polymer is probably poly(vinyl chloride) (PVC), and a lot of papers dealing with its 'antiplastification'[25] demonstrate, in my opinion, the formation of stable nuclei that cannot, however, grow further, due, again, to the inherent diphility of this polymer and their being either less ordered or smaller than the corresponding nuclei in PVA. Hence, in this case the 'antiplastification' is (probably) the result of the kinetic stimulation of nucleation.

4.2 'Real' antiplastification: ordered structures that can be formed only in the presence of solvent or diluent

There exists, however, another class of ordered polymer–solvent systems where the order itself exists only in bicomponent systems. Probably the first data of this kind (though not quite typical: the effect is 'damaged' by an 'admixture' of the primitive effect described in the preceding subsection) are found in a relatively early paper of Kargin *et al.*[26]. They observed the formation and disappearance of isotactic polyacrylic acid spherulites on alternating dilution and evaporation or heating and cooling. The essential fact in their study was that the spherulites of the same size and shape appeared in the same places, as registered by polarizing microscope photographs.

Also in early work with Shaltyko and Agranova[27] I observed very similar behaviour in poly-γ-benzyl-L-glutamate (PBG)–solvent systems. There was, however, an essential difference in this case from the preceding one, namely, the spherulites were disappearing at polymer contents approximately 85 to 90 per cent (w/v), and reappearing at the same places—and in general being *the same ones*—on isothermal dilution with dioxane which is known as a helix-promoting solvent for PBG.

On a later reconsideration of this work we proceeded in the same manner using as a solvent the dichloroacetic acid (DCA) which is well known as a promotor of the helix–coil transition in PBG. However, in this case we

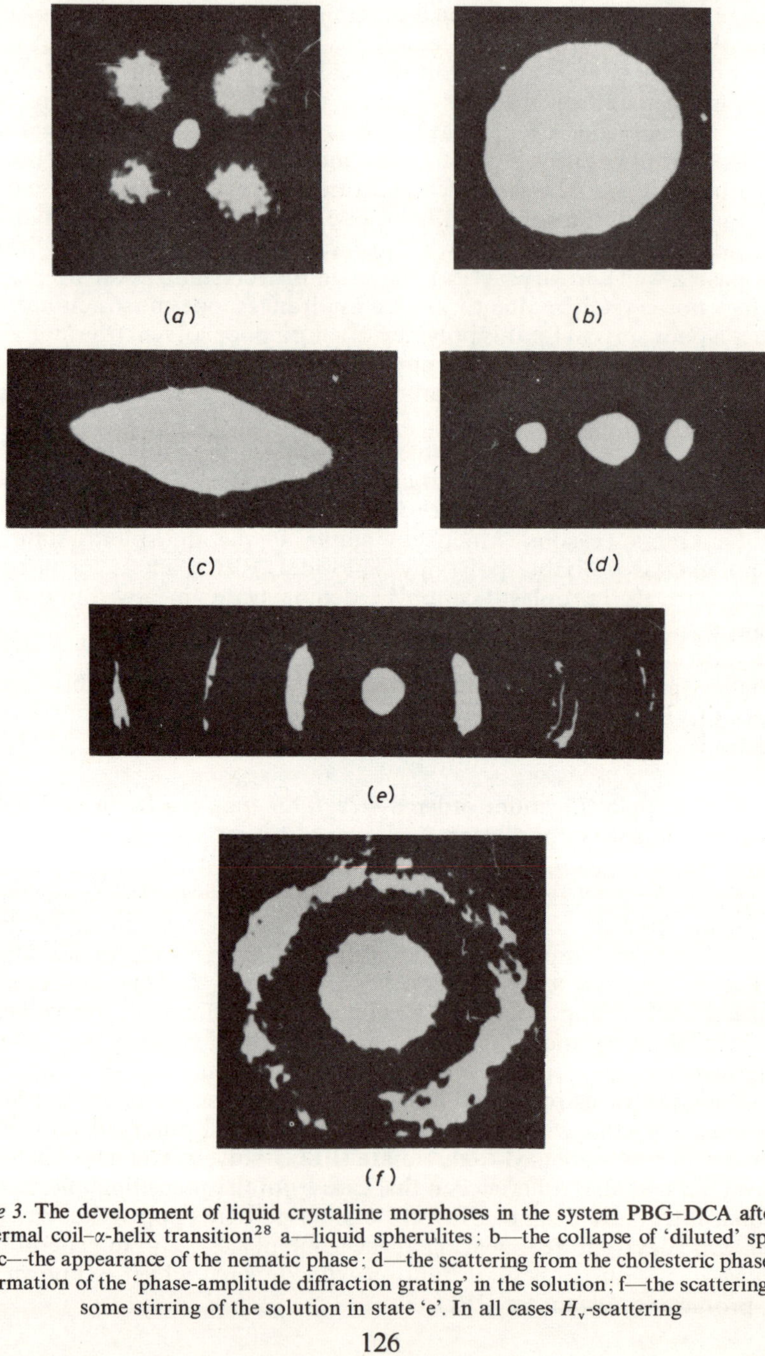

Figure 3. The development of liquid crystalline morphoses in the system PBG–DCA after the isothermal coil–α-helix transition[28] a—liquid spherulites; b—the collapse of 'diluted' spherulites; c—the appearance of the nematic phase; d—the scattering from the cholesteric phase; e—the formation of the 'phase-amplitude diffraction grating' in the solution; f—the scattering after some stirring of the solution in state 'e'. In all cases H_v-scattering

observed very different events at different concentration ranges. At a low concentration, just as one should expect, the solution was isotropic, and the PBG molecules existed in the random coil conformation as revealed by i.r. spectroscopy measurements. In the intermediate concentration range (the crucial concentration being dependent on molecular weight and temperature) an isothermal transition coil–α-helix takes place (checked by the amide I band), after which a whole set of liquid crystalline morphoses develops gradually (*Figure 3*, polarized small angle light scattering data, H_v-patterns). It should be mentioned that some similar morphoses but not as abundant were observed earlier by Robinson[29] who used polarizing microscopy only. At yet higher concentrations of PBG the situation with dioxane is repeated: in the 75 to 85 per cent PBG range liquid (?) spherulites appear, differing in shape, size and amount of order from those formed immediately after the coil–helix transition, but as DCA evaporates (without heating!) the spherulites collapse. If a drop of DCA or dioxane is inserted on the film with collapsed spherulites they reappear in the same place.

Though this study was described in detail in several papers[28,30,31], I should like to call to mind one specific feature of the most ordered morphose (*Figure 3e*) which was interpreted as optically equivalent to a phase-amplitude diffraction grating (the similarity of this polarized light diffraction pattern to wide angle x-ray diffraction patterns of highly oriented polymers also deserves some attention). This morphose displays two distinct temperature induced order–disorder transitions, each of them being a specific orientational melting, following J. Frenkel's[1] terminology (*Figure 4*). First to disappear is the higher range of order (cholesteric) as measured by the intensity of the first order reflection. However, in the course of a decrease of intensity the spacings

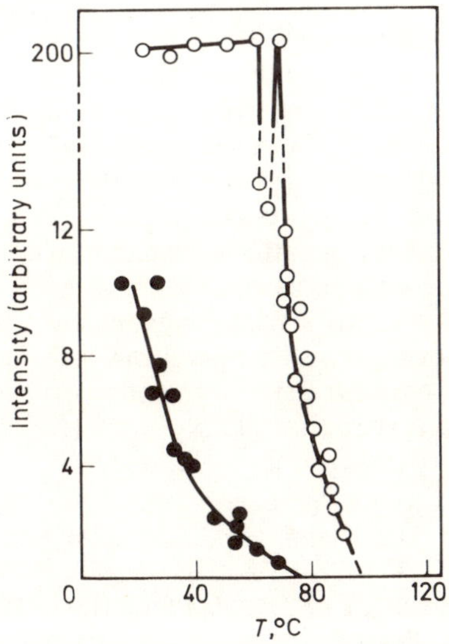

Figure 4. Orientational melting in the system PBG–DCA with the high degree of order (state 'e' of *Figure 3*). Black circles—1st reflection: open circles—the central spot

also decrease gradually, as if a desolvation of DCA proceeds[30] leading to an improvement of the liquid crystalline frame (quite an opposite behaviour was observed with dioxane as solvent[32]). With decreasing intensity of the first reflection the intensity of the central spot increases and splashes abruptly at the temperature of the first transition (71.5°C). Hence, the scattering intensity of the higher range of order is 'pumped over' into that of the lower one, which is probably simply nematic. This order persists, on further heating, until at 95°C a second orientational melting takes place, and the solution becomes isotropic, the macromolecules pertaining to the α-helical conformation as i.r. data indicate.

Thus a somewhat contradictory behaviour is displayed by PBG in different concentration ranges, DCA acting as a 'real' antiplastifier in the high concentration range only. However, an element of contradiction appears also within the liquid crystalline range, the higher type of order simultaneously improving (as the decrease in the 'interplanar distances' show) and gradually melting. Probably this observation together with the above-mentioned 'pumping over' is of general importance showing the possibility of temperature induced transitions proceeding in opposite directions (in respect to order) on different levels of structural (the Russian adopted term is 'supermolecular'[33]) organization. Moreover these data are a non-direct evidence of the coexistence of *two* (or more) liquid crystalline phases, as predicted by Flory's general treatment (see for details Miller's lecture[42]).

Returning to the isothermal concentration induced coil–helix transition followed by a phase transition from an isotropic solution to the liquid crystalline state, one must interpret it according to the Flory–DiMarzio[34, 35] theory of formation of a nematic phase by semiflexible chains. However, a simple interpretation in terms of a constant flexibility parameter f (being the molar fraction of 'flexible' bonds in the chain) and the corresponding critical concentration is here impossible. Again, though discussed elsewhere, it should be remembered that the same PBG sample ($M_w \approx 10^5$ with a narrow molecular-weight distribution) displays in dioxane quite another behaviour, starting to form liquid crystalline morphoses even at 7 to 8 per cent. At this same concentration, however, in DCA the macromolecules are still statistical coils with a supercritical value of f (i.e. $f > 0.63$) and thus incapable of forming a nematic phase at any concentration. What makes the phase transition possible is an abrupt change of the conformation of PBA leading to a change of the f value near to zero.

As was first pointed in our work with Elyashevich[36] and further discussed in my short communication at this Symposium[37] this means that the change of χ_1 at a specific PBG concentration φ_2^1 (and correspondingly DCA concentration $\varphi_1^1 = 1 - \varphi_2^1$) is such that the exponent $(\Delta\chi_1/kT)\varphi_1^1$ in the formula for the resulting flexibility coefficient

$$f = f_0 \exp\left(-\frac{\Delta\chi_1}{kT}\varphi_1^1\right)\Big/\left\{1 - f_0 + f_0 \exp\left(-\frac{\Delta\chi_1}{kT}\varphi_1^1\right)\right\} \tag{4}$$

is close to infinity thus turning f to zero, f_0 being the initial supercritical value of f. Such a catastrophic change (I was tempted to give this set of events the name of 'conformational catastrophe') may be due only to a desolvation—just of the same kind as described by Doty[38] for dilute PBG solution in

DCA–dichloroethane mixtures on heating (in a certain composition range).

The peculiar behaviour of PBG in a higher concentration range is probably connected with thermokinetic rather than thermodynamic reasons, the rodlike helical macromolecules being incapable of rearranging themselves in a crystalline array without the aid of some external field that will induce a new α–β conformational transition. Thus it is possible that the addition of DCA to a dried film on a supporting glass plate stimulates a series of local α–β transitions, the necessary energy being provided by surface tension, and the β-conformation being more suitable for packing of the chains into three-dimensional crystalline units which in their turn are packed in a spherulite or a similar morphose. In this respect (the energy is provided by surface tension) the reaction of the film on dioxane is interpreted similarly.

It should be pointed that this last interpretation is still hypothetical, contrary to that concerning the real liquid crystalline state of PBG. However, DCA in different concentration ranges produces undoubted intramolecular disorder–order transitions which are followed by transitions of related kind on higher structural levels. The source of all these effects is the specific and concentration dependent PBG–DCA interaction, most probably of a 'solvation–desolvation' kind, thus supporting the ideas presented in the lecture of Dr Kang-Jen Liu[13].

Quite recently a very similar behaviour of poly(2,6-dimethyl-1,4-phenylene oxide) (PPO) spherulites was reported by Horikiri and Kodera[39]. These authors were able to show directly that after the collapse of spherulites due to solvent elimination the dried and solvent free films were amorphous as checked by different methods. The addition of solvent led to an intense spherulitization showing just as in the preceding case that the presence of solvent (in the strict sense of the word—not 'diluent' or 'plastifier') is necessary not only to stimulate crystallization, but—which is far more important—*to stabilize the crystalline phase.*

Though the term 'antiplastification' arises from technological slang, if we still adopt it and want to ascribe it some physical meaning, we must denote by antiplastification a specific mechanism of stabilization of the crystalline (or other ordered) state in the presence of solvent.

This brings us back to another study made by us in the sixties[40] where real x-ray detected crystallization was possible only in the presence of solvents, each solvent (formic or acetic acid, or phenol) producing a somewhat different unit cell. The polymer was poly(octamethylenedibenzimidazole) (POMBI) with the repeating unit

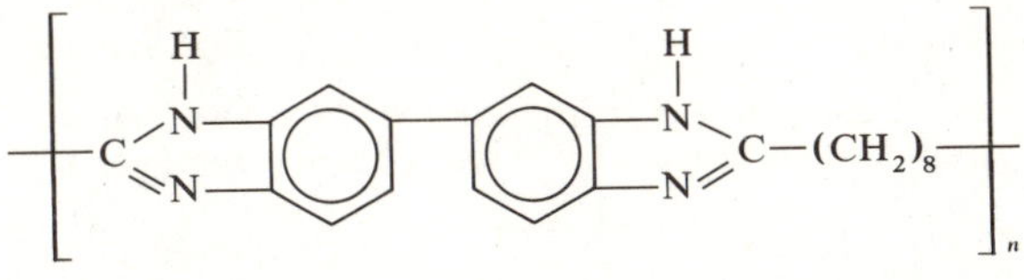

In this case it is probably possible to obtain the f-value by direct calculation: however, it is beyond doubt that on shortening of the methylene sequence f will decrease. The tetra-analogue of POMBI gave a rich x-ray pattern: there were some difficulties in obtaining the hexa-analogue in the crystalline state (in this case the action of formic acid being a mixture of kinetic stimulation and 'real antiplastification'), and with POMBI it was quite impossible to obtain a dry crystalline film or fibre, though in the latter case one could observe some traces of a not very developed unidimensional order.

Hence, simple logic leads to a conclusion that the increased flexibility of oligomethylene links between the bulky dibenzimidazole groups makes an ordered packing impossible. Again we must turn to equation 4 and suggest that some specific interaction of the solvent (usually, formic acid), with the polymer will this time be a kind of solvation (thus, a 'positive' interaction, as compared with PBG–DCA system in the intermediate concentration range). The solvated POMBI chain at some polymer:solvent ratios behaves as a complex with increased rigidity making it similar in respect of crystallizability to the $(CH_2)_4$-analogue. To account for this change of flexibility one must calculate the changes of free energy for the amorphous and crystalline states of the polymer–solvent system. The calculation leads for the free energy in the crystalline state

$$\Delta F_{cr} = \chi_c \varphi_1 + (C\varphi_2 - \varphi_1) \ln \frac{C\varphi_2 - \varphi_1}{C\varphi_2} + \varphi_1 \ln \frac{\varphi_1}{C\varphi_2} + \Delta F^0_{\varphi_1 = 0} \quad (5)$$

Here χ_c is the Flory–Huggins parameter in the bicomponent crystalline lattice, C is the stoichiometric number of solvent molecules that can be bound with one chain repeating unit, and $\Delta F^0_{\varphi_1=0}$ is the free energy of the polymer with the same lattice in absence of the solvent. Equation 5 holds for

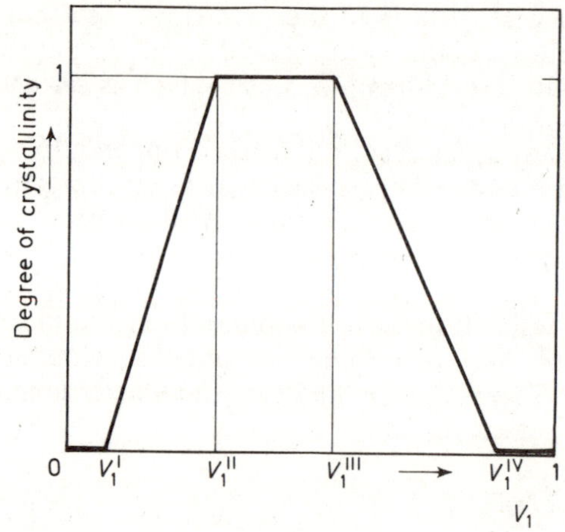

Figure 5. The coexistence diagram of crystal-solvates[36]. The volume fraction of the solvent is plotted on the abscissa (φ_1 in the notation of this paper)

$\varphi_1 < C\varphi_2$. At $\varphi_1 = C\varphi_2$ the lattice is completed, and this concentration corresponds to the minimum of free energy for the crystalline state, though crystallization starts actually at lower solvent concentrations. At $\varphi_1 > C\varphi_2$ the solvent can no longer enter the lattice without damaging it. *Figure 5* illustrates the situation. It can be seen, again, that the crystalline state is wholly stable only in an intermediate range of polymer:solvent ratios, dissolving on dilution and gradually 'melting' at concentrations $\varphi_2 \geqslant 1 - \varphi_1^{\mathrm{II}}$.

As to the mechanical properties of POMBI fibres, in the crystalline state they are sufficiently poor. However, after thermal treatment, elimination of solvent and loss of three-dimensional order the tensile strength improves essentially attaining values as high as 70 kg mm^{-2}. Hence, one should conclude that the crystals observed are plastic in nature. The same is most probably the case for PBG at low DCA concentrations and for the collapsing PPO spherulites. It follows that the choice between notions 'plastification' or 'antiplastification' is in this case a linguistic rather than physical problem.

5 THE NATURE AND SOME DYNAMIC PROPERTIES OF POLYMERIC LIQUID CRYSTALS

Much attention has been devoted to liquid crystals in this Symposium, both in the main lectures and in short communications. Two distinctly different types of ordered mesophases were discussed: one, containing rod-like or at least sufficiently rigid chains (with subcritical values of Flory's parameter: $f \leqslant 0.63$), and the other formed by block-copolymers. We shall discuss very briefly both kinds paying more attention to their common properties than to differences.

5.1 Tactoid and related systems

These systems obey in general the laws following from the theories of Flory and DiMarzio[34,35]. However, as can be seen from the data already presented, it is not necessary to treat their nature as essentially entropic[42], and purely energetic factors, such as specific polymer–solvent interactions or external influences on the chain conformation (see below) should be taken into account.

In this subsection I shall try to apply some of the recent criteria developed for ordinary liquid crystals to the polymeric ones, and in particular to the problem of their domain structures, the boundaries between domains being formed by disinclinations.

As an example of a 'classical' Flory–DiMarzio polymeric liquid crystal, the poly(*p*-benz)amide (PBA) has been chosen, which was studied in detail by Papkov and co-workers in the USSR[43–45]. Statistical mechanical considerations in Flory's[46] manner lead to a conclusion that in the case of *trans*-configuration of the amide bond the PBA chain

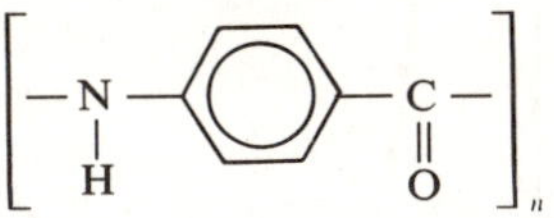

must be fairly rigid ($f \to 0$) in spite of the possibility of the rotation of the benzene ring. The behaviour PBA displays in dimethylacetamide with 3 per cent lithium chloride added is characteristic of rodlike macromolecules both in the high and medium dilution range. For instance, the critical concentration of the formation of the tactoid phase

$$\varphi_2^* = \frac{8}{p}(1 - 2/p) \tag{6}$$

is distinctly observed, the axial ratio p being of course proportional to the molecular weight: further, the intrinsic viscosity is connected with p or M_w by an exponential formula of the usual type $[\eta] = K_\eta M_w^a$ but with a value of a close to 1.7, which, according to Simha, is the limiting value of the exponent for rodlike particles[47].

However, most striking are the absolute viscosity data. *Figure 6* shows the

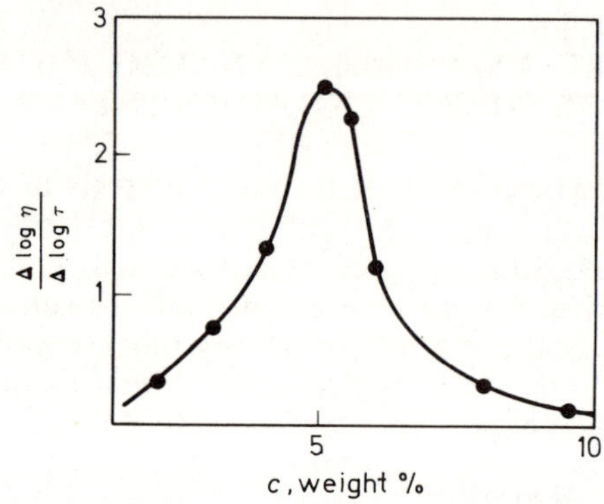

Figure 6. The concentration dependence of the 'viscosity anomaly' of PBA

viscosity shear stress relation as a function of the concentration of PBA, and *Figure 7* shows the zero shear viscosity of the same sample as a function of temperature. We shall omit different rheological data concerning the

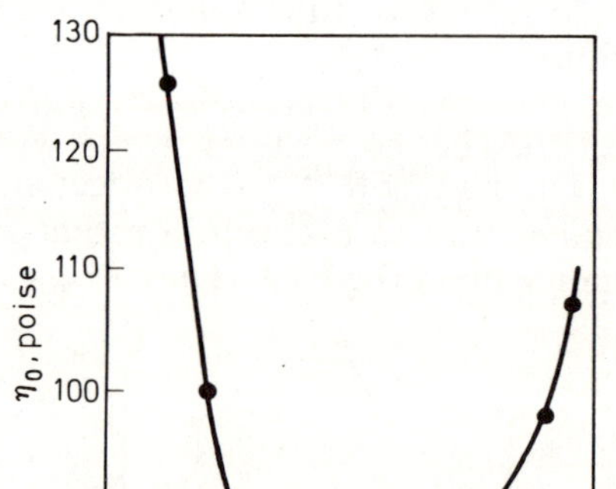

Figure 7. The temperature dependence of the viscosity of a 5.5 per cent solution of PBA

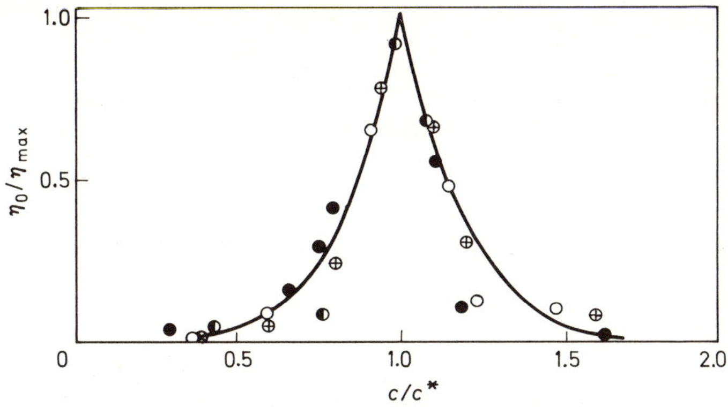

Figure 8. The dependence of the reduced absolute viscosity on the reduced concentration of PBA

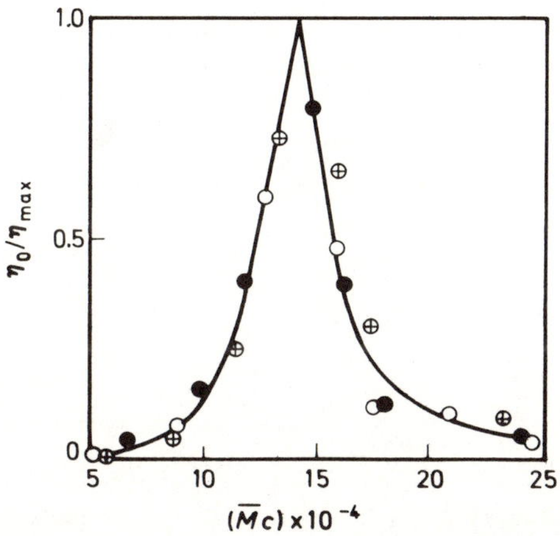

Figure 9. The reduced absolute viscosity of PBA as a function of the product $\overline{M}c$

master curves and related forms of representation, but some intermediate plots should be mentioned (*Figures 8* to *10*). There are distinct singularities in the dependences of reduced absolute viscosities on reduced concentrations or on the product $\overline{M}c$, the critical value of the latter being in all cases 1.3×10^5 at 20°C (c expressed in weight per cent).

All those singularities indicate that at a given molecular weight and concentration the PBA solution undergoes a phase transition from the isotropic to anisotropic state, the critical values on the abscissae allowing us

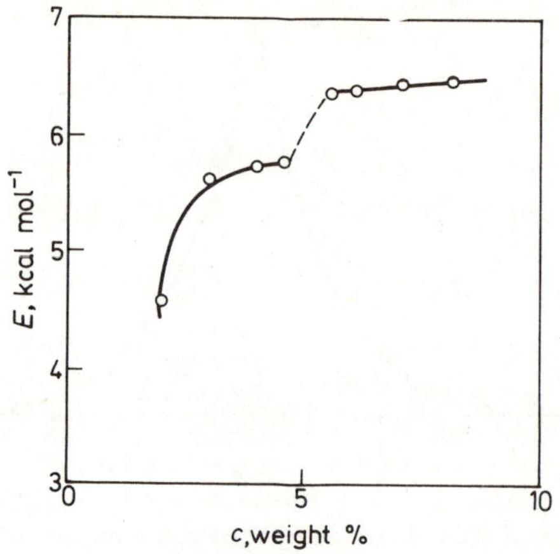

Figure 10. The concentration dependence of the energy of activation of viscous flow of PBA

to check the position of the binodal (*Figure 11*: the investigated temperature range is shown). Finally, x-ray data show a gradual decrease of interplanar distances with PBA concentration, these distances becoming constant after a specific (for the given sample) concentration is attained.

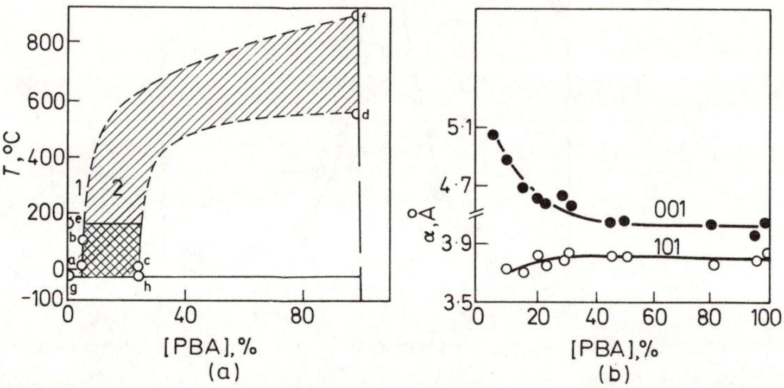

Figure 11. a—A part of the phase diagram of PBA. The hatched region (2) corresponds to the liquid crystalline state; the region actually evaluated is marked by cross-hatching. b—The concentration dependence of interplanar distances in PBA. The reflections are noted above the corresponding curves

The abrupt fall of the viscosity after the intersection of the binodal can be explained in one manner only: the mechanism of flow itself undergoes a change, the real viscous flow (with all anomalies peculiar to polymers) being changed into a slip along the disinclinations or the domain boundaries.

One rather important observation, this time of a technological nature, should be added. Turning to the American data[48, 49] obtained with the same polymer or with poly(phenyleneterephthal)amide, very similar both in structure and properties, or, with a 100 per cent *para*-aromatic polyamido-hydrazide which displays a very similar behaviour (these polymers being connected with the so called 'Fiber B' and 'Fiber X'), one can draw the conclusion that on wet spinning the solutions of these polymers allow the attainment of very high modulus and tensile strength in these cases only when one proceeds with spinning from the *anisotropic* state. The overall draw ratio in this case, as a rule, does not exceed 150 per cent. The most probable explanation of these results is that some sort of 'stocks' of the future fibre preexist in the solution, these stocks being the unperturbed domains. The mechanical longitudinal field during spinning packs the domains one along another in such a manner that the internal orientation in the domains coincides with the overall orientation. This statement is equivalent to the suggestion that the domains disappear, the whole sample attaining the state of a 'liquid single crystal'. That it really can be a case is shown by the data on the behaviour of the PBA solution in a strong magnetic field which I presented in one of the short communications at this Symposium[37] (see also *Figure 12*). In this case again the external field

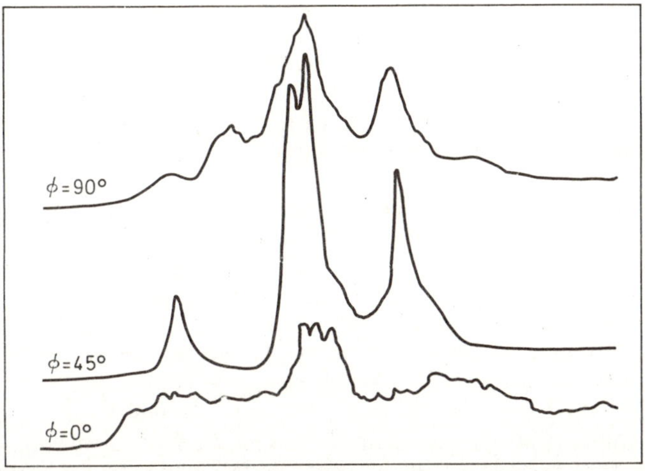

Figure 12. The anisotropy of p.m.r. spectra of dimethylacetamide at PBA concentrations higher than 8 per cent (by weight). ϕ—the angle between the orientation of the solution (attained before the rotation of the cell and persisting a short time after the turn) and the new orientation of the magnetic field (in the co-ordinate system of the sample)

reorganizes the overall structure on its highest (on the scale of dimensions) level promoting a very specific spatial uniaxial orientation, the solvent molecules participating in the organized macromolecular frame. If one changes the direction of the magnetic field or turns at the same angle the cell with the solution the latter reorients as a whole following the new field direction, the time of structural relaxation (or rather retardation) being approximately 30 minutes. Here also critical parameters exist, one of them being the concentration (or product $\overline{M}\varphi_2$), and the other the field intensity that must be of the order of 10^4 Oersted units. At lower fields no such effects are observable. The field-induced spatial nematic order is retained to some extent after switching the field off, the relaxation time (of reestablishment of the initial domain structure) attaining several days or even a week.

An effect similar in nature was previously observed by the Japanese for PBG[50].

However, in our case the structure and mechanical behaviour of a PBA film obtained in these conditions were estimated. The film has a practically zero tensile strength in the direction normal to that of the force lines of the magnetic field, fibrillating spontaneously along these lines. This is just the behaviour one should expect from a flat nematic solid single crystal. For further discussion, the analogy of this fibrillization in the rigid-chain. wholly-oriented system with that occurring in superoriented flexible chain polymer films should be kept in mind.

5.2 The mesophases in binary and sandwiched block copolymers

Turning to this type of liquid crystal one can immediately see a striking difference in morphology as demonstrated in the main lecture of Dr B. Gallot[51]. All periodic structures presented belong to higher levels of organization, the lower level being formed by an assembly of crystallizable (or non-crystallizable!) blocks and having an *ordinary* crystalline (or amorphous) structure. These primary units are assembled in their turn in one or other type of supermolecular organization characterized by x-ray large periods. This is the case where higher order structons appear quite definitely. One can mention at least two levels of packing. The first is formed by segregated blocks, whose aggregation leads to fairly monodisperse superunits, not necessarily crystalline. However, the mere fact that their dimensions are equal (due, of course, to the extremely narrow distribution of block length) makes possible their further packing in a fully ordered array which has every 'right' to be called crystalline, as has any fixed geometrically-ordered spatial distribution of equivalent units.

The diphility and resulting specific block–solvent and block–block interactions play in this case the role of a 'driving force' of a purely thermodynamic nature, which 'triggers' and further promotes the whole ordering process. It is a possibility (though needing more direct confirmation) that in the case of cylinders they behave as Flory's tactoids following the general scheme of formation of a tactoid (nematic) phase. This possibility is to some extent substantiated by the fact that under specific flow conditions (i.e. when the system is subjected to the action of an external field) the ordering may proceed at elevated temperatures *in the absence* of solvent, as was shown in the very fine study of the Bristol group (A. Keller *et al.*, short communication presented at this Symposium). Thus, the thermodynamic driving force may be substituted by a mechanical one, leading to the conclusion (which is preliminary. still needing more direct support) that the systems under consideration enter into the frame of the equivalence principle discussed below (Section 6). There are several other features common to the systems ordered by segregation phenomena and to those formed by rigid macromolecules.

(*i*) As a rule, the macromolecules able to form liquid crystalline mesophases are diphilic—either inherently or in the explicit form. All block copolymers present an extreme case of diphility, each block having its 'own' preferential or selective solvent which usually behaves differently towards the other type of blocks. As is well known, in dilute solution this diphility results in the formation of molecular micelles[52, 53] which are visualized as insoluble 'nuclei' still kept in solution by the soluble block or blocks. Without doubt, if the insoluble block is crystallizable and flexible it will form a folded crystal nucleus, the nucleation being monomolecular. Thus the growth of lamellar crystals from blocks of only one kind is inevitable due to secondary nucleation proceeding in the same manner when the overall concentration is increased. The lamellae are still held in solution or gel, by the soluble blocks, thus forming one of the special cases of liquid crystals. The styrene–ethylene-oxide block copolymers are an example.

If we return to PBG the very same type of behaviour is observed if the affinity of the solvent for the side groups is sufficiently greater than for the polypeptide backbone. The solubility of helical PBG as compared to the *in*solubility, for instance, of helical poly-L-alanine may be explained in the same terms as the formation of molecular micelles. The α-helical backbone is *essentially desolvated*, forming some kind of intramolecular 'linear-crystalline' structure (about the classification of intramolecular structures see Ref. 54) which one can visualize as an insoluble spring still held in solution by the side groups. This model accounts fairly well for the desolvation effects discussed in the preceding section.

The behaviour of PBA could be explained in a similar manner, though the discussion of the details would lead us far aside from the main item.

(*ii*) The diphility and the connected segregation effects, especially in the presence of selective solvents, may give rise to crystallization or liquid crystallization mechanisms of the same type as discussed above (variation of solvent leading to sufficiently strong variation of χ_{1A}, the second subscript indicating the crystallizable block in the case of block copolymers). It should also be kept in mind that the non-crystallizable block in the case of partial compatibility may affect the glass temperature of the crystallizable block, thus providing better kinetic conditions for crystallization.

(*iii*) In the case of *in*compatibility of two blocks both being soluble in a common solvent at least one of them may attain, as a result of segregation, a subcritical f value which is followed as concentration is increased by formation of nematic or cholesteric domains.

One could proceed with (*iv*), (*v*) . . . etc.; however, any attempt at a further, even qualitative analysis will be premature, for the preceding discussion was held mainly on a hypothetical, though not improbable, basis. Nevertheless, the concept of diphility and resulting 'partial solubility' as necessary conditions for the formation of liquid crystals or other specific bi- and tri-component ordered states in which the solvent participates *actively* is not a hypothesis and may turn out to be the 'common denominator' for all known types of polymeric liquid crystals. If this is true, then Flory's formalism in the modification with the variable flexibility parameter and incorporation of higher order structons (tactoids formed by aggregated blocks) will also hold for all cases of ordered structure formation. In the next section we shall try to show that the same is true in the presence of external fields.

6 NUCLEATION AND CRYSTAL GROWTH IN THE PRESENCE OF EXTERNAL FIELDS AND THE EQUIVALENCE PRINCIPLE OF THERMOKINETICS

Now we return to the thermokinetic treatment of the formation of ordered structures in dynamic conditions. One should immediately note that Zachmann's treatment[55] is wholly applicable to these conditions when the external field drastically affects both the long range diffusion of chain segments and the probability of incorporation of a new chain in the growing crystal. However, this is a problem of orientational crystallization that will be treated in the end: just now we shall consider the direct application of the equivalence principle to systems subjected to longitudinal flow.

6.1 Stream–thread transition and the 'orientational catastrophe'

Our earlier work[22,23] that was reconsidered recently consisted in

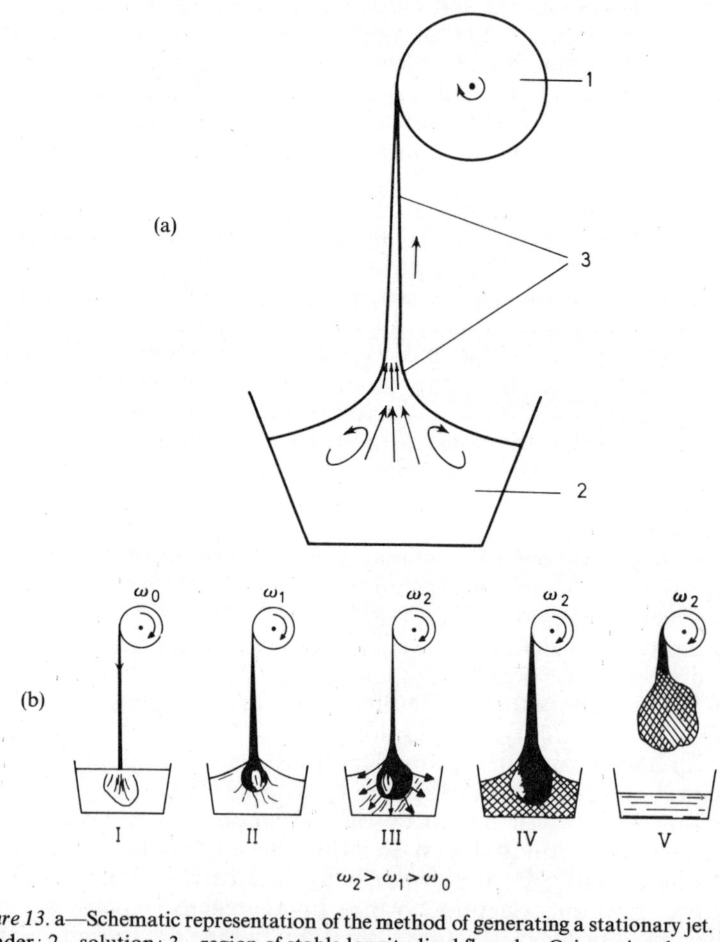

Figure 13. a—Schematic representation of the method of generating a stationary jet. 1—Rotating cylinder; 2—solution; 3—region of stable longitudinal flow. b—Orientational catastrophe with the fibroin solution. The arrows show the direction of propagation of the perturbation wave

generating a stable liquid filament by means of a rotating cylinder (*Figure 13*) the height of which above the level of the polymer solution and the angular velocity ω could be varied. The solutions studied were PMMA in dimethylformamide and silk fibroin *Bombyx mori* in acetone–lithium bromide solution with some admixture of water. Only visual observations of the stream–thread transition were made, some of them substantiated later by means of polarized light scattering with a laser as a light source.

As was described in 1965, the transformation in PMMA at increasing ω starts with the appearance of strong birefringence followed by opacity. The stream ejects macroscopic drops of the solvent (probably not pure solvent; see below), and at very high ω or, which is the same, longitudinal velocity gradients $\dot{\gamma}$, a kind of fog is observed around the stream.

The fibres thus formed still contain approximately 30 per cent of solvent and are strongly birefringent. Being kept in a free state they gradually lose the solvent and contract, the birefringence disappearing and thus showing the relaxation of chain orientation. However, under isometric conditions, when the fibre cannot contract, an internal stress of the order of 25 to 50 kg mm^{-2} develops, and at some ultimate moment depending on the remaining solvent concentration the fibre 'annihilates' being not merely torn but 'exploded' into powder by internal stress. This first kind of orientational catastrophe strongly resembles the well known 'annihilation' of inorganic glasses on spontaneous crystallization.

The second type of orientational catastrophe observed with fibroin is even more spectacular. In this case one can follow the propagation of the crystallization (or, rather, solidification wave) in a direction opposite to the direction of flow, the solution changing in colour from light yellowish-brown to dark brown. This wave (*Figure 13*) attains the level of the liquid and then expands in the bulk of the solution, which is finally pulled out of the

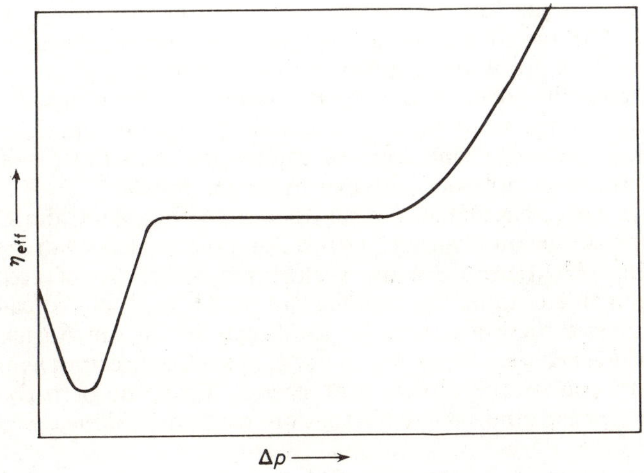

Figure 14. The dependence of the effective viscosity on the longitudinal velocity gradient. (In the actual conditions of the experiment[22] the pressure decrement Δp was measured in the jet viscometer instead of $\dot{\gamma}$: Δp is proportional to $\dot{\gamma}$)

vessel as a gelated bulb. An estimate of the tensile strength of the filament needed for this process gives again a figure of approximately 50 kg mm^{-2}.

Knowing the structure of fibroin and the ability of its crystallizable peptides to display the $\alpha \rightarrow \beta$ transformation, one can explain this catastrophic transition in the stretched solution by just such a transformation induced by stretch. *Figure 14*, showing the so called complex flow of fibroin solutions, substantiates this conclusion. The minimum of effective viscosity η_{eff} occurs as a result of competition between ordinary (shear) and longitudinal flow[23]. After the minimum is attained the latter predominates, a positive feedback being established between longitudinal viscosity $\bar{\eta}$ and the rate of strain. The first increase of η_{eff} is due to uncoiling, and the plateau region corresponds to a kind of steady state where sufficient energy must be accumulated to produce the $\alpha \rightarrow$ coil $\rightarrow \beta$ transformation of the molecular level. The second rise of $\bar{\eta}$ (the shear viscosity being now negligible) is due to the abrupt increase of multilateral interaction of the extended chains, this interaction proceeding as a formation of interchain hydrogen bonds characteristic of the β-conformation of polypeptides.

To explain at least qualitatively the phenomena described one must accentuate their main features. There are:

(i) The strong birefringence of solvent containing PMMA fibres and their collapse on loss of solvent;

(ii) The occurrence of a crystallization or perturbation wave moving in the direction opposite to that of flow;

(iii) The ejection of solvent at increasing angular velocities during the transformation.

Features (i) and (iii) resemble the collapse of PBG or PPO spherulites on isothermal evaporation of the solvent, and the desolvation phenomena in the PBG–DCA system leading to formation of liquid crystals, respectively. Feature (ii) needs deeper consideration. The initial state and the state preceding the expansion of the perturbation wave in the volume of fibroin solution present just the case discussed in a general manner in Section 3 (*Figure 2*). The orientational catastrophe with the fibroin solution being transformed to a cross-linked gel is a *very typical spinodal phase separation* resulting from the 'deformation of the binodal'. The stress induced conformational change leads to an immediate increase of the Flory–Huggins parameter χ_1, strong desolvation of chains and substitution of polymer–solvent contacts by polymer–polymer hydrogen bonds.

It should be remembered further that according to Cahn's theory of spinodal phase decomposition verified for polymers by van Aartsen and co-workers[17–19], there are some preferential directions of growth of the composition fluctuations. As pointed out by Cahn[56], in the case of metal alloys or glasses these are some lattice directions. In van Aartsen's studies there were no such directions, but in our case such a direction again appears, and it is, of course, the direction of stretch. Thus, the perturbation wave observed represents the front of fluctuations expanding along this preferential direction.

If we turn to the F–T formalism, then the situation is presented as follows (*Figure 15*). The stable solution of the crystallizable polymer (fibroin) may exist at a given concentration in two states—crystalline and amorphous.

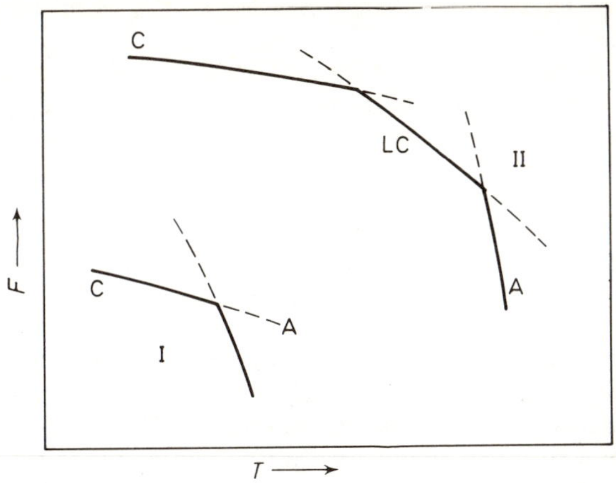

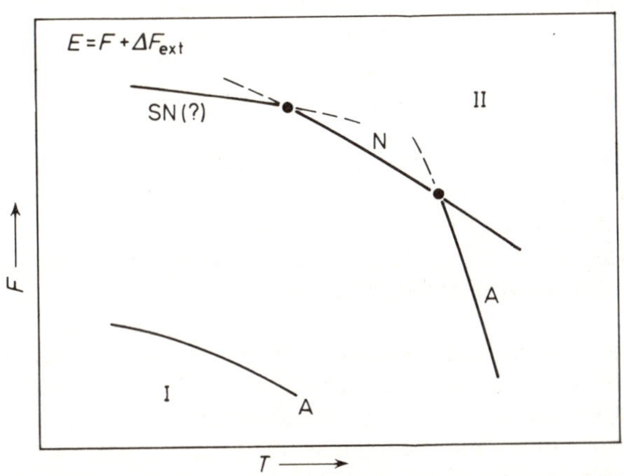

Figure 15. The illustration of the stream–thread transition on the F–T diagram. I—unperturbed system. II—a stretched system. a—fibroin. b—PMMA. Notations of states: A—amorphous; C—crystalline: LC—liquid crystalline (any kind): N—nematic, SN—solid nematic. The interrogation sign denotes the doubtfulness of its stability: see text

with a transition point (*Figure 15a*). A hypothetical full analogue of fibroin differing from the latter only by the value of the flexibility parameter, f being subcritical, will display a different behaviour (the upper set of curves on *Figure 15a*), existing in three possible states—crystalline, nematic, and isotropic. However, by subjecting the solution to the action of a mechanical field we *in fact changed* the initial value of the flexibility parameter f_0 to a new one

$$f = f_0 \exp(-\Delta\varepsilon/kT)/\{1 - f_0 + f_0 \exp(-\Delta\varepsilon/kT)\} \qquad (7)$$

$\Delta\varepsilon$ being the part of the external field energy ΔF 'pumped over' in the internal energy of the chain limiting its freedom of rotation, the other part of this energy affecting the configurational entropy of the solution with chains already extended.

Hence, from the point of view of Flory–DiMarzio theory there is *in fact* no difference between the behaviour of an unloaded solution of inflexible molecules and a loaded one of flexible molecules. The equivalence comprises not only the kinetics and morphology of phase separation, but also its thermodynamics, for desolvation *in fact* occurs and the additional energetic term $\Delta\varepsilon$ can be substituted by $\Delta\chi_1\varphi_1$ from equation 4.

However, this equivalence may also be treated as just a special consequence of another general principle advanced by Katchalsky and Kuhn—the teinochemical principle[57,58].

The situation with PMMA is very similar in its general features (*Figure 15b*); however, due to the fact that it cannot form an organized phase at rest only one state is possible in the untreated solution. But when this solution is stretched, again, a spinodal phase separation occurs, the more dilute phase being ejected from the stream and the more concentrated rearranging itself into a nematic phase. Just at this point an essential difference between fibroin and PMMA should be noted. In fibroin the ordered state, either nematic or crystalline, is fixed due to hydrogen bonding. In PMMA the ordered state can be sustained only on the condition that some external force equilibrates the rubber-like internal entropic stress which leads in the free state to contraction. In addition, the nematic state needs some amount of solvent for its stabilization, the elimination of solvent leading to multiple fracture ('explosion'). Of course, the cause of the orientational catastrophe with PMMA is the high amount of (mechanical) energy absorbed by the system and not released as heat of crystallization.

Discussing the stream–thread transition in thermokinetic terms and using the equivalence principle one must not forget about the purely mechanical side of events and the existence of the Poisson coefficient adding a 'mechanical term' to the desolvation energy $\Delta\chi_1\varphi_1$. In addition, in the course of desolvation (or solvent ejection, which is the same at first approximation), both $\Delta\chi_1$ and φ_1 change in different directions, but this does not affect the general qualitative physical treatment presented.

At the end of this subsection I would like to make two remarks not directly connected with the main item but following from the binodal deformation and corresponding *F–T* formalism.

(i) Many unsuccessful attempts to obtain fibres from globular proteins are due to the fact that their diphility towards moisture is ignored. One can proceed with ordinary spinning using ordinary processing for these proteins, but their behaviour will be the same as for PMMA filaments though for other reasons. The so-called Fisher diagram[59] (*Figure 16*) shows a general correlation between the linear dimensions of proteins in water and their hydrophobic amino acid content. All natural protein fibres are formed by proteins situated in the upper left side of the diagram, the globular proteins being spread along the right side of the regression curve. The parameter *L* on Fisher's diagram formally corresponds to a χ_1-dependent value of the flexibility parameter. One *can* extend the chains of the globular proteins and obtain fibres, but in contact with water due to the product $\Delta\chi_1\varphi_{H_2O}$ the effective *f*-parameter attains at a high rate its supercritical value, and the

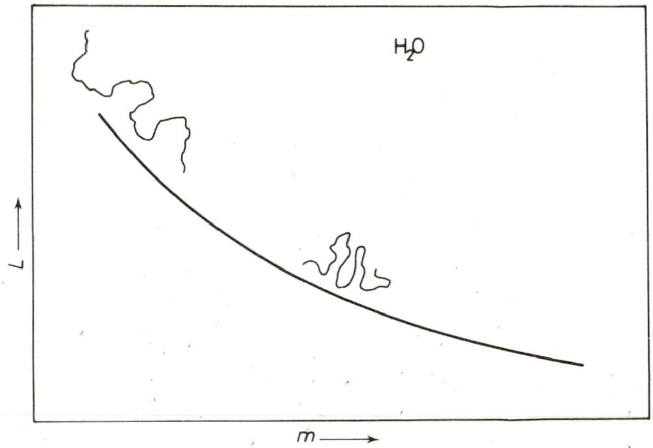

Figure 16. The Fisher diagram for proteins. *L* denotes linear dimensions which increase on decrease of the effective *f*-value. *m*—Fraction of hydrophobic units in the chain. Above the regression curve the corresponding conformations are schematically presented

resulting internal entropic stress may be sufficient not only to produce severe contraction of the fabric but even its 'annihilation'.

(ii) Under some flow conditions one observes pulsations which are ascribed to 'rheological anomalies' of polymer solutions. However, under conditions close to turbulence all can be explained by periodic binodal deformations, this time a negative feedback being provoked in the system. Due to turbulence some chains start to uncoil and fall out of solution; the forming of a smooth cluster of molecules extinguishes the turbulence, and the corresponding local region of solution returns to the unperturbed state, which is followed by the back dissolution of the fluctuation induced. This is illustrated by *Figure 17* which needs no further comments.

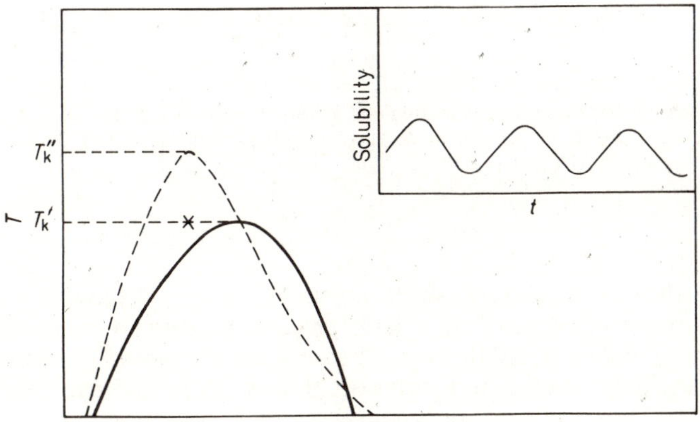

Figure 17. The interpretation of flow pulsations in terms of 'binodal deformation'

6.2 Orientational crystallization of concentrated flexible chain polymer solutions

This case is very similar to that discussed above for fibroin, the only difference consisting in the fact that only simple uncoiling takes place, without intramolecular transitions and additional fixation due to hydrogen bonds. Since the orientational crystallization was discussed in more detail in a special paper presented at this Symposium[60], I shall remind you just of the main items.

In any external field that uncoils and orients the macromolecules at some intermediate field strength a competition begins between mono- and multi-molecular nucleation, the rate of both being given by a formula of the usual type:

$$\dot{N} = N_0 \exp(-\Delta F^*/kT) \tag{8}$$

the free energy term ΔF^* taking into account the degree of molecular orientation induced by the field. If one plots ΔF^* as a function of molecular orientation (*Figure 18*) one can see that there exists some critical orientation

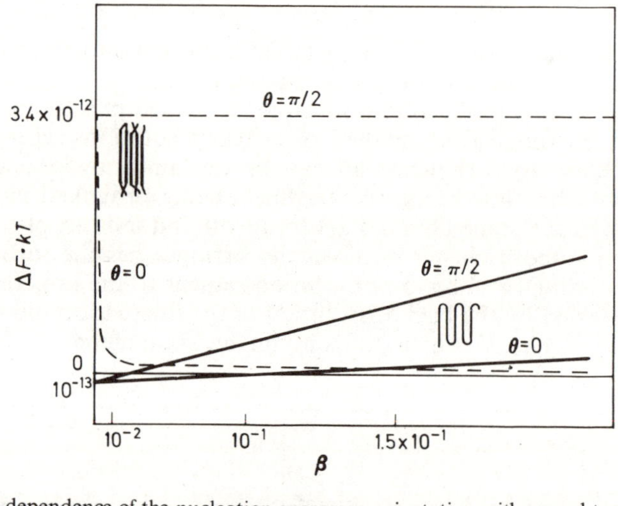

Figure 18. The dependence of the nucleation energy on orientation with regard to the stretching field (orientation angle θ) and on the mode of nucleation (schematized near the corresponding curves). β—Degree of coiling; when β tends to unity, f tends to zero

at which the formation of multimolecular nuclei becomes preferable. Turning from orientations to the fields producing them we may state that at intermediate field strength the 'non-thermodynamic' mechanism of Zachmann will act leading to formation of large loops and irregular folding; however, at still higher fields the general thermokinetic equivalence principle must lead to the possibility of coexistence of three phases (*Figure 19*). Then the fate of the growing crystals will depend on the position of nucleation on

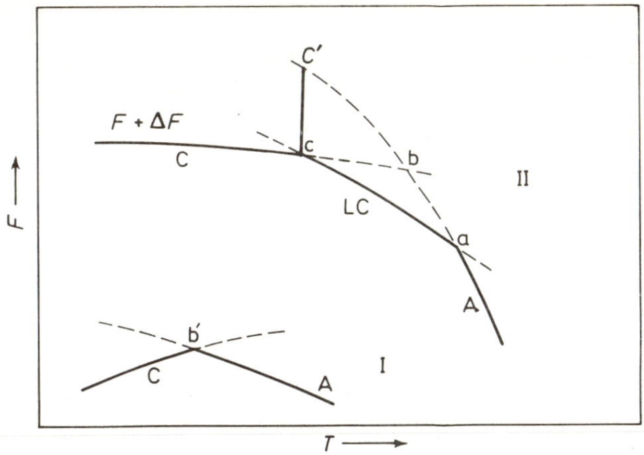

Figure 19. The analysis of orientational crystallization with aid of an F–T diagram. For details see text, for notation see *Figure 15*

the free energy–temperature diagram. Nucleation from the nematic state will lead inevitably to extended chain crystals while the less probable nucleation from the metastable isotropic state will still supply the system with folded chain crystallites. The energy difference between the nucleation points C' and C corresponds to the ΔF^* differences in *Figure 18* for the same degrees of orientation. One more nucleation type of the 'unperturbed' origin (point b′ on *Figure 19*) may occur if due to crystallization and subsequent local stress relaxation some parts of the solution become unstretched.

All three types of crystals may in principle, be observed by means of low angle x-ray scattering. The coexistence of folded and extended chain crystals is revealed readily by the fact that the average length of crystallites (calculated in the usual manner) *exceeds* the length of the large periods.

Processes of the nature discussed can play a part both in melts and solutions, the general treatment being the same, for a concentrated solution is just a swollen or diluted melt. Of course there are some minor differences, such as ejection of solvent in stretched solutions, the shifts of transition points due to dilution [in reality the $(F + \Delta F)$–T diagrams should be three-dimensional, the third axis being kept for concentration] but all these particularities are essentially unimportant corrections to the general picture.

Experimental evidence for extended chain crystallization under stretch or shear is provided by Porter's[61] and Pennings's[20,21] studies, as well as by some other data. Very characteristic is the fact that Pennings obtained from stirred solutions polyethylene films with tensile strength as high as 100 kg mm^{-2} and with a yarn-type morphology, i.e. strongly fibrillated. Hence, one can again see the striking similarity in behaviour of essentially oriented films of rigid and flexible chain polymers both displaying properties of a solid nematic crystal.

7 CONCLUDING REMARKS

The advantages of the thermokinetic approach to structural processes in polymers as compared to the purely thermodynamic or purely kinetic consists not only and not so much in the fact that it allows us to describe conveniently or to predict the behaviour of a complex polymer system subjected to different kinds of influences. More important, from my point of view, is the possibility of interpreting at the same time processes which would seem very different (spontaneous ordering at subcritical f values and formation of ordered structures in external fields) or morphologies (liquid crystalline systems formed by rodlike macromolecules and by flexible chain block copolymers)—in other words, to reduce them to a 'common denominator', which was, strictly speaking, proposed to the author to do in this lecture. This 'common denominator' is represented by the equivalence principle; its use allows us to avoid complicated mathematical formalisms and to reduce different modes of growth of organized structures in polymeric solutions to classical variants (playing the role of lemmas) elaborated by Flory and his school and by DiMarzio. Such a 'reduction principle' is a natural consequence of the equivalence principle, and if in cases of common systems the expediency of its application may be disputed, in the case of polymers with their many-level structural organization the convenience of this principle is unquestionable.

Together with the thermokinetic approach I used widely the concept of diphility. The inherent or explicit diphility are responsible practically for all the types of ordering in solutions or gels that were included in the programme of this Symposium. The part diphility plays in the phenomena of intra- or intermolecular ordering was discussed in sufficient detail and hardly needs additional comments. As to nucleation phenomena, the diphility equally may be either a hindrance for them (as, for instance, in case of PVA), or can promote crystallization. One should note, however, that in solutions and/or gels the nucleation phenomena are frequently complicated by the relative autonomy of different levels of the structural organization. Hence even the term 'nucleation' itself needs to be used with care, and it is necessary to define accurately the nucleation of what particular structure is discussed.

Significant in this respect are the phenomena of the 'kinetic memory' (see for example Ref. 62) which we met already when considering the collapse and restoration of spherulites. However, if in the cases mentioned above the presence of the solvent was necessary to stabilize the spherulites, now I refer to common systems, for instance, solutions of polypropylene in decalin. On repeated melting–crystallization treatment the crystallization rate (and, consequently, the primary nucleation rate) gradually increases, attaining at the fifth cycle a value approximately five times higher than the rate during the first crystallization. After this the rate changes no more, and the spherulites melt 'normally' but reappear on cooling in the same places, wholly restoring their size and internal morphology.

This result was interpreted as the persistence of 'stocks' of spherulites: the chains forming the spherulitic radii and the radii themselves retain after the fifth treatment their spatial distribution, so that the melting process was located only at the lower level of the structural organization, namely, the crystallites. But the persistence in an implicit form of the higher levels hindered the diffusion controlled disordering of the chains forming the crystallites; hence the nucleation also was non-diffusion controlled and the respective activation energy was consequently very low. Of course this habit is also of a thermokinetic nature, the increase in temperature in the melting thermostat or the prolongation of time of incubation leading to the 'effacement' of memory, i.e. to degradation of the stocks.

It is of interest that the autonomy of levels once occurring in solution or melt persists to some extent also in the solid polymer, the kinetic memory being transformed into a 'genetic' one: again, in definite time–temperature ranges even at relatively high elongations (400 to 500 per cent) one can still observe correlation regions resembling the interspherulitic boundaries, though the spherulites themselves were long ago transformed into macrofibrils with a quite different internal structure[63].

In the solutions the presence of many levels predetermines transitions proceeding in opposite directions of the type discussed in Section 4; correspondingly, the nucleation of one type of ordered structures may proceed as a result of degradation of another type or level of organization. In combination with diphility this leads to a great abundance of different ordered morphoses in solutions: using Flory's formalism (connected with free energies) one can take into account the change of free energy on variation of solvent and/or temperature in terms of χ_1 and the related parameter f and then with aid of F–T diagrams *reduce* this variety of morphoses to the ordinary polymorphism treated, for instance, in Mandelkern's monograph[7]. It is remarkable that in his analysis of polymorphism Mandelkern uses F–T diagrams having the same appearance and the same physical sense as, for example, *Figure 19*. However, there is no reason to be astonished: the nucleation in presence of an external field is not simply 'reduced' to polymorphism but creates a kind of *topological* polymorphism in the true sense of the word. As was shown, this polymorphism due to external fields is possible in cases of amorphous polymers also: however, the stabilization of ordered morphoses in the absence of ordering interactions with the solvent is hardly possible; on the contrary, in crystallizable polymers the parallel packing of extended chains is fairly stable, the energy 'pumped over' from the external field being compensated for by the heat of crystallization.

Thus, from the thermokinetic viewpoint the behaviour of liquid and three-dimensional crystals, at least during the formation of corresponding phases, is of an essentially similar type—which, strictly speaking, is just what one should expect from phase transitions similar in order.

One more contribution to the 'common denominator' is provided by the structon concept. The introduction of higher order structons allows us to encompass within the frame of the equivalence principle the behaviour of block-copolymers (see p. 136–137). Moreover it accentuates the 'polymorphic' treatment, especially in cases of phase reversal as discussed by Gallot[51]. Finally one should once more mention the very likely similarity in behaviour of simple tactoids and anisometric aggregates of homodisperse blocks.

I avoided in this lecture a detailed discussion of polymer complexes though implicitly this problem was treated in Section 5. It should be mentioned that complexes discussed by Liu in his main lecture[13] and also in the paper Ref. 64

may be equally interpreted either as nuclei of ordered structures or as interactions leading to mutual crystallization hindrances but increasing the solubility of mixed systems. Finally, the specific complex formation considered in the main lecture of Chapman[65] being obviously connected with effects of diphility shows that selective interaction can produce not only organized structures but also a specific sort of 'organized behaviour' (the membranes acting as triggers). This aspect of macromolecular interactions in solutions connected with the teinochemical principle gives a basis for the incorporation of a whole system of feed-backs in polymeric ensembles and, correspondingly. for the elaboration of molecular cybernetics (see Ref. 66 though the term belongs to Ch. Sadron).

REFERENCES

1 J. Frenkel, *Kinetic Theory of Liquids*, Dover Publ., Inc., New York (1955).
2 M. L. Huggins, *Brit. Polymer J.*, **4**, 465 (1973).
3 R. Hoseman and S. Bagchi, *Direct Analysis of Diffraction by Matter*, Amsterdam (1962).
4 S. Frenkel and G. Elyashevich, *Vysokomolek. Soedin.*, **A13**, 493 (1971).
5 S. Frenkel and G. Elyashevich, in *Relaxation Phenomena in Polymers*, chap. 4, pp. 229, 234, 240, Khimia, Leningrad (1972).
6 R. Brout, *Phase Transition*, New York, Amsterdam (1965).
7 L. Mandelkern, *Crystallization of Polymers*, Chap. 5. McGraw-Hill: New York (1964).
8 S. Ya. Frenkel, L. V. Kuchareva, B. M. Ginsburg, K. A. Gasparyan and V. I. Vorob'ev. *Biophysica*, **10**, 735 (1965).
9 L. V. Kuchareva, B. M. Ginsburg, S. Ya Frenkel and V. I. Vorob'ev, *Biorheology*, **7**, 37 (1970).
10 O. V. Klenina, V. I. Klenin and S. Ya. Frenkel, *Vysokomolek. Soedin.*, **A12**, 1277 (1970).
11 O. V. Klenina, V. I. Klenin and S. Ya. Frenkel, *Vysokomolek. Soedin.*, **A14**, 2192 (1972).
12 V. G. Baranov, T. I. Volkov and S. Ya. Frenkel, *Dokl. Akad. Nauk USSR*, **172**, 849 (1967)
13 Kang-Jen Lui, *Pure Appl. Chem.* **38** (1–2), 65 (1974).
14 V. N. Zvetkov, V. E. Eskin and S. Ya. Frenkel. *Structure of Macromolecules in Solution*. Moscow, Nauka (1964).
15 Sz. Frenkel, *MTA Kém. Tudoményok osztalyanak Köleményei*, **14**, 63 (1960). (Hungarian).
16 M. Gordon, *Pure Appl. Chem.* **38** (1–2), 97 (1974).
17 J. J. van Aartsen, *Europ. Pol. J.*, **6**, 9195 (1970).
18 J. J. van Aartsen and C. A. Smolders, *Europ. Pol. J.*, **6**, 1105 (1970).
19 P. T. van Emmerik and C. A. Smolders, *Europ Pol. J.*, **9**, 293 (1973).
20 A. Pennings, C. J. H. Scholten and A. M. Keil, *J. Polymer Sci.*. **C38**, 167 (1972).
21 A. Pennings and A. Keil, *Kolloid-Z.*, **205**, 160 (1965).
22 S. Ya. Frenkel, V. G. Baranov, Yu. N. Panov and T. I. Samsonova, *Ukrainian Phys. J.*, 282 (1967).
23 S. Ya. Frenkel, V. G. Baranov, Yu. N. Panov, S. A. Agranova, V. G Aldoshin, L. N. Korzhavin and T. I. Samsonova. In the special issue of *Colloid Journ.* (*USSR*): *Mechanisms of Processes of Formation of Films from Polymer Solutions and Dispersions*, Moscow, p. 22 (1966).
24 P. Flory, *Principles of Polymer Chemistry*, Ithaca, N.Y. (1953).
25 N. Kimo and T. Nakagawa, *Polymer J.*, **4**, 143 (1973).
26 V. A. Kargin, V. A. Kabanov, S. Ya. Mirlina and A. V. Vlasov. *Vysokomolek. Soedin.*, **3**, 134 (1961).
27 S. Ya. Frenkel and L. G. Shaltyko, *Abstracts of the 1st All-Union Biochemical Congress*. Ed Acad. Sci. USSR (1963).
28 S. Ya. Frenkel, L. G. Shaltyko and G. K. Elyashevich. *J. Polymer. Sci., Part C*, **30**, 47 (1970).
29 C. Robinson, J. C. Ward and R. B. Beevers, *Disc. Faraday Soc.*, **25**, 29 (1958).
30 L. G. Shaltyko, A. A. Shepelevsky and S. Ya. Frenkel. *Molecular Biology*, **2**, 29 (1968); *Vysokomolek. Soedin.*, **10B**, 639 (1968).
31 S. Ya. Frenkel, Appendix I (by Editor) to the Russian translation of F. Geil, *Polymer Single Crystals*, Moscow, Khimia, p. 478 (1968).
32 V. G. Baranov, T. I. Volkov and S. Ya. Frenkel, *Dokl. Akad. Nauk USSR*, **162**, 836 (1965).

33 V. A. Kargin and G. L. Slonimsky, *Short Essays on the Physical Chemistry of Polymers*, 2nd Ed., Moscow, Khimia (1967).
34 P. Flory, *Proc. Roy. Soc.*, **A234**, 73 (1956).
35 E. DiMarzio, *J. Chem. Phys.*, **35**, 658 (1961).
36 G. K. Elyashevich and S. Ya. Frenkel, *Vysokomolek. Soedin.*, **A9**, 323 (1969).
37 S. Frenkel, Short communication at this Symposium, topic 4.
38 J. Yang and P. Doty, *J. Amer. Chem. Soc.*, **79**, 761 (1957).
39 S. Horikiri and K. Kodera, *Polymer J.*, **4**, 213 (1973).
40 S. Frenkel and B. M. Ginsburg, *J. Pol. Sci.*, **22C**, 813 (1969).
41 S. Ya. Frenkel, G. K. Elyashevich and Yu. N. Panov, in *Progress of the Chemistry and Physics of Polymers*, Moscow, Khimia, p. 87 (1970).
42 W. G. Miller, *Pure Appl. Chem.* **38** (1–2), 37 (1974).
43 S. P. Papkov, V. G. Kulchihin, A. Ya. Malkin, V. D. Kalmikova, A. V. Volohina and L. I. Gudim, *Vysokomolek. Soedin.*, **14B**, 244 (1972).
44 S. P. Papkov, M. M. Iovleva, L. P. Milkova, A. V. Volohina, V. D. Kalmikova and G. I. Kudryavzeva, *Vysokomolek Soedin.*, **15B**, 757 (1973).
45 V. G. Kulchihin, A. Ya. Malkin, S. P. Papkov, O. N. Korolkova, V. D. Kalmykova, A. V. Volohina and O. V. Semenov, *Vysokomolek, Soedin.*, **16A**, 169 (1974).
46 P. Flory, *Statistical Mechanics of Polymer Chains*, New York (1969).
47 R. Simha, *J. Phys. Chem.*, **44**, 250 (1940); *J. Chem. Phys.*, **13**, 188 (1945).
48 *US Pat. No.* 3 671 542 (1972).
49 J. Preston *et al. Preprints Symposium High Modulus Wholly Aromatic Fibers*, Americal Chemical Society: Boston (1972).
50 S. Sobajima, *J. Phys. Soc., Japan*, **23**, 1070 (1967): Ch. Joshida, Jo. Uematsu and I. Uematsu. *Rept. on Progr. Polymer Phys. Japan*, **15**, 645 (1972).
51 B. Gallot, *Pure Appl. Chem.*, **38** (1–2), 1 (1974).
52 A. A. Korotkov, L. A. Shibaev, L. M. Pyrkov, V. G. Aldoshin and S. Ya. Frenkel, *Vysokomolek. Soedin.*, **1**, 443 (1959): see also Ref. 15
53 Ch. Sadron, *Ann. Fac. Sci. Univ. Clermont, Phys.*, No. 15, 33 (1966).
54 S. Ya. Frenkel, *Introduction to the Statistical Theory of Polymerisation*, Moscow–Leningrad. Nauka, chap. 2. (1965)
55 H. G. Zachmann, *Pure Appl. Chem.*, **38** (1–2), 79 (1974).
56 J. W. Cahn, *J. Chem. Phys.*, **42**, 93 (1965).
57 A. Katchalsky, *Progr. Biophys. Biophys. Chem.*, **4**, 1 (1954).
58 W. Kuhn, A. Ramel, D. Walters, G. Ebner and H. Kuhn, *Fortschr. Hochpolym. Forsch.*, **1**. 540 (1960).
59 H. Fisher, *Proc. Natl Acad. Sci USA*, **51**, 1285 (1964).
60 R. S. Zurabyan, V. G. Baranov and S. Ya Frenkel, Short communication at this Symposium.
61 I. H. Southern, N. Weeks, R. S. Porter and R. G. Crystal *Macromol. Chem*, **162**, 19 (1972).
62 S. Ya. Frenkel, V. G. Baranov and T. I. Volkov, *J. Polymer Sci.*, **C16**, 1655 (1967).
63 V. G. Baranov and K. A. Gasparyan, *J. Polymer Sci.*, **A2–8**, 1015 (1970).
64 H. Z. Liu and K.-J. Lui, *Macromolecules*, **1**, 157 (1968).
65 D. Chapman, *Pure Appl. Chem.*, **38** (1–2), 59 (1974).
66 S. Frenkel, *Molekularkybernetik, Ideen des exakten Wissens*, DVA, Stuttgart, No. 9. 529 (1970).

SPECIFICITY OF INTERMOLECULAR ACTION IN SYSTEMS INCLUDING POLYELECTROLYTES AND IONS OF ORGANIC PHYSIOLOGICALLY ACTIVE SUBSTANCES

G. V. Samsonov

Institute of High Molecular Compounds
Academy of Sciences of the USSR
Leningrad, USSR

ABSTRACT

Thermodynamic analysis of the sorption laws of quaternary ammonium bases with aromatic and aliphatic radicals, of other organic bases and organic acids, and ions of antibiotics in particular, has shown that a combination of electrovalency interaction and weak interaction plays a decisive part in the specificity of interaction. This combination depends both on the polyelectrolyte structure and on properties of organic counterions. The increase in the entropy of the system during sorption of a number of complex organic ions, the existence of the compensation effect of enthalpy and entropy for many systems may be interpreted as a coexistence of a multiplet additional interaction, i.e. the occurrence of a comparatively large number of microstates for each ion bound with a polyelectrolyte, or as the hydrophobic effect. It is quite probable that both effects are operative.

In the systems under investigation an additional interaction becomes apparent as a cooperative effect which is revealed by an increase in the free energy of bonding with the increase in the quantity of organic ions bound with polyelectrolyte.

The introduction of large amounts of organic substances into a polyelectrolyte gel changes the dielectric properties of the medium and shifts the usual electrovalent interactions. Thus, in cation exchangers, a shift of the hydrogen–sodium exchange occurs and the organic ion–sodium binary system is gradually transformed into the organic ion–sodium–hydrogen ternary system. Sorption of bipolar ions (amino acid proteins) also occurs with electrochemical recombination of sorbate, i.e. its protonation.

A study of the interaction of amino acids, polypeptides and proteins, especially enzymes, with carboxylic cation exchangers, enabled us to predict the necessary structures of crosslinked polyelectrolytes for reversible specific high capacity sorption of many enzymes. A new class of carboxylic cation exchangers 'Biosorb' has been synthesized and with their aid many proteases, pectinases, amylases and other enzymes have been isolated and purified. Thus, the microbe protease was completely separated from amylase of almost the same molecular weight with a close isoelectric point. The process is carried out at a very high enzyme–sorbent ratio exceeding by several hundred times the corresponding value for gel filtration. The columns are only 3 to 5 cm high.

A unique feature of the new group of crosslinked carboxylic polyelectrolytes is the combination of high permeability, low variation of hydration with changing pH of the medium and high selectivity of the sorption of organic ions. Separation with the aid of these gel sorbents is determined by the molecular weight of the proteins and by their electrochemical properties including isoelectric points and local bipolarity.

A special class of systems includes interacting soluble polyelectrolytes, in particular linear polyelectrolytes and ions of organic physiologically active substances. Under certain conditions, the modification of biopolymers, such as enzymes or protein hormones, which takes place in these interactions leads to stabilization of their structure, maintenance, inhibition or increase of biological activity and shows paths for synthesizing new drugs. A method for protecting insulin from the action of proteolytic enzymes and of acid medium with retention of hormonal activity is based on these principles. A systematic investigation of the interaction of pancreatic ribonuclease with polyelectrolytes in solution led to preparation of stabilized and activated enzymes or to inhibition of nuclease. Not only the structure of the polymer support, but also its molecular weight is of great importance for the different effects on nuclease. The cooperative effect of intermolecular interaction contributes to the effect of inhibition whereas random interaction favours the effect of enzyme activation.

Preparation and study of crosslinked and soluble polyelectrolytes specifically binding ions of organic physiologically active compounds, such as antibiotics, hormones, enzymes or alkaloids is of considerable interest in connection with methods of separation and fine purification of these compounds as well as their modification and stabilization. It is also important to analyse theoretically general relationships governing the behaviour of supermolecular biologically active polyelectrolytic systems.

While considering ion exchange systems including ions of organic substances of a relatively low molecular weight, the sorption phenomena for quaternary ammonium bases, antibiotics and to a certain extent alkaloids have been studied in greatest detail. When the interaction of synthetic polyelectrolytes with biopolymers was investigated, the greatest attention was directed to the binding of proteins, in particular enzymes.

1. PRINCIPLES OF ION EXCHANGE SORPTION OF QUATERNARY AMMONIUM BASES AND ANTIBIOTICS

Many essential principles of the intermolecular action of complex ions of physiologically active substances with linear and crosslinked polyelectrolytes are observed in the sorption of ions of quaternary ammonium bases on ion exchange resins[1–9]. These systems may be considered as model substances although they themselves are of interest, e.g in investigating the mechanism of action of some psychotropic preparations.

Various antibiotics were considered as the next group of more complex organic ions. It was taken into account that in accordance with their degree of complexity they resemble either simple symmetric ions or more complicated ions with many functional groups[10–22]. It should be noted, however, that ion exchange sorption of antibiotics is of great practical importance

Table 1. Selectivity of sorption for ions of quaternary ammonium bases on crosslinked sulpho-polyelectrolytes

Cation exchanger	Main monomer unit	Second counterion	$(CH_3)_4N^+$ K	$\Delta\phi$	ΔH	$T\Delta S$	$(CH_3)_3C_2H_4OHN^+$ K	$\Delta\phi$	ΔH	$T\Delta S$	$(C_2H_5)_3C_6H_5CH_2N^+$ K	$\Delta\phi$	ΔH	$T\Delta S$
Dowex 50 × 2	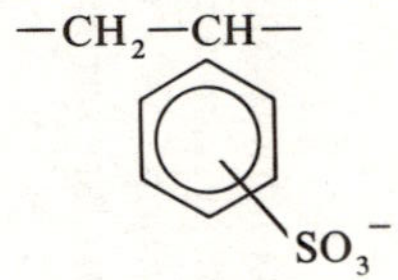 + DVB	Na^+	1.26	−0.14	−0.41	−0.27	1.61	−0.3	−1.6	−1.3	4.8	−1.2	−1.2	0
KU-5A	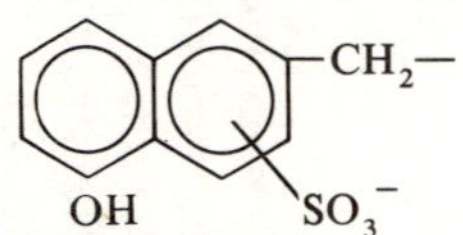	Na^+	2.9	−10.6	−1.3	−0.7	3.64	−0.8	−2.0	−1.2	20.0	−2.1	−0.9	1.2
KU-1	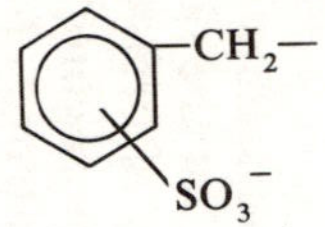	Na^+	6.9	−21.1	−1.4	−0.3	9.1	−1.3	−2.7	−1.2	162	−32	+2.6	+5.8
Sulphosephadex SE-25	dextran	H^+	1.0	0			1.0	0			1.0	0		
Sulphosephadex SE-50	dextran	H^+	1.0	0			1.0	0			1.0	0		

K is the ion exchange constant; $\Delta\phi$, ΔH, $T\Delta S$ in kcal mol^{-1}.

for developing industrial methods of separation and purification of these physiologically active substances.

We shall consider the relationships of ion exchange with participation of quaternary ammonium bases taking into account the complex structure of these ions and the peculiarities of the structure of crosslinked polyelectrolytes. *Table 1* shows thermodynamic constants of ion exchange characterizing the selectivity of sorption of these ions in their competition with sodium or hydrogen ions. The latter two types of ions exhibit more or less the same affinity for the investigated sulpho cation exchangers. For the first three cation exchangers the selectivity of sorption increases in the following order: tetramethylammonium, trimethyloxyethylammonium, trimethylbenzylammonium. The introduction of a polar hydroxyethyl group and of a still more polar benzyl group which are capable of additional interaction with ion exchangers on the basis of a dipole–dipole interaction or some other type of interaction increases the specificity of the binding of ions of quaternary ammonium bases. The increase in the complexity of structure of crosslinked polyelectrolyte also leads to similar results. The replacement of a benzene ring by a naphthalene ring or the appearance of a phenyl radical (i.e. the introduction of groups increasing additional interaction with radicals with sorbed ions) leads to a considerable increase in the selectivity of sorption of organic ions. In this connection we should point out the extremely low selectivity of sorption for all ions of quaternary ammonium bases on sulpho-Sephadexes. It is well known that in the absence of ionogenic groups the matrices of these gels interact very slightly with organic ions. This peculiarity was used for the development of gel chromatography in which a weak reversible interaction is observed (for example with ions of biopolymers), and substances are separated according to their molecular volumes with a very small effect from their functional groups on their interaction with Sephadexes.

Table 2. Selectivity of ion exchange sorption of antibiotics

Counterions	Ion exchanger	$\Delta\phi$	ΔH	$T\Delta S$
Penicillin–chlorine	FAF	5.5	10.0	11.0
Erythromycin–sodium	KFUKh	60.0	3.5	5.9
Chlorotetracycline–sodium	SBS	123	1.7	4.4
Oxytetracycline–sodium	SBS	148	−3.6	−0.77
Novobiocyn–chlorine	FAF	950	4.0	8.3

$\Delta\phi$, ΔH, $T\Delta S$ in kcal mol^{-1}.

In considering thermodynamic functions in this phenomenon of ion exchange it is of particular interest that the positive entropy sign appears for systems with the highest selectivity of organic ion sorption. As a rule, ions of antibiotics, which are of much more complex structure, are sorbed by a mechanism leading to an increase in entropy (*Table 2*).

Coming back to the problem of participation of the matrix of synthetic polyelectrolyte in the formation of additional bonds with organic ions, we shall consider the influence of ionogenic and non-ionogenic monomers on the selectivity of sorption. *Figure 1* shows the thermodynamic constant of the ion exchange of ions of quaternary ammonium bases with hydrogen

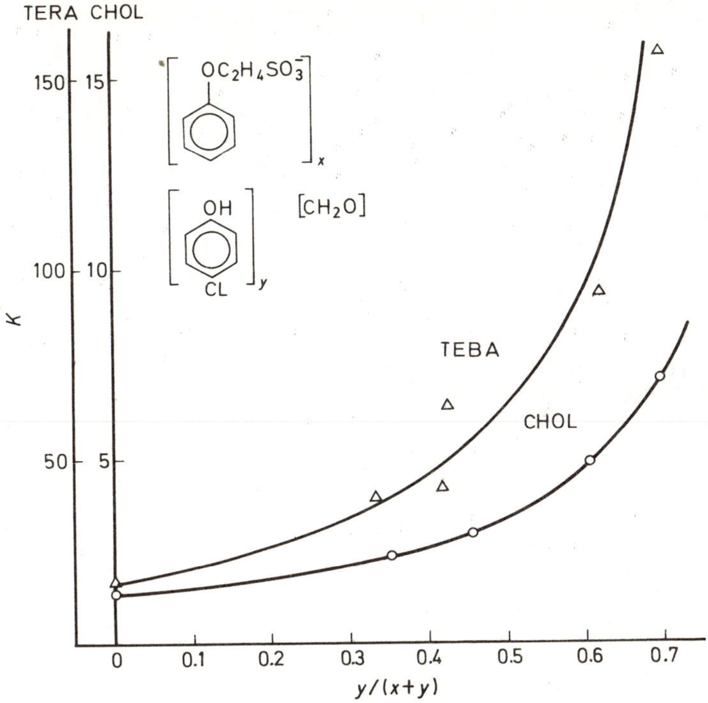

Figure 1. Effect of monomer ratio in the resin on the sorption selectivity of quaternary ammonium ions. TEBA—$(C_2H_5)_3CH_2C_6H_5N^+$; CHOL—$(CH_3)_3C_2H_4OHN^+$. $K = \int_{x=0}^{x=1} K_s dx$; $K_s = m_{org}C_H/m_H C_{org}$.

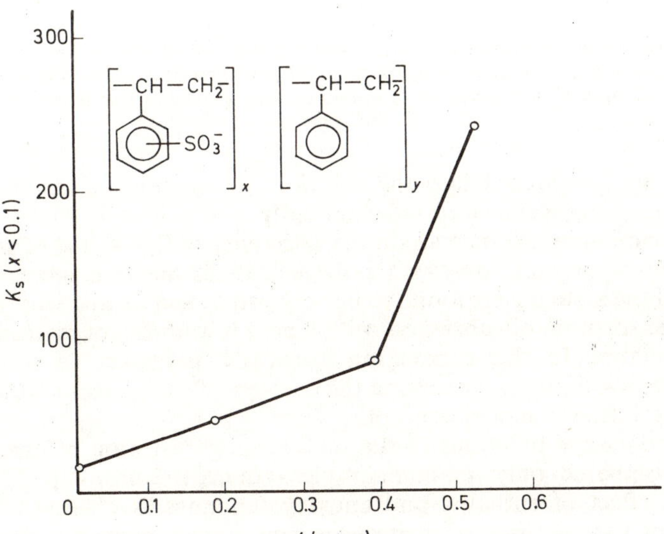

Figure 2. Effect of the degree of desulphonation of sulpho cation exchanger on the selectivity of ion exchange sorption of oxytetracycline. The second counterion is hydrogen.

ions versus the content of non-ionogenic monomers (*p*-chlorophenol in acid solutions used in the experiment was not ionized). In both systems the selectivity of sorption markedly increases with decreasing content of ionogenic groups. The selectivity increases especially sharply when the molar fraction of non-ionogenic groups exceeds 0.5. A similar phenomenon was observed in the sorption of oxytetracycline on partly desulphonated exchangers of the type of Dowex-50 (*Figure 2*). The increase in the content of desulphonated monomers in the exchanger leads to a drastic increase in the selectivity of sorption of antibiotics.

The increased selectivity of sorption related to the hydrophobization of polyelectrolyte and the simultaneous increase in the entropy of the system suggests that in some systems including organic ions and polyelectrolytes an additional hydrophobic interaction may play a certain part. Indeed, *Figure 3* shows that when substances capable of destroying the aqueous

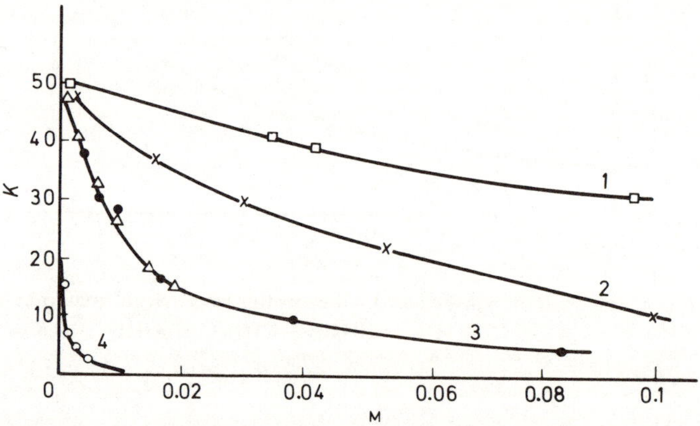

Figure 3. Changes in ion exchange constant with decrease in hydrophobic interaction of oxytetracycline (Dowex-50). *M* is the molar fraction of the component introduced into the solution. 1 Methanol; 2 ethanol; 3 butanol–urea; 4 sodium dodecylsulphate (the second counterion is hydrogen).

structure are introduced into the solution, the sorption selectivity of, for example, oxytetracycline ions falls markedly.

At the same time the increase in the selectivity of the course of additional interaction of organic ions with polyelectrolytes and the increase in the entropy of the system were found to be very widely spread and were observed even in the sorption of substances with a very low ability of rearranging the water structure. In this connection it became necessary to consider an alternative possibility of increasing the entropy of the system and, hence, to investigate a second model concept.

Thermodynamic functions of the ion exchange sorption of organic ions may be compared only for comparable systems (*Figures 4–6*). The compensation effect of enthalpy and entropy functions is observed both for sorption of various ions of quaternary ammonium bases on one sodium form of sulpho cation exchanger and for exchange of oxytetracycline with hydrogen on sulphostyrene polyelectrolyte with different content of divinyl-

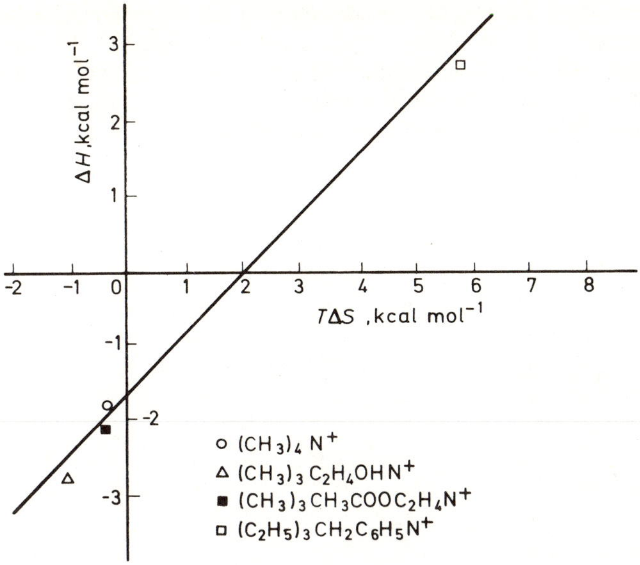

Figure 4. The compensation effect in ion exchange of quaternary ammonium bases on sulphoresin KU-1 (the second counterion is sodium).

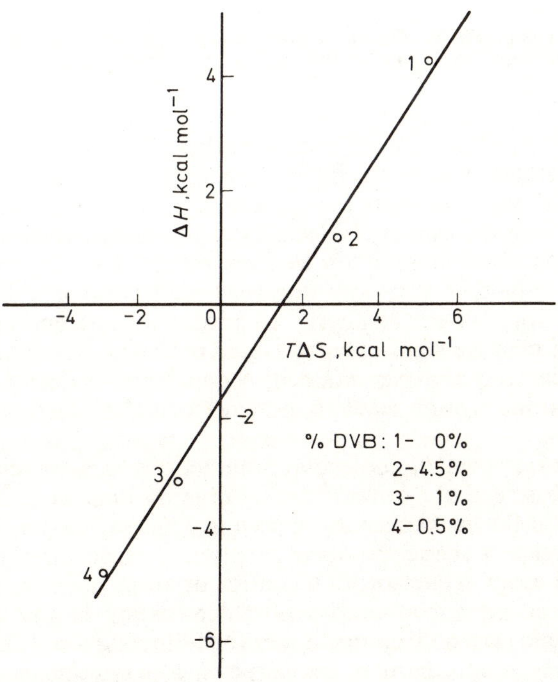

Figure 5. The compensation effect in sorption of oxytetracycline on sulphoresins Dowex-50 with different content of divinylbenzene (the second counterion is hydrogen).

benzene (up to its complete absence in a linear polyelectrolyte). A similar phenomenon occurs when erythromycin is sorbed by different cation exchangers (*Figure 6*). Proceeding from the undoubted importance of additional interactions for the selectivity of sorption of complex organic ions and varying values and even signs of functions of entropy and enthalpy,

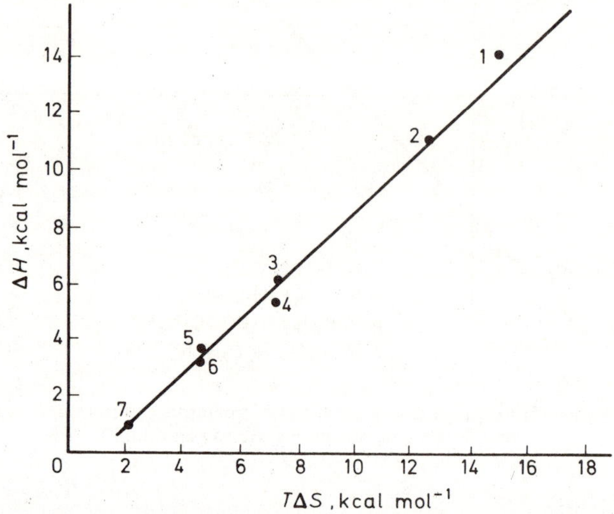

Figure 6. The compensation effect in sorption of erythromycin on different cation exchangers (the second counterion is sodium). 1—Dowex 50 × 1; 2—Dowex 50 × 2; 3—Amberlite TRC-50; 4—KFUKh; 5—SBS-3; 6—KFSKh-3; 7—SBST-3.

it is possible to advance a rational hypothesis that several (sometimes even very many) states exist for the interacting system organic ion–polyelectrolyte in contrast to one microstate for the interaction of ions of hydrogen, sodium or chlorine with the same polyelectrolyte. These microstates occur when the energy of additional interaction leading to the appearance of an additional microstate is neither very low nor very high compared to the energy of thermal motion. When the energy of additional interaction becomes sufficiently high, positive signs of entropy and enthalpy are replaced by negative signs and high selectivity is retained. A considerable decrease in the energy of additional interaction leads to a decrease in the selectivity of sorption of organic ions.

A complex system of molecular interactions is also revealed when we consider the selectivity of sorption of organic ions as a function of their molar ratio in the exchanger. As shown in *Figures 7* and *8*, we may observe either a decrease in the selectivity of sorption of organic ions due to a gradual exclusion of more active sorption centres or an increase in the selectivity of their sorption, i.e. a cooperative effect consisting of the influence of the sorbed organic ions on the sorption of the following ions. This phenomenon is particularly pronounced in the sorption of oxytetracycline (*Figure 8*) in which the cooperative effect is observed only when there are many closely spaced ionogenic groups in the exchanger (high exchange capacity and

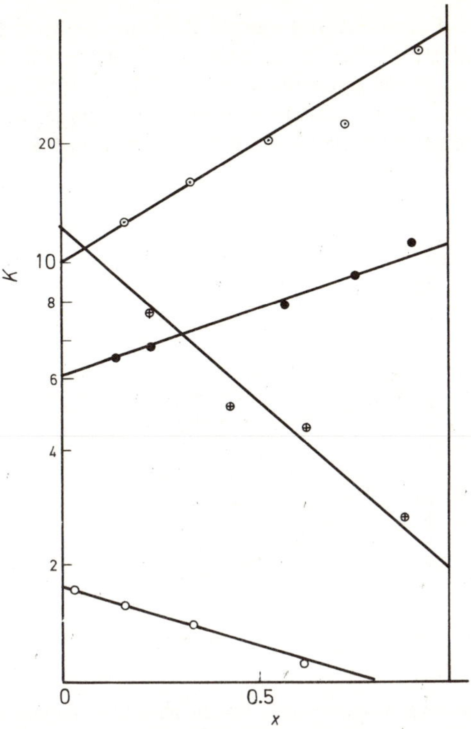

Figure 7. Sorption selectivity of oleandomycin versus its molar ratio in the exchanger (x). K is the ion exchange constant (the second counterion is sodium).

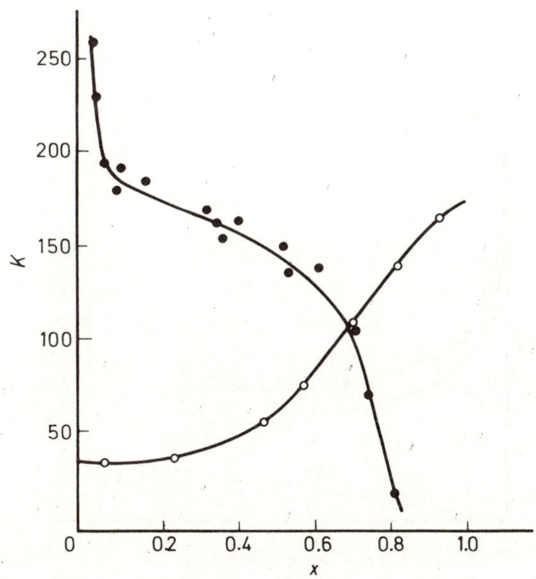

Figure 8. Soeprion selectivity of oxytetracycline versus its molar ratio in the exchanger (x). The second ion is hydrogen. ●, Dowex-50 × 1, exchange capacity m_0 = 2.3 meq g^{-1}; ○, Dowex 50 × 2, exchange capacity m_0 = 5.0 meq g^{-1}.

increased amount of crosslinking agent). An increase in the distance between sorbed ions completely eliminates the cooperative effect. Desulphonated exchanger which is highly selective at a low degree of filling of the exchanger with oxytetracycline cannot compete with an exchanger of high capacity and low selectivity after the sorbed ions exhibit cooperative interaction—when their molar fraction exceeds 0.7.

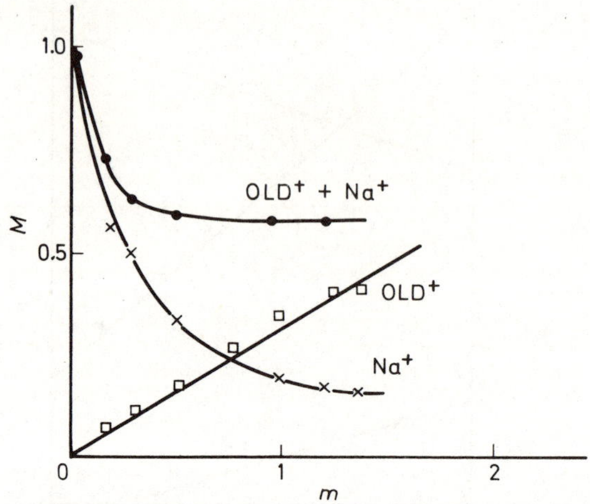

Figure 9. Transformation of a one-component system into a three-component system in oleandomycin sorption. M is the molar ratio of ions in the sorbed state, m is the quantity of sorbed oleandomycin meq g^{-1} of the carboxylic resin KFUKh.

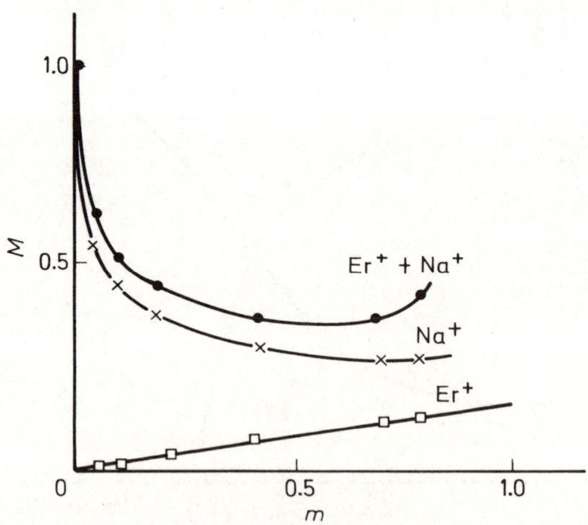

Figure 10. Transformation of a one-component system into a three-component system in erythromycin sorption. M is the molar ratio of ions in the sorbed state, m is the quantity of sorbed erythromycin (meq g^{-1}).

Additional interaction is of great importance and determines the high selectivity of sorption of organic ions which for specially synthesized ion exchangers attains a tremendous value, of an order of magnitude of 10^2–10^3. Nevertheless, as shown by the term itself, this interaction is only an additional effect operative when the electrovalency interaction appears. In crosslinked polyelectrolytes in the course of organic ion sorption electrovalency interaction acquires particular properties determined by changes in the dielectric permittivity of the medium. *Figures 9* and *10* show that sorption of antibiotics —oleandomycin and erythromycin—on the Na-form of carboxyl resins proceeds with simultaneous desorption of a more than equivalent amount of sodium as a result of changes in hydrogen–sodium exchange and of progressive replacement of sodium ions by hydrogen ions in the exchanger. although the process occurs at neutral pH. Changes in the degree of ionization of carboxyl groups in crosslinked polyelectrolytes in the presence of organic ions was also proved by direct titration of ion exchanger by organic bases.

2. INTERACTION OF CROSSLINKED POLYELECTROLYTES WITH PROTEINS (SORPTION AND CHROMATOGRAPHY)

Interaction of bipolar ions with polyelectrolytes, first of all with polyacids of the crosslinked type, i.e. with ion exchangers, has been studied in detail. Many relationships in these systems, in particular the part played by the additional interaction, the increase in entropy at high selectivity of sorption. the appearance of the cooperative effect and the shift of equilibrium for mineral ions by sorption of organic ions, are to a certain extent similar to general dependences for which the sorption of quaternary ammonium bases and ions of antibiotics serve as examples.

Monoaminomonocarboxylic amino acids at pH nearly 7 are sorbed by cation exchangers as cations, i.e. the formation of cations from bipolar ions occurs in the course of their sorption[23]. This type of process is more difficult for dipeptides, tripeptides and even more so for tetrapeptides; as a result they are less sorbed by the salt forms of cation exchangers. These dependences were applied to proteins for which it was also shown that sorption by cation exchangers is a specific phenomenon determined by the isoelectric state of the proteins and the state of their local dipolarity[24]. The possibility of the specific sorption of proteins including enzymes permitted the development of methods for their separation by using high capacity sorption (up to many hundreds of milligrams of proteins per gram of sorbent). These processes might be carried out not only analytically but also on a large scale and even industrially. For solving these problems it was necessary to abandon the inert matrices which are used in Sephadexes, biogels and their ion exchange variants. It is known, however, that the sorption of proteins by ion exchangers usually proceeds at a low sorption capacity owing to insufficient permittivity of the network of the ion exchanger, and is irreversible or occurs with partial or complete denaturation resulting from strong intermolecular action. For this reason, if we wished to develop sorption processes of high capacity for protein separation, it was necessary to synthesize and to use highly permeable crosslinked polyelectrolytes capable of reversible interaction with the protein macromolecules. In this work we proceeded from the assumption that the irreversible interaction and denaturation of proteins in this case are largely due to a combination of high free energy of interaction with the ability of the polyelectrolyte network to undergo deformation. An important stage of this work was the preparation and use of crosslinked polyelectrolytes undergoing little deformation and at the same time highly permeable and accessible to the diffusion of macromolecules. The use of a small amount of a crosslinking agent could not give good results since the networks of slightly crosslinked polyelectrolytes undergo great deformation when the pH of the solution and the ionic strength are changed, i.e. under usual working conditions. We achieved the best results by using triazine, that is a monomer containing three vinyl radicals, as a crosslinking agent. In combination with acrylic acids we obtained crosslinked polyelectrolytes of the carboxylic type which undergo little deformation even when the degree of ionization of carboxyl groups changes[25–28]. As shown in *Figure 11*, for protease of

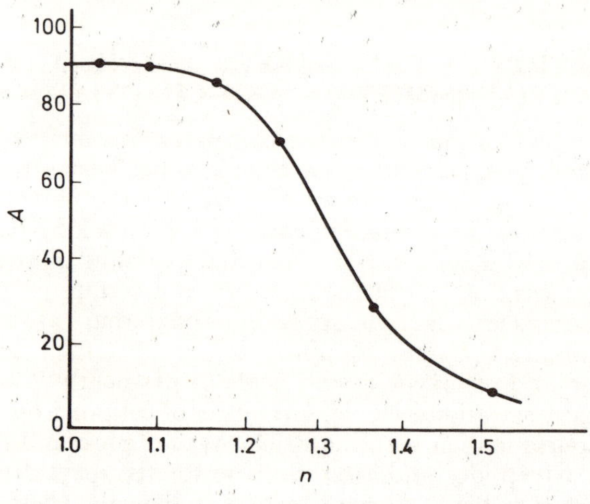

Figure 11. Desorption of proteolytic enzyme (terrylitine) from Biosorb-KMT. *A* is enzyme desorption (%); *n* is the ratio of swelling of the Na-form of Biosorb-KMT to swelling of the H-form.

microbic origin the degree of desorption from such a crosslinked polyelectrolyte essentially depends on the ratio of the degrees of swelling in the Na- and H-forms. Only when this value tends to unity is an almost quantitative yield in enzyme attained during desorption. Common highly swelling and slightly crosslinked carboxylic cation exchangers are characterized by the ratio of swelling exceeding 2 or even 3. Under certain conditions the yield of the polyelectrolyte used dropped below 10 per cent. Naturally, the transition to more inert matrices decreases the degree of irreversible sorption of proteins and at the same time decreases the selectivity and capacity of sorption in the presence of competing ions.

Conditions for separating proteins on a carboxylic crosslinked polyelectrolyte undergoing slight deformation are determined by the dependence of the sorption capacity on pH and the ionic strength of solution (*Figures 12* and *13*). The left-hand branch of the sorption curves in *Figure 12* is determined

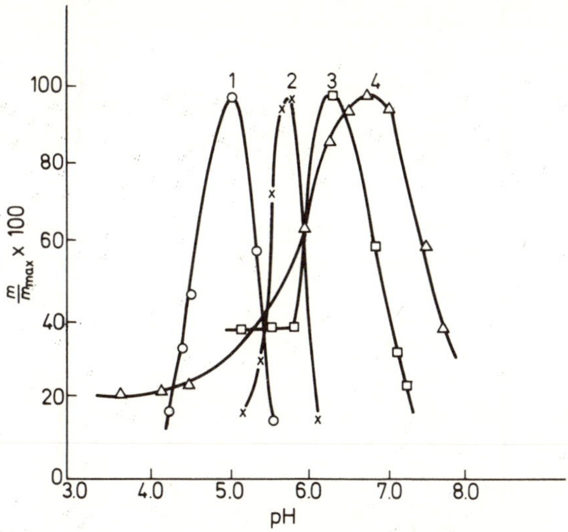

Figure 12. Sorption capacity of proteins on Biosorb-KMT versus pH of the equilibrium solution. m/m_{max} is the ratio of the sorption capacity to the maximum capacity: 1. thermolysine, $m_{max} = 650$ meq g^{-1}; 2. serum albumin, $m_{max} = 580$ meq g^{-1}; 3. chymotrypsinogen, $m_{max} = 330$ meq g^{-1}; 4. ribonuclease, $m_{max} = 230$ meq g^{-1}.

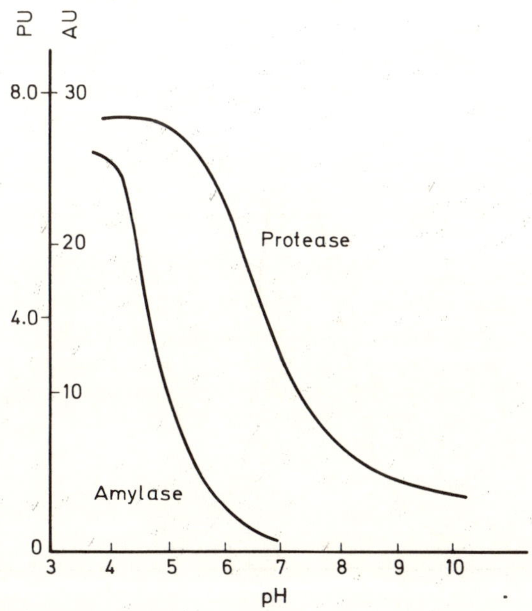

Figure 13. The right-hand branches of sorption curves for protease and amylase from *Asperg. terricala* on Biosorb-KMT. PU Proteolytic units: AU amylolytic units.

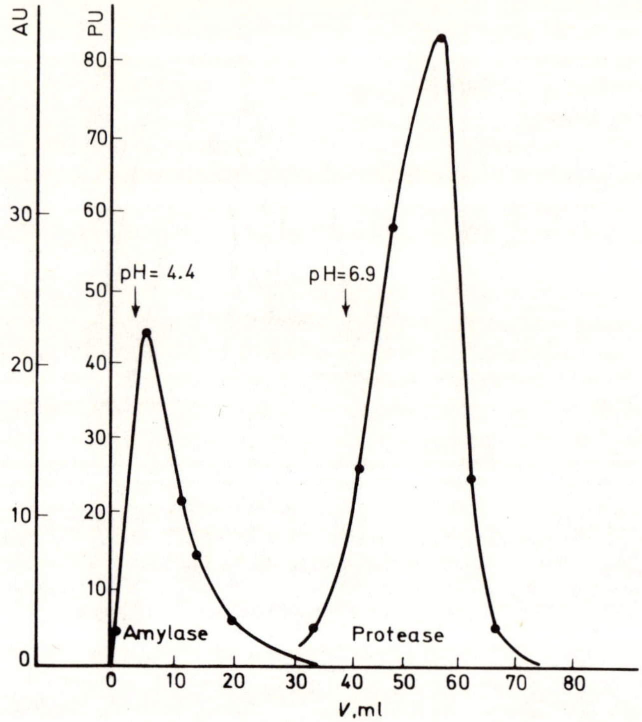

Figure 14. Fractionation of protease and amylase on Biosorb-KMT. PU Proteolytic units; AU amylolytic units.

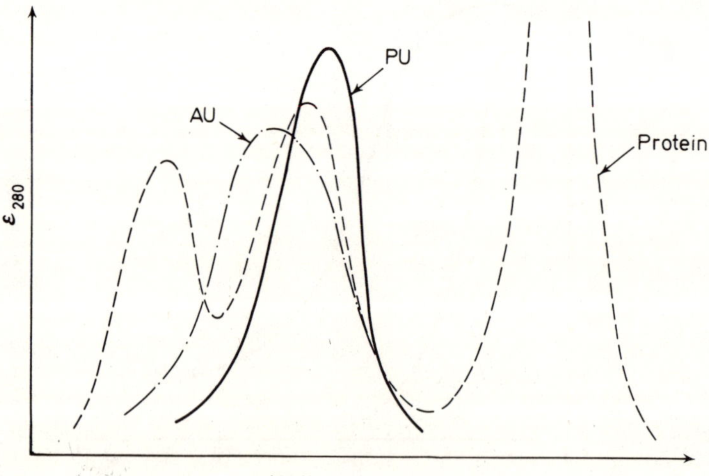

Figure 15. Gel chromatography of protease and amylase on Sephadex G-50. PU Proteolytic units: AU amylolytic units.

mainly by the competition with hydrogen ions and often lies in the region of protein denaturation. For selecting the conditions of separation it is advisable to use the right-hand branch: its run is governed by the competition with cations in solution, in our case with sodium ions. Two microbic enzymes—protease and amylase—with very little difference in molecular weight (M = 26 800 and 29 000, respectively) and in their isoelectric points (pI = 4.6 and 4.2) exhibit quite different sorption curves (*Figure 13*). This is due to differences in their interaction with polyelectrolytes because this interaction depends not only on the charge of the macromolecules but also on local bipolarity and on all the combined effects of additional interaction.

All these sorption properties made possible complete separation of enzymes on a carboxylic crosslinked polyelectrolvte undergoing little deformation (*Figure 14*) whereas the standard methoc using Sephadex G-50 led only to partial separation (*Figure 15*). It should also be noted that the Sephadex column is of 1 m height while the height of the carboxylic polyelectrolyte in

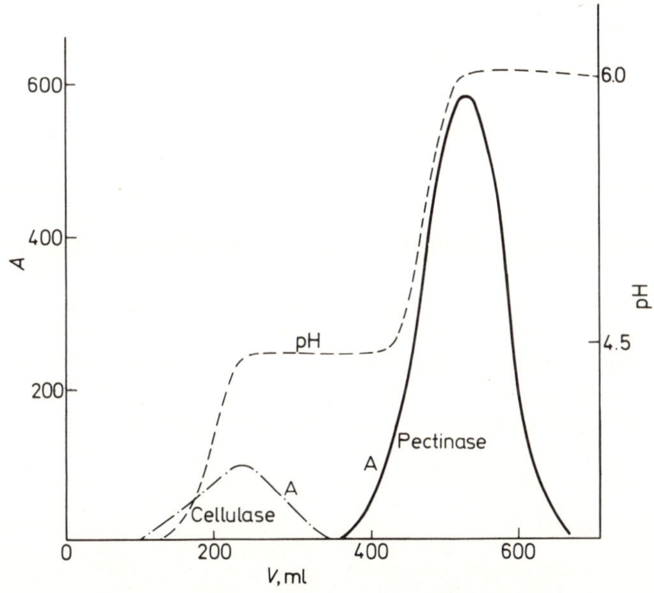

Figure 16. Chromatographic separation of cellulase and pectinase from *Asp. avamori* on Biosorb-KMT. *A* activity (units ml^{-1}). Sorption capacity—40 mg of pectinase on 1 g of sorbent.

the column is only 5 cm. It is of particular importance that this small column separates a 10-fold amount of enzymes compared to the first column.

Hence, the efficiency of the sorbent operating on the principle of selective sorption is hundreds of times greater than the efficiency of gel chromatography. Finally, the time of experiments is also several times shorter. We used this principle of protein separation for various enzymes. Thus, *Figure 16* shows the separation of cellulase and pectinase.

The second possibility of reversible selective ion exchange sorption and

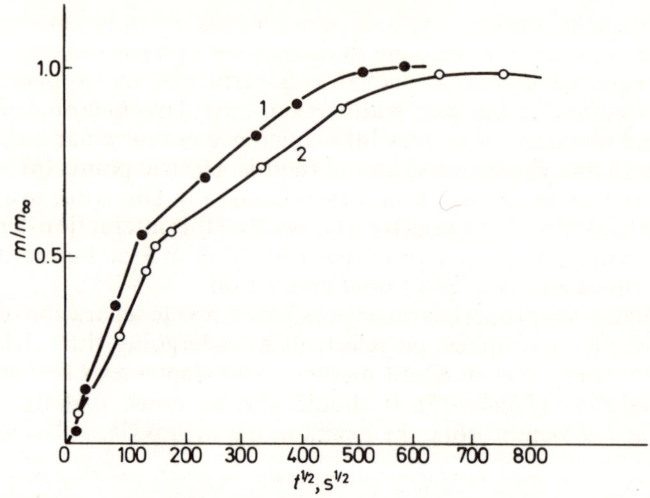

Figure 17. Kinetic curves of insulin (1) and lysozyme (2) sorption on macroporous sulpho cation exchanger KU-23. m/m_∞ ratio of sorption capacity to limiting capacity; t time (s).

protein chromatography consists in using macroporous resins obtained by copolymerization of monovinyl monomers in the presence of large amounts of of a divinyl compound in an organic solvent. Our main experiments were carried out with a sulphostyrene copolymer containing 30 per cent of divinylbenzene and showing a channel structure in the dehydrated state[29, 30]. This sorbent exhibits high internal porosity and an internal surface of about $70 \ m^2 \ g^{-1}$ with a pore diameter of up to 80 Å. For many macromolecules protein sorption occurs in two stages: the first stage proceeds at a high rate and the internal surface is filled with a monomolecular layer of macromolecules (*Figure 17*). In this stage the diffusion coefficient is $10^{-9} \ cm^2 \ s^{-1}$. Then the sorption of proteins proceeds less intensively and the diffusion coefficient

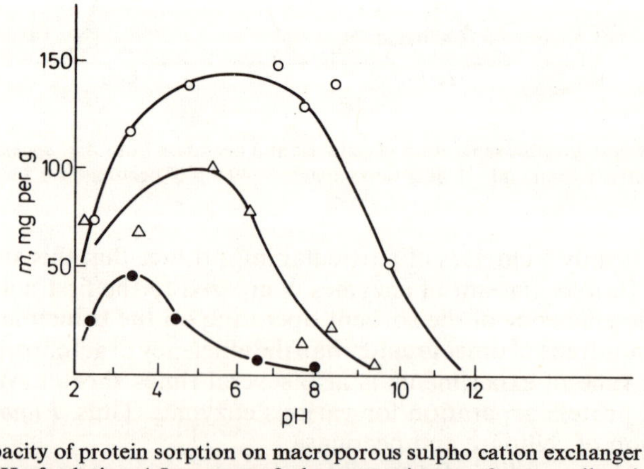

Figure 18. Capacity of protein sorption on macroporous sulpho cation exchanger KU-23 versus pH of solution. 1 Lysozyme; 2 chymotrypsinogen; 3 serum albumin.

decreases to 10^{-10} to 10^{-11} $cm^2 s^{-1}$. The selectivity also decreases; this is indicated by a decrease in the slope of the sorption isotherm. The upper point of the linear portion of the isotherm (*Figure 17*) agrees reasonably with the sorption capacity calculated for monomolecular filling of the sorbent proceeding from the value of internal surface.

For these resins the dependence of the sorption capacity on the pH of the solution is represented by a bell-shaped curve just as for carboxylic cross-linked rigid polyelectrolytes (*Figure 18*). The one-stage process of column sorption and desorption of insulin on a macroporous sorbent with pancreatic extract as the initial material yields a highly purified product (*Figure 19*). Crystallization of insulin may be carried out directly from the eluate[31].

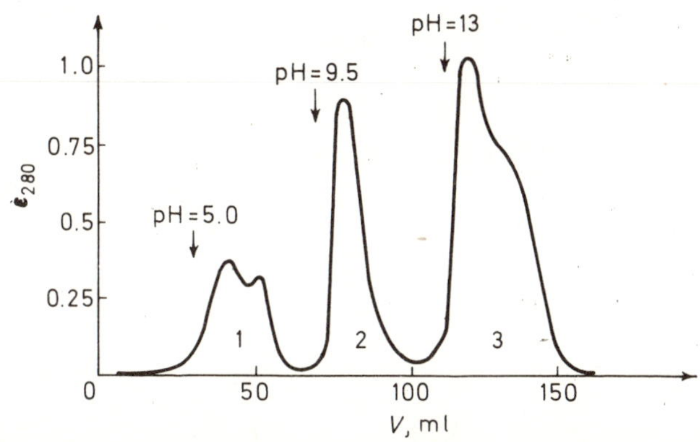

Figure 19. Chromatographic isolation of insulin from pancreatic extract on macroporous sulpho cation exchanger KU-23. 2 Insulin; 1,3 impurities. Elution by buffer solutions, pH = 5.0; 9.5 and 13.

3. REVERSIBLY DISSOCIATING COMPLEXES OF PROTEINS WITH SOLUBLE SYNTHETIC POLYELECTROLYTES

Ion exchange interactions in solutions of systems including physiologically active substances and synthetic polyelectrolytes are not only of theoretical significance. They can be used in practice for modification of physiologically active substances, for their stabilization (which is very advantageous, e.g. for enzymes and protein hormones) and for preparation of new types of medicines, such as those administered orally or having a prolonged action. Enzymes modified with the aid of synthetic polymers including soluble polymers are also used in industry.

Protection of insulin against the action of proteases and the acidity of the medium with retention of hormonal activity is made possible by the formation of a complex with polymethacrylic acid[32, 33]. Gel chromatography of this complex after using pepsin and trypsin for degrading (*Figure 20*) shows that one native component in the complex is retained while incubation of unprotected insulin with the same enzymes under the same conditions results in a marked degradation of the hormone. A direct determination of the activity confirms this statement. The formation of reversibly dissociating complexes of trypsin with soluble polybases leads to stabilization of the

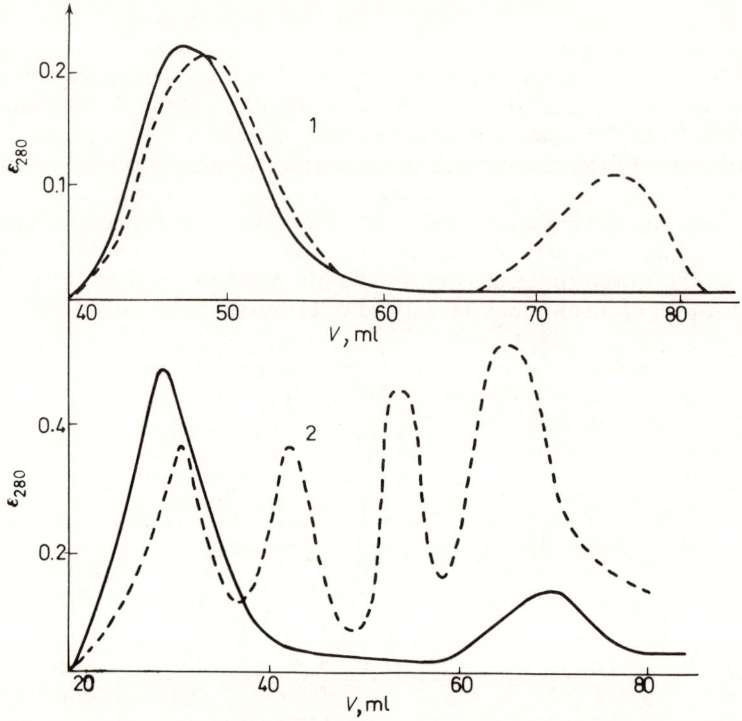

Figure 20. Gel chromatography of hydrolysis products for insulin–polymethacrylic acid complex by trypsin (1) and pepsin (2). Insulin, -----; complex, ———.

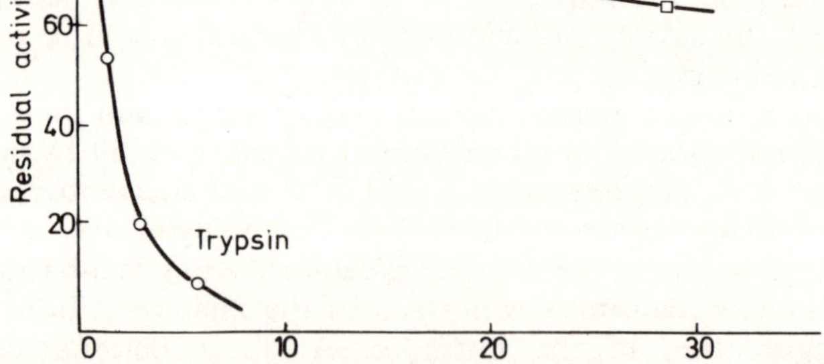

Figure 21. Stability of a trypsin complex with poly(vinyl-*N*-imidazole).

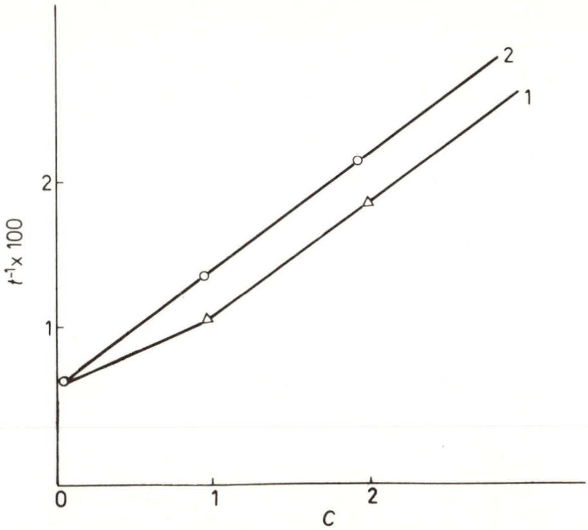

Figure 22. Rate of thrombin dissolution by trypsin in the presence of polybases. *t* Time of dissolution of standard thrombi in min; *c* concentration of polybases of two types (1 and 2) in mg ml^{-1}. 1. Poly(vinyl-4(S)-imidazole); 2. poly(vinyl-*N*-imidazole).

enzyme (*Figure 21*). The stability of the pH value is also observed for these complexes. The value of the optical activity of the complex is retained over a wider pH range than in the initial trypsin (*Figure 22*). Moreover, this complex dissolves standard thrombi at a greater rate than native trypsin (*Figure 23*).

The peculiarities of reversible interaction in the formation of soluble molecular complexes (including polymer complexes) are usually studied by hydrodynamic, electrophoretic and chromatographic methods and also by investigating the membrane equilibrium or the kinetics of diffusion through membranes[34–40]. In particular, quantitative chromatographic analysis

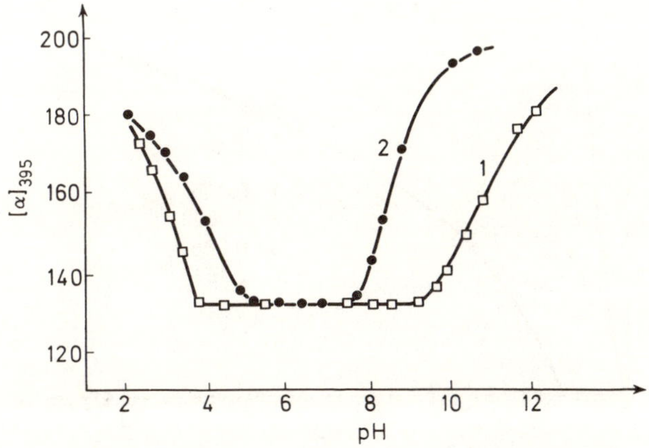

Figure 23. Optical activity of trypsin complex with poly(vinyl-*N*-imidazole) versus pH. Ionic strength 0.01; $T = 22°C$. 1. Polymer + trypsin; 2. trypsin.

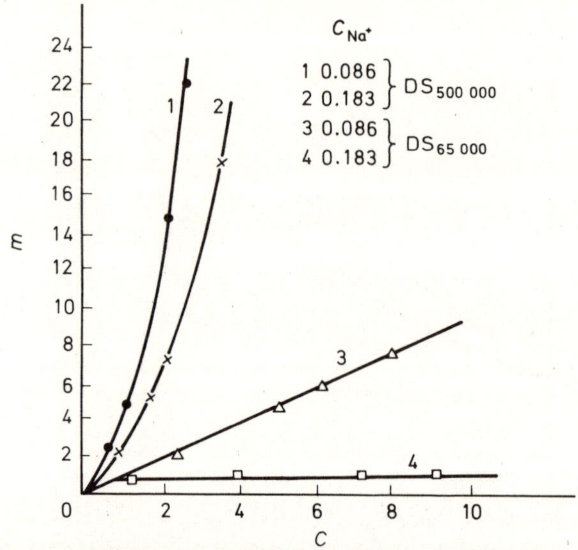

Figure 24. Isotherm of formation of a complex of ribonuclease with dextran sulphates, *m* is the number of moles of RNAase in the complex per mole of dextran sulphates; *c* is the concentration of RNAase in a free state in solution (mol $ml^{-1} \times 10^5$) in initial concentration of dextransulphates (1 mg ml^{-1}).

enables us to calculate both the equilibrium constants and the molecular ratios of the components of the complexes[41–43].

It is of interest to study inhibition, stabilization and activation of pancreatic ribonuclease and desoxyribonuclease by soluble polymeric dextransulphates. As shown in *Figure 24*, one molecule of high molecular weight dextransulphate

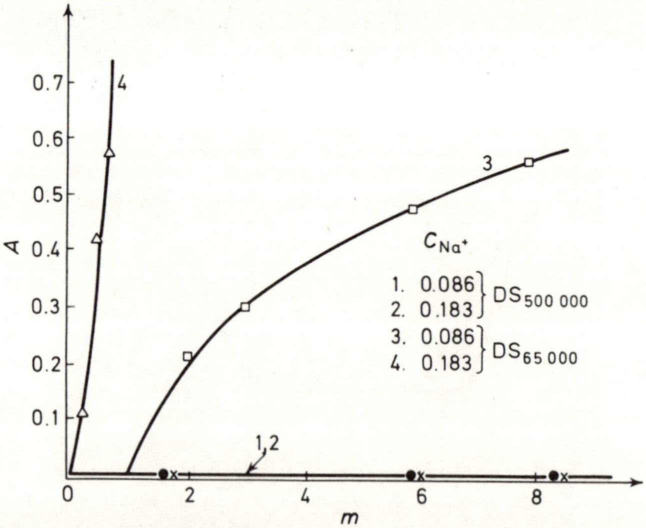

Figure 25. Specific activity *A* of RNAase bonded in a complex with dextransulphates. m = [RNAase]/[DS]; substrate is uridine 2′3′-cyclophosphate.

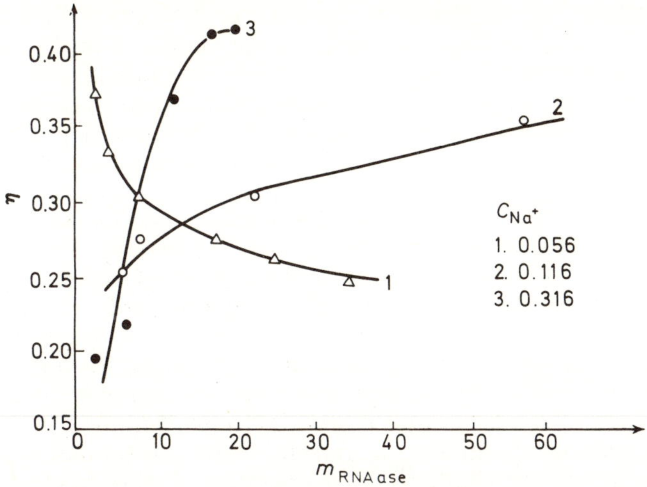

Figure 26. Viscosity of solutions of complexes of RNAase with dextransulphates. (M.W. 500 000); m_{RNAase} = [RNAase]/[DS]; T = 20°C; pH = 7.0.

(M = 500 000) binds up to 20 molecules of enzyme whereas a dextransulphate of a relatively low molecular weight (M = 65 000) binds a relatively small quantity of ribonuclease molecules, particularly when the ionic strength increases. In the first case the shape of isotherms for binding a large quantity of enzyme (the increase in the slope of the isotherm indicates the cooperative effect of binding) shows mutual interaction of molecules of RNAase fixed on

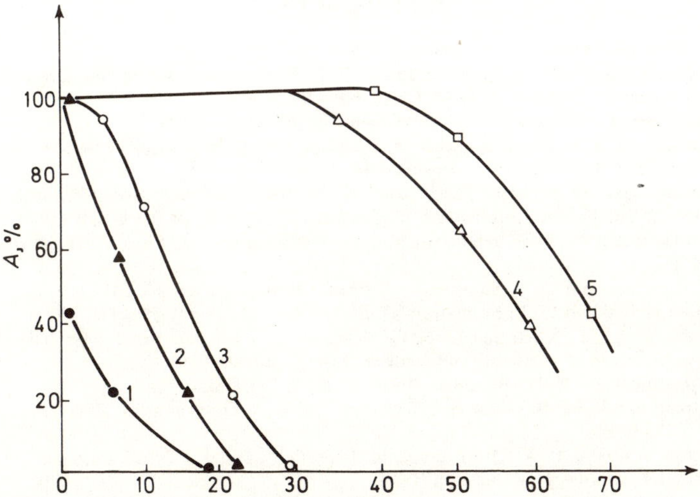

Figure 27. Stabilization of deoxyribonuclease by dextransulphates. A is the activity of DNAase (measured for DNA from calf thymus). 1, DNAase + $DS_{500\,000}$ (17% S); 2, DNAase; 3, DNAase + $DS_{50\,000}$ (17% S); 4, DNAase + $DS_{50\,000}$ (11% S); 5, DNAase + $DS_{50\,000}$ (8% S).

dextransulphate. The use of high molecular weight dextransulphate exhibiting the cooperative binding effect leads to the inhibition of RNAase, while the binding of enzyme by dextransulphate of a lower molecular weight results in an increase in the enzyme activity (*Figure 25*). It is also possible to develop conditions for its activation with an increase in the activity to 130 per cent with respect to the activity of native enzyme. Successive binding of RNAase by dextransulphate under conditions of cooperative interaction leads to a decrease in the viscosity of the solution whereas non-cooperative or slightly cooperative interaction increases the viscosity (*Figure 26*). The effect of the molecular weight of the polymer and of the number of ionogenic groups in polyelectrolytes on the peculiarities of interaction with enzymes is also apparent for desoxyribunuclease (*Figure 27*). Dextransulphates of relatively low molecular weight with a limited number of sulpho groups exhibit the stabilization of enzyme whereas high molecular weight polyelectrolytes with a large amount of sulpho groups inhibit DNAase. Both factors, i.e. the increase in the number of sulpho groups and the increase in the molecular weight of dextransulphate contribute to the cooperative effect of binding DNAase and its inhibition. In contrast, independent fixation of this enzyme on single active centres of polyelectrolyte leads to the stabilization of very labile pancreatic DNAase.

ACKNOWLEDGEMENT

The work described in this paper was carried out with the participation of L. V. Dmitrenko, B. V. Moskvichev, R. B. Ponomareva, L. K. Shataeva, N. N. Kuznetsova, D. I. Ostrovsky, A. A. Selezneva and R. V. Koltsova.

REFERENCES

[1] T. R. E. Kressman and J. A. Kitchener, *J. Chem. Soc.* 1208 (1949).
[2] J. R. Millar, D. G. Smith, N. E. Marr and T. R. E. Kressman, *J. Chem. Soc.* 2740 (1964).
[3] A. Schwarz and G. E. Boyd. *J. Phys. Chem.* **69**. 4268 (1965).
[4] G. E. Boyd and O. V. Larsen, *J. Amer. Chem. Soc.* **89**, 6038 (1967).
[5] G. V. Samsonov, B. V. Moskvichev, V. A. Pasechnik, C. B. Musabeckov, B. S. Chakyna and N. M. Smirnova. *J. Phys. Chem.* (Russian) **44**. 2589 (1970).
[6] G. V. Samsonov, B. V. Moskvichev and N. N. Kuznetsova, in *Ion Exchange and Ion-exchangers*. p. 90. Science publication of Academy of Sciences of the USSR (1970).
[7] B. V. Moskvichev, N. N. Kuznetsova and G. V. Samsonov, *J. Phys. Chem.* (Russian) **45**, 1873 (1971).
[8] B. V. Moskvichev and G. V. Samsonov, *J. Phys. Chem.* (Russian) **45**, 1317 (1971).
[9] B. V. Moskvichev and G. V. Samsonov, *J. Phys. Chem.* (Russian) **45**, 1371 (1971).
[10] G. V. Samsonov, S. F. Klich and D. N. Gulaeva, in *Ion Exchangers and Ion-exchange*, p. 91. Science publication of Academy of Sciences of the USSR (1965).
[11] G. V. Samsonov and R. M. Bakaeva, *Antibioticks*, 737 (1968).
[12] G. V. Samsonov, R. M. Bakaeva, S. S. Urusova and L. A. Domracheva. *Chem. Pharmaceutic J.* (No. 7) 28 (1969).
[13] G. V. Samsonov and L. V. Dimitrenko, in *Thermodynamics of Ion Exchange*, 178, Minsk, Science and Technology publication (1968).
[14] G. V. Samsonov, E. B. Trostianskaya and G. E. Elkin, *Ion Exchange Sorption of Organic Substances* Science publication of Academy of Sciences of the USSR (1969).
[15] G. V. Samsonov. V. G. Vorobyeva and M. V. Kravetz. *Colloid J.* 427 (1970).
[16] L. V. Dmitrenko. A. D. Morozova. V. S. Pyrogov and G. V. Samsonov, *High Molec Comp.*, **14**, 353 (1972).

[17] L. V. Dmitrenko, A. D. Morozova and G. V. Samsonov, *Izvest. Akad. Nauk SSSR*. 1563 (1972).
[18] A. V. Michalev and G. S. Libinson, *J. Phys. Chem.* (Russian) **43**, 1333 (1969).
[19] A. V. Michalev and G. S. Libinson, *J. Phys. Chem.* (Russian) **44**, 2887 (1970).
[20] A. V. Michalev and G. S. Libinson, *J. Phys. Chem.* (Russian) **45**, 2699 (1971).
[21] A. Sh. Genedi and G. V. Samsonov. *Kolloid J.* (Russian) **31**, 674 (1969).
[22] L. V. Dmitrenko, A. D. Morozova and G. V. Samsonov, *Izvest. Akad. Nauk SSSR*, 1563 (1972).
[23] G. V. Samsonov and N. P. Kuznetzova, *Dokl. Akad Nauk SSSR*, **115**, 351 (1957).
[24] G. V. Samsonov and R. B. Ponomareva, *Biochem.*(Russian) **24**, 63 (1959).
[25] L. K. Shataeva, A. A. Selezneva, O. V. Orlyevskaya, D. I. Ostrovsky, V. V. Korshunov, R. B. Ponomareva, N. N. Kuznetzova, K. P. Papukova and G. V. Samsonov, in *Synthesis, Structure and Properties of Polymers*, 225, Science publication of the Academy of Sciences of the USSR (1970).
[26] A. A. Selezneva, G. A. Babenko, O. V. Orlyevskaya and G. V. Samsonov, *Microbiol. Industry*, 14 (1972).
[27] Pat. USSR, N322332, 17.7.1970. *Bull. Izobret.* N36 (1971).
[28] O. V. Orlyevskaya, L. K. Shataeva and G. V. Samsonov, *Appl. Biochem. Microbiology*. (Russian) 7, 1971 (1971).
[29] D. I. Ostrovsky, A. D. Morozova, L. V. Dmitrenko, P. D. Kozlova and G. V. Samsonov, *Izvest. Akad. Nauk SSSR, ser. chem.* 1214 (1972).
[30] L. V. Dmitrenko, D. I. Ostrovsky and G. V. Samsonov, *High Molec. Comp.* **14**, 859 (1972).
[31] Pat. USSR, N213256, 10.5.1963, *Bull. Izobr.* N10 (1968).
[32] M. V. Glykina, N. N. Kuznetzova, I. A. Bolotina and G. V. Samsonov, *Molec. Biology*, **4**, 692 (1970).
[33] M. V. Glykina, N. P. Kuznetzova, S. V. Koltzova and G. V. Samsonov, *Biochemistry* (Russian), **36**, 258 (1971).
[34] G A. Gilbert, *Proc. Roy. Soc.* **A250**, 377 (1959).
[35] G. A. Gilbert, *Nature*, **212**, 296 (1966).
[36] R. Towner, J. Winterbottom and S. W. Timasheff, *J. Amer. Chem. Soc.* **82**, 3161 (1960).
[37] H. Noguvh, *Biochem. Biophys. Acta*, **22**, 459 (1956).
[38] G. K. Ackers and T. E. Tompson, *Proc. Nat. Acad. Sci., USA*, **53**, 342 (1965).
[39] D. J. Winsor and H. A. Scheraga, *Biochemistry*, **2**, 1263 (1965).
[40] G. K. Ackers, *J. Biol. Chem.* **242**, 3026 (1967).
[41] G. V. Samsonov and R. B. Ponomareva, *Biofizika*, **13**, 213 (1968).
[42] G. V. Samsonov, R. B. Ponomareva and R. G. Luchko, *Biofizika*, **14**, 634 (1969)
[43] G. V. Samsonov and R. B. Ponomareva, *Studia biophysics*, 24/25, 399 (1970).

SURFACE TRANSFORMATIONS OF POLYMERS

O. Wichterle

Institute of Macromolecular Chemistry
Czechoslovak Academy of Sciences
Prague, Czechoslovakia

ABSTRACT

Polymeranalogous transformations due to the treatment of a compact polymer phase with liquid or gaseous agents lead—depending on the ratio of the diffusion penetration rate of the agent to the reaction rate—to more or less inhomogeneous products having a higher conversion on the surface than in the lower polymer layers.

In some cases the above finding can be used for the preparation of laminated structures having certain advantages compared to similar systems obtained by coating the polymeric substrate with a film of a different polymer.

The most important group has so far been represented by transformations of this type so chosen as to obtain thin layers exhibiting a specific chemical or biological activity on an inert substrate whose function is purely mechanical. Besides substitution transformations of hydrocarbon polymers by which e.g. ionizable groups are introduced, these reactions also involve chemical transformations of typical functional groups attached to a stable polymeric matrix. The foremost position in the latter category is held by acrylic and methacrylic polymers and copolymers in which it is possible to alter either the functions derived from the carboxylic group or functional groups attached by means of the ester or amide link.

The introduction of strongly hydrophilic groups into the surface of hydrophobic linear polymers may lead to their release from the substrate. This may be prevented by introducing crosslinks via polyfunctional reaction components simultaneously with the hydrophilization transformation. Such a precaution is not necessary for crosslinked polymers whose crosslinks resist chemical agents. Among these polymers a special role is played by macroporous gels, the large internal surface of which may be modified in many ways by chemical reactions of the functional groups.

When solid polymers react with agents which do not dissolve the original polymer or its reaction product the polymer surface remains macroscopically undamaged and the chemical transformation of the polymer proceeds continuously from the surface into the bulk. The thickness of the layer transformed, or more exactly the dependence of conversion on the distance from the surface, is determined by the rate of diffusion of the agent, the rate of its reaction with the polymer and the time of action of the agent. In any case, however, the transition from maximum conversion on the polymer surface to zero conversion in the unaffected bulk of the polymer is a continuous one. In this way the laminar structures obtained are fundamentally different from those arising by coating the polymeric substrate with foreign polymer layers. Obviously, such operations make sense only for polymers which have been subjected to definitive shaping, or for polymer fibres and films.

The scope of this Symposium envisages only such surface transformations which may be called 'chemical transformations of functional groups'. We therefore shall not be dealing here with by far the most extensive, and perhaps most important group of surface polymer treatments consisting in the grafting of a foreign polymer onto a polymeric substrate. It would not fit in with the conventional terminology to designate sites on which the graft starts growing as functional groups, and to classify the grafting itself as a chemical transformation of these groups. Moreover, the nature of grafting consists rather in the addition of a new polymer than in a functional transformation of the original polymer.

I shall here make a marginal remark about the transformations of polymers which do not contain any typical chemical functional groups and in which such groups only just start being formed due to treatment with extremely strong agents. Thus, for instance, even the very inert polytetrafluoroethylene may be modified on its surface by treating it with alkali metal solutions in liquid ammonia, whereby fluorine atoms are eliminated from the surface molecules and conjugated double bond sequences are formed owing to the zip character of the reaction[1]. The surface of a polymer thus prepared is then characterized by a greater adhesion to other materials and to surface-deposited metals. Polyethylene may be surface hydrophilized by treating it with chlorosulphonic acid[2] which introduces into the surface layer sulphonic, sulphoester, hydroxyl, carbonyl and carboxyl groups and double bonds[3]. Similarly, chromosulphuric acid introduces into the surface of polyethylene hydroxyl and carboxyl functions[4]. If the latter reaction is carried out at room temperature and by a short immersion, then—although the transformation of the extremely thin superficial layer may hardly be detected analytically—the adhesion of polyethylene treated in this manner to the hydrophilic films coated on it strikingly improves[5]. The surface layers of polymeric hydrocarbons may also be modified by the incorporation of highly reactive electron-deficient torsos, such as carbenes and nitrenes[6]. The substitution reaction of aromatic hydrocarbon polymers may also be restricted to their surface layers if the penetration of the agent into the polymer is slowed down. A comparative study concerning the localization of sulpho groups in the copolymer styrene–divinylbenzene according to the reaction conditions will also be presented at this meeting[7].

A continuous transition of the chemically modified surface layer into the unchanged substrate allows us to obtain mechanically stable structures by the transformation of typical functional groups even in the case of a very drastic change of properties of the surface layer. A typical example can be seen in the surface hydrolysis of poly(methyl methacrylate) combined with the crosslinking reesterification of a polyhydric alcohol[8]. The final product of this transformation is a loosely crosslinked polymer of methacrylic acid, which, particularly in the form of a neutral alkaline salt, is swollen by water to several times its dry value. In spite of this, such a strongly swollen surface layer remains safely connected with the completely unswollen poly(methyl methacrylate) substrate even in the neutral aqueous medium; if the surface layer is not much thicker than about 100 μm, it resists even comparatively brutal abrasion. It is not possible to find a sharp boundary between the hydrolysed and undamaged substrate by removing the maximally swollen surface layer, although the transition between the fully hydrolysed and unchanged fraction is concentrated in a comparatively thin intermediate layer. Owing to the low reactivity of the polymethacrylic ester group the above reaction requires very drastic conditions: to attain a thickness of tens or hundreds of micrometres of the hydrolysed layer, it is necessary to treat the surface of poly(methyl methacrylate) during e.g. one minute by a mixture of concentrated sulphuric acid and glycerol in the ratio 1:4 at 100–115°C.

Shaped objects made from polyacrylic polymers may be modified similarly, but under considerably milder conditions. A special importance is assigned to the surface hydrophilization of the block copolymers of acrylonitrile and acrylamide obtained by the zip hydrolysis of polyacrylonitrile[9]. These polymers, which if swollen with water are elastomers of excellent mechanical properties, may be further hydrolysed by immersing them in a mixture of sulphuric acid and polyhydric alcohol for a short time with simultaneous loose crosslinking, thus obtaining extremely hydrophilic surface layers of the polyanionic gel[10].

If cellulose esters, e.g. triacetylcellulose films, are treated with sulphuric acid, a narrow range of experimental conditions may be found under which there is no visible corrosion of the surface, but the surface is nevertheless strongly hydrophilized, probably as a consequence of the reesterification substitution of the acetoxyl groups with acidic sulphoester groups. This is reflected in a marked antifogging effect and affinity to dyes[11].

Three-dimensional polymeric systems are particularly suited for such surface substitutions if their crosslinks remain unperturbed during the reaction. If, for example, shaped objects made from the hydrophilic loosely crosslinked 2-hydroxyethyl methacrylate gel are exposed on drying to the vapours of sulphur trioxide, the hydroxyl groups in the thin surface layer are transformed into sulphoester groups[12]. On equilibrium swelling in water the resulting enhanced hydrophilic nature and negative ionic charge of the surface layer produce, for example, reduced friction in contact with living tissues.

A special group of the surface reactions of polymers is represented by their reversible transformations which allow the regeneration of the original polymer and whose value consists in an interruption of the ordering of macromolecules in the surface layer, thus enabling a new order to be established during regeneration. An example can be seen in the treatment of a polyamide fibre with iodine followed by the regeneration of macromolecules with thiosulphate in a different crystalline arrangement, which is supposed to give the fibres an increased adhesion to deposited metals[13]. The surface crystalline structure of polyamide fibres may also be disturbed by a short-term exposure to the vapours of strong acids, e.g. hydrogen chloride, or to solutions of strong acids in nonpolar solvents[14]. The crystalline structure can then be regenerated in a modified form by neutralization; the fibres which were in contact in the acidic cycle remain strongly connected and as if welded after neutralization. This fact could be used for the preparation of some textile structures.

Typical functional changes concerning exclusively the polymer surface are represented by chemical bonds of biologically active compounds on polymeric carriers. Despite the great variety and importance of such surface reactions we shall not be dealing with them now, since they were dealt with at a recent conference in Prague. Let us only recall that it may also be useful to direct to the polymer surface those transformations which lead to carrier synthesis. It should be borne in mind that the homogeneous distribution of ligand groups or their precursors need not always be in accordance with the final requirements of the carriers of biological polymers. What we require from these carriers are two more or less opposite properties: (i) high modulus of elasticity, i.e. a rigid structure which is in the first place determined by a limited degree of swelling in the medium used, and (ii) the highest possible mobility of the specifically reactive groups attached to the surface which is above all aided by the high degree of swelling of their surroundings. Better than by applying a compromise, i.e. a uniform carrier structure, this conflict may be resolved by the maximal possible fulfilment of these requirements in two domains of the carrier, i.e. in its rigid skeleton and in its surface, the chemical functions of which may be varied without jeopardizing the good mechanical properties of the whole system to any essential degree. In this respect, there seems to be a wide scope of applications of macroporous gels which, besides possessing a widely adjustable porosity and good mechanical properties, also contain chemical functions able to undergo a number of surface chemical transformations. Macroporous glycol methacrylate gels may be cited as an example of such widely modifiable prefabs for the preparation of specific carriers. If they are treated in the dry or slightly swollen state with agents in which gels do not swell, various substitution transformations of the hydroxyl groups may be carried out, preferentially on the internal surface of the macroporous gel. Thus, by treating the gel with gaseous phosgene or solutions of the latter in nonpolar solvents, chloroformate groups may be formed on the surface of this gel which can become the site of further chemical transformations[15]. By treating the gels with phosphorylchloride and pyridine in vapours or inert solvents, it is possible to achieve phosphorylation of the surface groups; by treating these groups with vapours of sulphur trioxide, they may be transformed into acid sulphuric acid esters. These and similar reactions do not lead to any changes in the internal rigid structure of the gel, although new functional groups arise on its surfaces, which if uniformly distributed throughout the bulk of the polymer could cause some undesirable swelling effects.

In conclusion, I should like to summarize the practical goals of the surface transformations of polymers. They are as follows:

(i) Modification of the surface of a plastic for a better adhesion of foreign polymeric films coated on it.

(ii) Creation of functional layers on polymer carriers; these functions may be mechanical (lowering of surface friction or drag reduction), sorptional (chromatography, antifogging treatment), increased affinity to dyes, fixation of biological materials or synthetic catalysts, improved tolerance of plastics to the biological medium.

It seems that the latter possibilities of application are the strongest stimuli for taking an interest in improving our mastery of this technique. We are particularly attracted by the hope that the surface treatment of some mechanically satisfactory materials would allow us to attain antithrombogenic properties which so far have been lacking in all the known synthetic materials. We have in mind the surface fixation of heparinoid structures which would prevent blood clotting in a similar way to surface-bonded heparin, but which in contrast with the bonded natural heparin would resist heparinase or other enzymes of the blood medium. The progress already achieved by the introduction of anionic structures with a high degree of swelling of the surface layer is encouraging, although we are fully aware how far we still are from the accomplishment which would allow the chemists to lay safe foundations for alloplasty of the blood system and organs bound to blood circulation.

REFERENCES

[1] E. I. duPont de Nemours, *Brit. Pat. No. 806 422* (23 Dec. 1958): *Chem. Abstr.* **53**, 1189e (1959).
[2] J. A. Nelson and W. K. Vollmer, *US Pat. No. 2879 177* (Mar. 1959); *Chem. Abstr.* 53, 11889e (1959).
[3] J. Hradil and J. Stamberg, *Coll. Czech. Chem. Commun.* **37**, 3868 (1972).
[4] J. Hradil, A. L. Kaganov, J. Štamberg and P. Synek, paper presented to this Symposium.
[5] A. L. Kaganov, P. Synek and J. Štamberg, in press.
[6] A. J. Osteraas and D. A. Olsen, *J. Appl. Polymer Sci.* **13**, 1537 (1969).
[7] J. Prokop and K. Setinek, paper presented to this Symposium.
[8] O. Wichterle, *US Pat. Appl.* 294 846 (1972).
[9] V. Stoy, paper presented to this Symposium.
[10] O. Wichterle and A. Stoy; Czech. Pat. Appl.
[11] O. Wichterle: *U.S. Pat. Appl. 326,419* (1972).
[12] O. Wichterle: *U.S. Pat. Appl. 294,845* (1972).
[13] I. A. Abu-Isa, *J. Appl. Polymer Sci.* **15**, 2865 (1971).
[14] S. Landa, O. Wichterle and M. Kolínský: *Fr. Pat. No. 875 600* (Sep. 1941).
[15] J. Čoupek, private communication.

OXIDATION–REDUCTION POLYMERS

GEORG MANECKE

Institute of Organic Chemistry, Free University of Berlin

and

Fritz-Haber-Institute of the Max-Planck-Gesellschaft, Berlin

ABSTRACT

This review summarizes our work on the synthesis and redox properties of polymers with various quinone–hydroquinone systems. Information is given on our attempts to increase the chemical stability of the redox resins by suitable substitution and by other methods.

The redox properties were thoroughly investigated (e.g. redox potentials, redox capacity, rate of the redox reaction). A theoretical treatment is given for the redox behaviour of tetra- and hexa-valent redox systems.

Redox resins can be used as insoluble, regenerable oxidants or reductants. By means of these resins some organic substrates could be dehydrogenated and several metal ions could be reduced or deposited.

For several years we have synthesized polymer redox systems and have investigated their electrochemical properties. Polymer redox systems are known by the names oxidation–reduction polymers, redox polymers, redox resins and electron-transfer polymers. Water-swellable polymers which can reversibly be oxidized or reduced are usually named redox resins. The redox components of these polymers can be links of the polymer chains or they can be substituents of the chains. The chemical structure of the redox systems can be varied widely. There exist several reviews[1–3] on this subject: those of Cassidy can be especially recommended. This paper is a selective review of investigations of redox resins containing hydroquinone–quinone redox systems which have been performed in our laboratories.

SYNTHESIS OF REDOX RESINS

Redox polymers may be addition polymers synthesized by polymerizing hydroquinone–quinone derivates which contain vinyl groups, or condensation polymers prepared by polycondensation of suitable compounds. They can also be prepared by polymer analogous reactions on preformed polymers.

Examples of addition polymers are the redox resins which were investigated by Cassidy and his co-workers[4]. They were prepared on the basis of vinylhydroquinone. As the hydroquinone–quinone system inhibits polymerization, protecting groups (e.g. ether, acetate, benzoate) are usually introduced, which can be quantitatively removed after polymerization. A copolymer can be obtained from vinyl hydroquinones, protected by acetylating the hydroxy groups, with divinylbenzene (DVB) as crosslinking agent and styrene as an additional comonomer. After sulphonation and regeneration of the redox system by splitting off the protecting groups a redox resin is synthesized which is swellable in water. This is a classical method of preparing redox resins by addition reaction. However, it was soon observed that to a certain extent irreversible oxidation of the polymer also takes place. That means the redox polymer is partially destroyed. This was believed to be caused by radical attack on the hydrogen atoms of the unsubstituted ring of the hydroquinone–quinone system during oxidation.

The hydroquinone–quinone systems can be stabilized by substituting the ring hydrogens with methyl groups[5,6]. So, we have synthesized several methylated vinylhydroquinone acetates and investigated their polymerizability. The synthesis is shown in the following scheme.

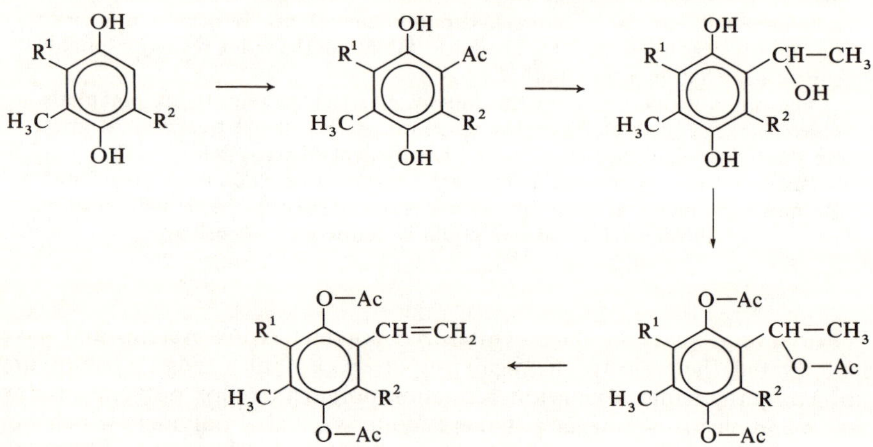

R^1	R^2
H	H
CH_3	H
H	CH_3
CH_3	CH_3

We found that monomers with methyl groups at position 5 (*meta* to the vinyl group) polymerize badly[7]. The polymerization was initiated by radicals from benzoyl peroxide or α,α′-azo-bis-isobutyronitrile in *n*-butanol.

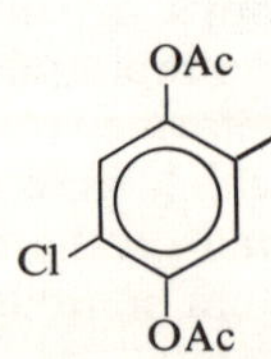

(++)

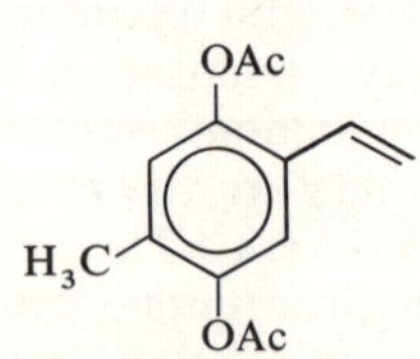

(++)

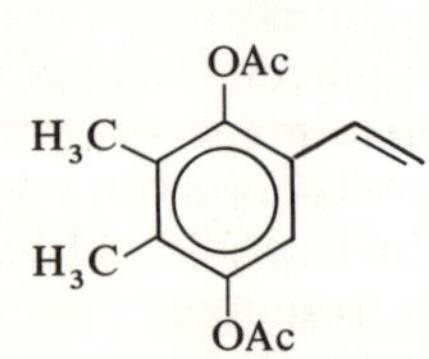

(+)

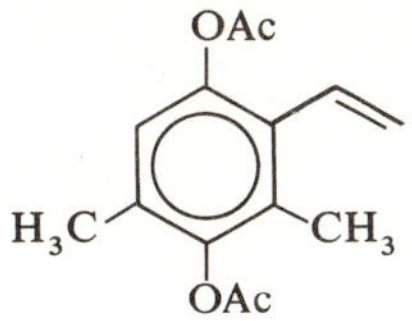

(+ +)

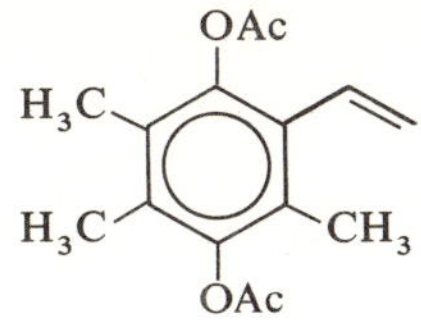

(+)

(+ +) = good polymerizability (80—90 % yield)
(+) = poor polymerizability (*ca.* 20 % yield)

Redox resins produced from 1-vinyl-2,4-dimethyl-3,6-diacetoxybenzene by the usual procedure had good chemical resistance[7]. Their redox capacity was *ca.* 2.1 meq g^{-1}, i.e. 1 g of dry resin reduces 2.1 mmol of Fe^{3+} to Fe^{2+}. As the resin contains SO_3H groups, the redox resin showed an ion-exchange capacity of 3.5 meq g^{-1}. By incorporating methyl substituents as electron-donating groups into the hydroquinone–quinone system the redox potential decreases. Therefore, depending on the substitution, redox resins can be produced with different redox potentials.

The hydroquinone–quinone redox systems can be stabilized by anellation of the quinone system with aromatic rings. 2-Methyl-3-vinylnaphthoquinone[8] was synthesized, which polymerized badly as hydroquinone diacetate. Therefore, the 2,3-dimethyl-5-vinylnaphthoquinone[9] was prepared. Unfortunately this monomer and its hydroquinone derivates were not polymerizable. It was found, however, that epoxidation of the quinone yielded a compound which could easily be copolymerized. The polymer quinone epoxide was converted into the corresponding poly(vinylnaphthoquinone) by using KI in aqueous acetic acid.

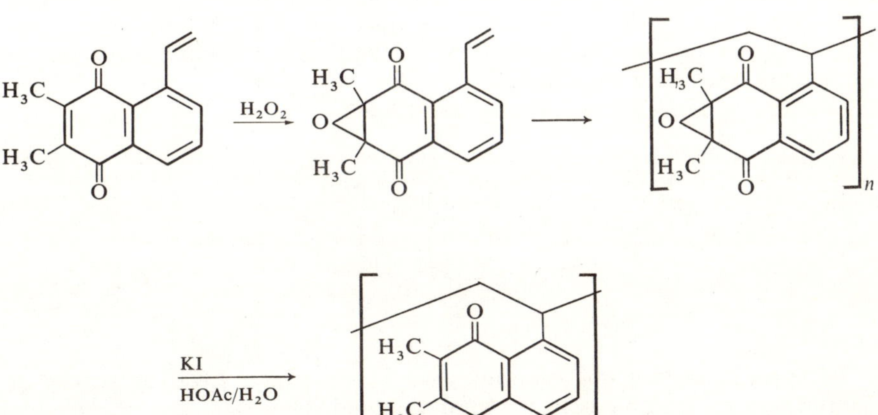

The following polymerizable vinylnaphthohydroquinone diacetates were also synthesized[10, 11].

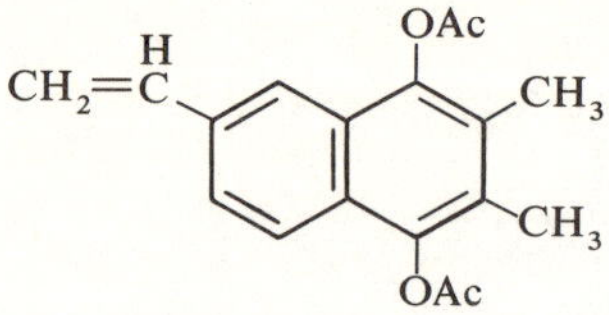

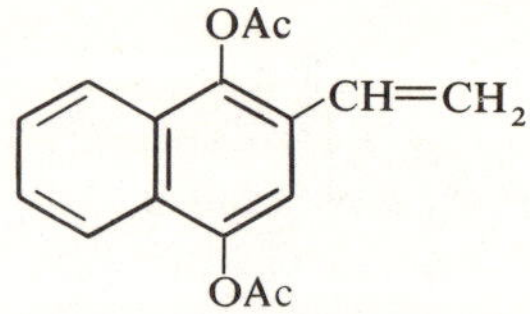

By annelating both sides of the benzoquinone the very stable anthraquinone system is obtained. Both vinylanthraquinone isomers were synthesized[12].

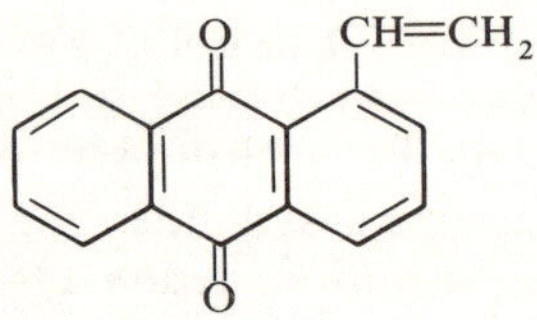

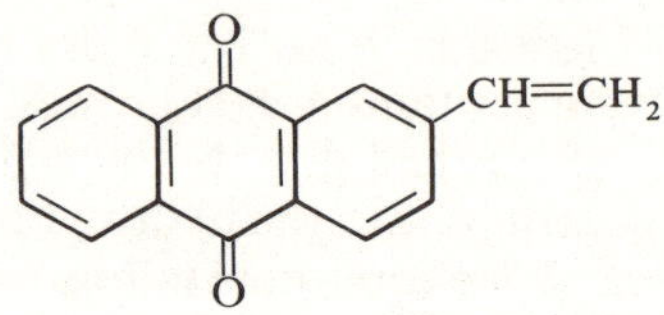

The 1-vinylanthraquinone polymerized very badly. The 2-vinylanthraquinone could be polymerized and copolymerized with styrene and DVB[13]. The copolymerization had to be performed in dimethylsulphoxide by thermal initiation, because the monomer is only slightly soluble. After sulphonation resins were obtained with redox capacities up to 4.76 meq g^{-1} and ion exchange capacities from 2.0 to 4.5 meq g^{-1}. The midpoint redox potential of this redox resin was *ca.* 127 mV (pH 1.09). Also macroreticular resins could be prepared with this monomer. Although the anthraquinone redox resin had a relatively good chemical resistance, it was noticed that hydrogen peroxide attacks the resin and decreases the crosslinking.

To improve the oxidation resistance of the polymer chain 2-isopropenylanthraquinone[14] was synthesized.

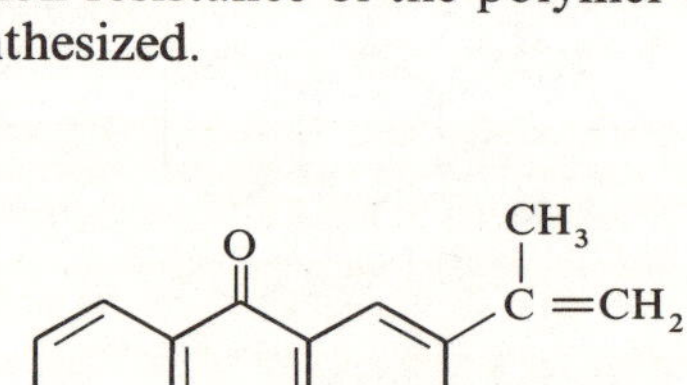

By copolymerizing this monomer with styrene and DVB or acrylic or methacrylic acid, stable resins[15] were obtained. They could be prepared in bead-form and also with a macroreticular structure. This could be done because the monomer had a good solubility. These redox resins are now commercially available[16].

Many model compounds in which hydroquinone groups and derivatives are linked by the sulphone function have been prepared in our laboratory[17–22].

The quinoid forms of many hydroquinone sulphones are unstable in aqueous media. Therefore, we had to use a special titration procedure[19,23]. Protecting the quinone nucleus by permethylation or anellation with aromatic rings resulted in compounds which were sufficiently stable to be potentiometrically measured in the acetic acid–water (1 : 1) system[24].

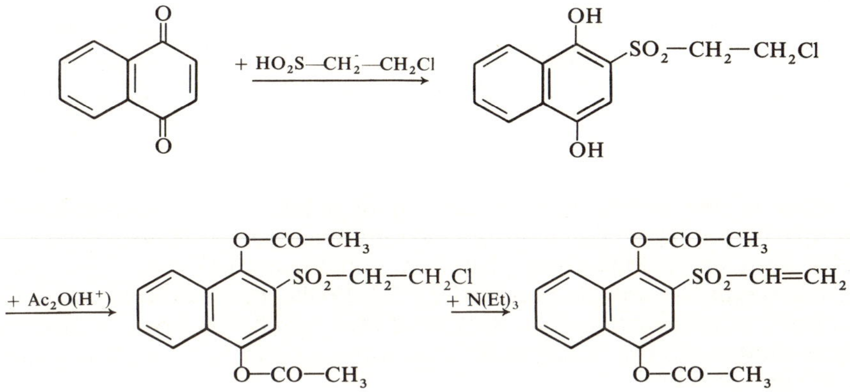

1,4-Diacetoxy-2-vinylsulphonylnaphthaline[22] was synthesized. This compound yielded no homopolymer by thermal or free-radical initiation, yet we were able to produce copolymers with either styrene or styrene–DVB. In order to improve swellability by means of introducing hydrophilic groups the preswollen copolymers were chlorosulphonated. The subsequent hydrolysis to sulphonic groups resulted in splitting off the protecting acetyl groups at the same time. The crosslinked copolymer had good mechanical properties and pretty good chemical stability. The redox capacity of 3.2 meq g^{-1} and the exchange capacity of 3.26 meq g^{-1} coincided with the calculated values.

Hydroquinonesulphonamides were also included in our investigations. We prepared a lot of model substances which were potentiometrically measured[23,25]. An increase in the hydrolysis stability was observed by substituting hydroquinone with the —NH—SO_2—R group instead of —SO_2—R (R = phenyl or NH—phenyl). Amongst other polymers we prepared a polymer hydroquinonesulphonamide by a polymer analogous reaction. It was stable and had good redox properties (4.7 meq g^{-1})[26].

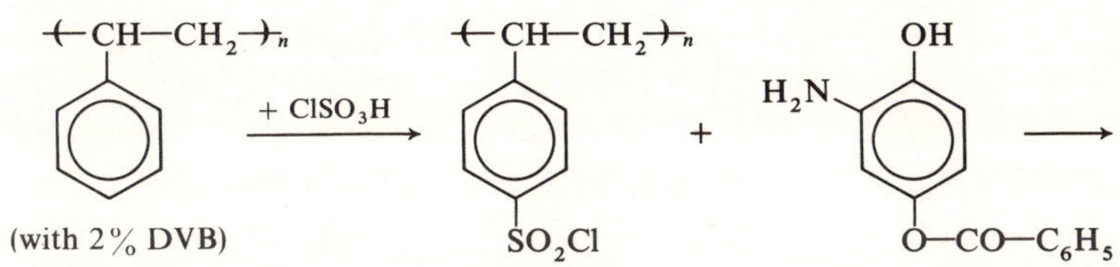

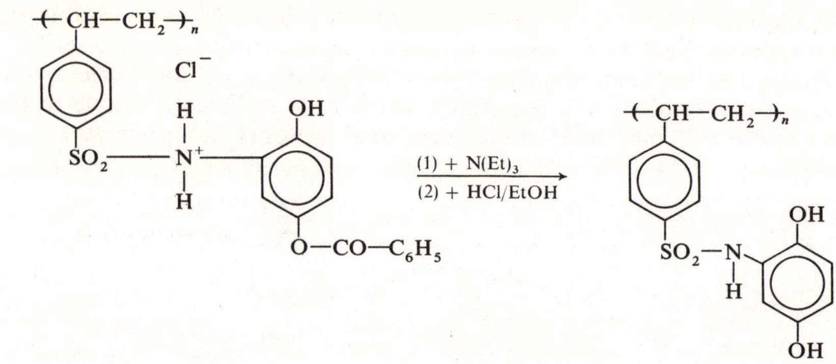

Most of the redox systems containing sulphone and sulphonamide groups did not have perfect properties. Our investigations have shown that substituted pyrazoloquinones yield very stable redox polymers[27–30]. Vinyl pyrazoloquinones can be prepared by 1,3-dipolar cycloaddition of diazo-propen-2 to suitable quinones. Because of their acid N—H pyrazole function these compounds can be sulphalkylated with excellent yields or *N*-vinylated with vinyl acetate.

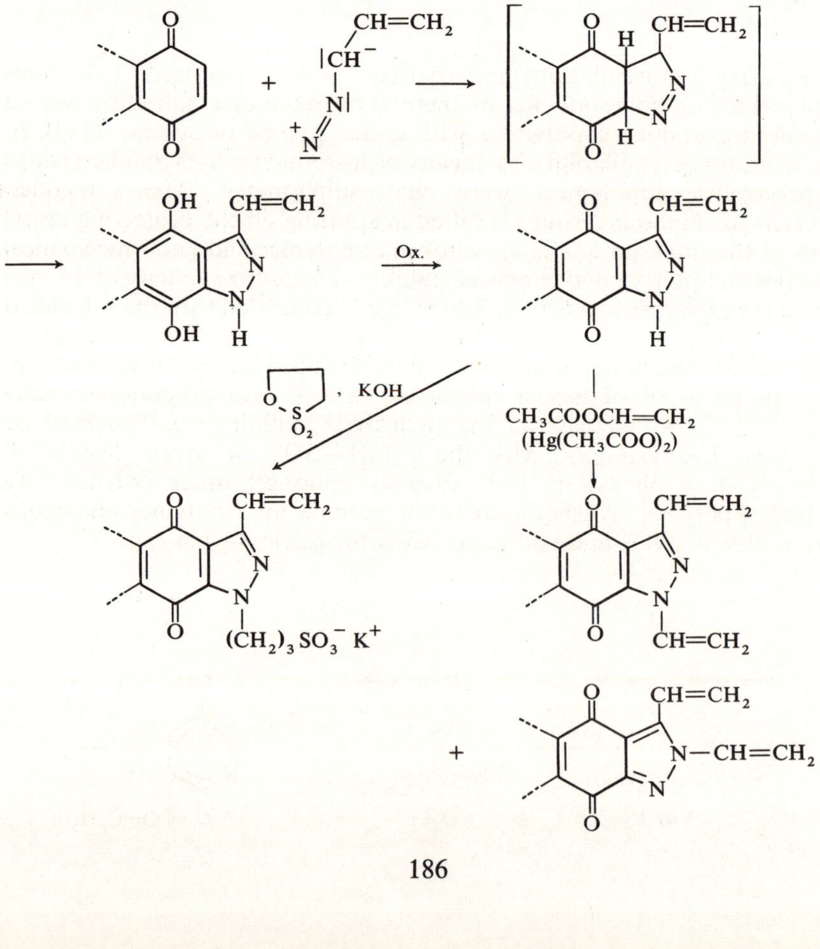

It is interesting to notice that most of the *N*-vinylated pyrazoloquinones cannot be polymerized by radicals unless an epoxy group is introduced into the quinone moiety. This epoxy group can be reduced later with KI in acetic acid.

Table 1. *N*-vinylpyrazoloquinone derivates and their polymerizability*

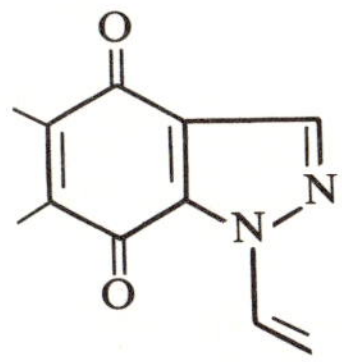

(−)

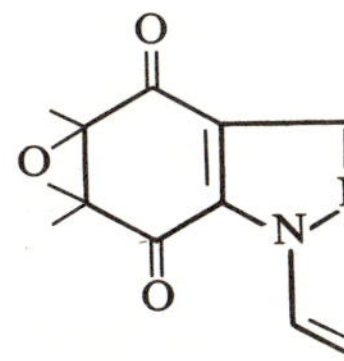

(++)

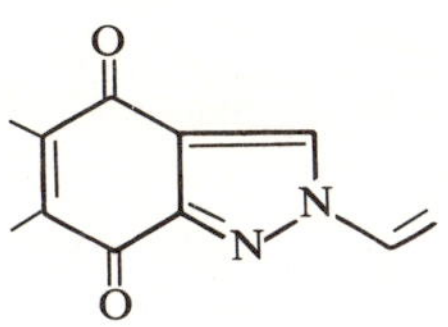

(−)

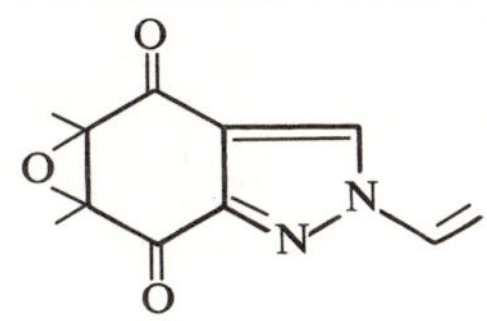

(++)

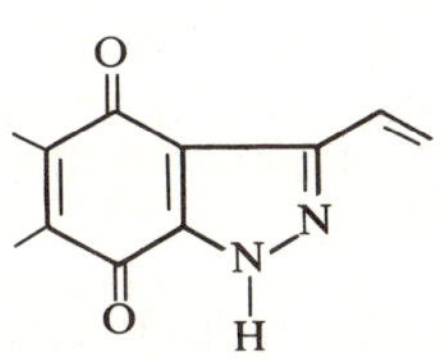

(+)

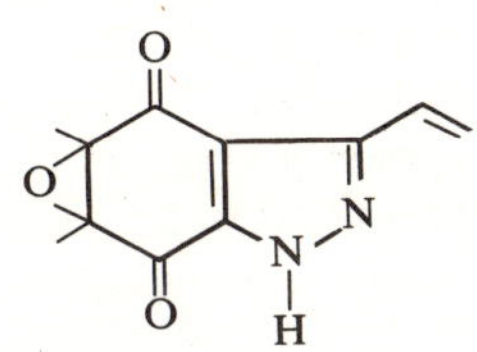

(++)

* as indicated in brackets

3-Vinyl-1H-benzo[*f*]indazole-4,9-dione and the crosslinking agent[30] 3,7-divinyl-1H,5H-pyrazolo[5,4-*f*]indazole-4,8-dione were copolymerized after being sulphalkylated. The swellability of the resulting product was too high, leading to poor mechanical stability. Resins with much better mechanical stability were obtained from these monomers performing the sulphalkylation *after* the copolymerization. (Redox capacity 5.0 meq g^{-1}, ion exchange capacity 2.68 meq g^{-1}, polarographic half-wave potential of the monomers *ca.* 130 mV).

Previously prepared polymers of this type had low redox potentials. Therefore, we tried to synthesize polymers with higher potentials[27,31]. For example 5,6-dichloro-3-vinyl-1H-indazole-4,7-dione could easily be polymerized after being reduced and acetylated to 5,6-dichloro-3-vinyl-indazole-hydroquinone-1,4,7-triacetate. Redox polymers were obtained by copolymerization of this monomer with the crosslinking agent 5,6-dichloro-1,3-divinyl-indazolehydroquinone-4,7-diacetate. To improve swellability the polymer was sulphalkylated.

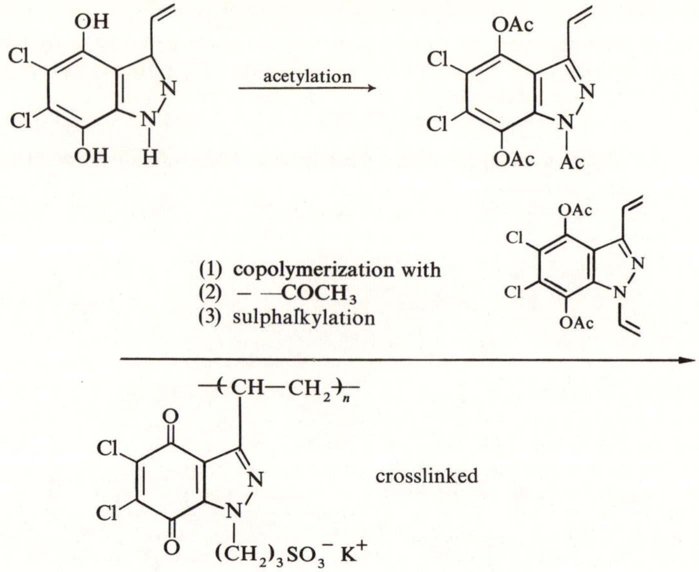

The swelling curve indicates that the swelling volume of the redox resin changes only slightly within a broad pH range. This is very favourable for column application. The redox capacity was 4.25 meq g^{-1} and the ion exchange capacity was 2.27 meq g^{-1}. The apparent standard redox potential E_0 was determined to be 476 mV (20°C).

Good results are achieved with some new nitrogen-containing redox polymers[27]. The quinone system inhibits the radical polymerization of some vinyl quinones. Introducing the aziridinyl group into certain vinylated quinones by adding aziridine to the vinyl group, monomers could be obtained which can easily be polymerized by cations. Vinyl quinones not having a completely substituted quinone nucleus can also add aziridine to the quinone nucleus by 1,4-addition[27,32]. Homogeneous products were obtained by reacting some vinylanthraquinones and vinylpyrazoloquinones with a large excess of aziridine in sealed tubes.

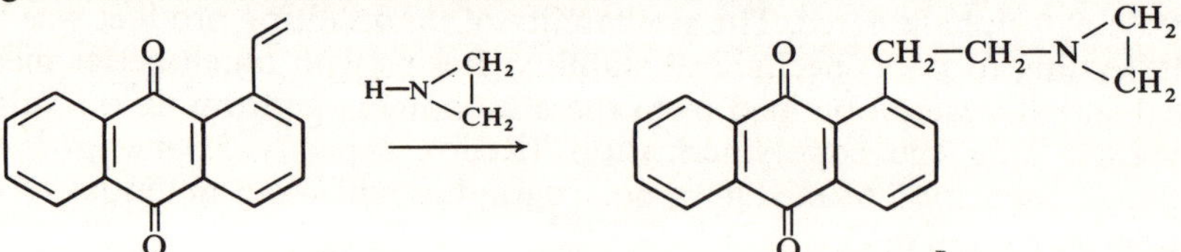

I

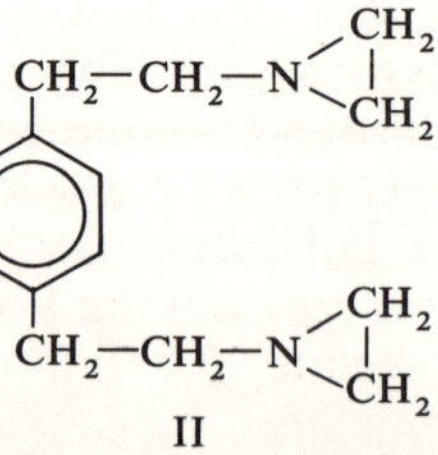

II

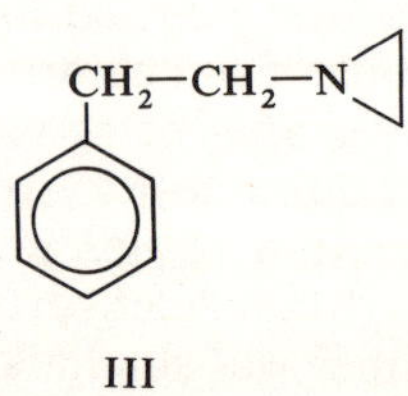

III

1-[2(1-Aziridinyl)ethyl]-anthraquinone (I) obtained from 1-vinylanthraquinone could be copolymerized with various amounts of the crosslinking agent 1,4-bis[2-(1-aziridinyl)-ethyl]-benzene (II) but the product was too soft. In order to obtain polymers with good mechanical properties, large amounts of crosslinking agent had to be used. This deteriorated the swelling properties of the products. By synthesizing a terpolymer with [2-(1-aziridinyl)-ethyl]-benzene (III), a polymer with both good strength and good swelling characteristics could be obtained[33].

Additional introduction of hydrophilic groups into these redox polymers was not necessary because of their content of tertiary amino groups. For practical applications it is important that these resins have good chemical stability and a high redox capacity (3.6–5.3 meq g^{-1}) in addition to good swelling and mechanical properties in acid solutions.

Redox resins are easily accessible by condensing formaldehyde with phenol and di- or tri-hydroxybenzenes[34, 35]. We have investigated with special accuracy the polycondensates of phenol–formaldehyde–hydroquinone which yielded rather stable spongy redox resins with good redox characteristics (redox capacity: 4.5–6.0 meq g^{-1}). Also polycondensates with hydroxylated naphthoquinones and anthraquinones were prepared[36, 37].

We also succeeded in synthesizing redox resins by a polymeranalogous reaction: Poly(α-methylstyrene) was reacted with phthalic acid anhydride under Friedel–Crafts conditions. By cyclization with polyphosphoric acid we obtained a polymer containing 8 mol % 2-isopropenylanthraquinone units[38].

THE REDOX BEHAVIOUR OF THE REDOX RESINS

For the characterization of redox resins their redox capacities and their redox potentials are the main data. Besides these, the velocity of their redox reactions is an important factor for their practical use. The redox capacities depend on the composition of the polymers and are measured in meq g^{-1} resin. For estimates of their possible application in oxidation–reduction reactions it is important to know their redox potentials. The potentiometric titration of redox systems is a convenient method to determine their redox potentials.

The redox process of the hydroquinone–quinone system can be described by the following equation:

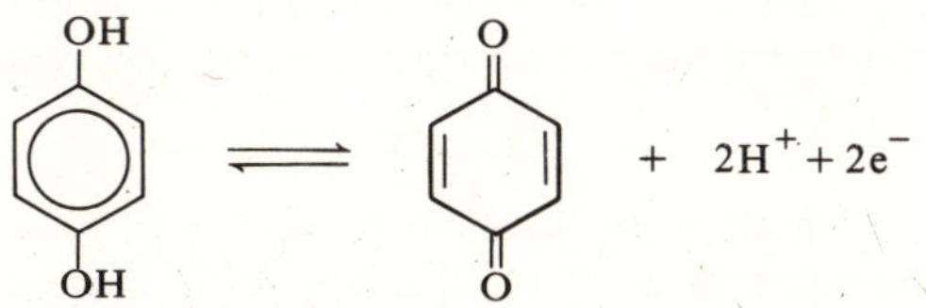

For better clearness we shall use this equation in the following abbreviated form:

$$H \rightleftharpoons Q + 2e^-$$

The potentiometric titration curve of this system can be described by the simple Nernst equation for processes consisting of a two-electron step.

$$E = E'_m + \frac{RT}{2F} \ln \frac{[Q]}{[H]}$$

E'_m is the pH-dependent midpoint potential which is measured at 50 per cent oxidation ($E'_m = E_0 - 0.059$ pH).

The shape of the titration curve can be characterized by the index potential E_i, the difference of the potentials at 50 and 25 per cent oxidation or 75 and 50 per cent oxidation[39]. A deviation of the limiting value of the index potential ($E_i = 14.1$ mV at 25°C) indicates the presence of redox species other than H and Q, e.g. of semiquinones.

The redox potential of a reversible redox system in solution can be measured at an inert electrode against the usual reference electrodes. If the redox system is insoluble, a direct determination of the redox potential is impossible. Therefore, it is convenient in most cases to add small amounts of a soluble redox system (called a mediator) with a standard potential similar to the polymer. The mediator should reach equilibrium with the redox resin. In this case the determined redox potential of the dissolved system equals the redox potential of the polymer. Since the equilibration process within the redox resin usually takes a long time, it is difficult to determine whether equilibrium is reached or not.

As was shown above the redox process also involves protons, leading to a pH-dependence of the redox potential. Because most of the redox resins contain ionogenic groups as hydrophilic parts (—SO_3H, —COOH, —OH, etc.) the distribution of the protons between the outer solution and the resin will be ruled by the Donnan equilibrium. This affects the potential of the resin in a complicated way, so that the E'_m of the redox resins are usually measured at definite conditions.

Potentiometric titration curves of the redox resins generally show higher midpoint potentials (E'_m) than the corresponding monomer units. The slopes of the curves show a steeper increase and sometimes the curves are unsymmetrical. As there are many possible explanations for this behaviour, soluble model compounds of definite structure were synthesized and their redox behaviour was investigated.

Tetravalent redox systems with the following structures have been synthesized and measured potentiometrically (cf. *Table 2*).

A tetravalent system may be described by the following abbreviation

$$H \sim H \rightleftharpoons Q \sim Q + 4e^-$$

For a four-electron step the titration curve would be very flat according to the Nernst equation with an E_i of 7.1 mV at 25°C. Actually the measurements always result in an $E_i \geqslant 14.1$ mV (25°C). So it can be assumed that the redox reaction consists of two divalent steps, inserting an intermediate I[42a, b].

$$H \sim H \rightleftharpoons I + 2e^-$$

$$I \rightleftharpoons Q \sim Q + 2e^-$$

Table 2 Potentiometric data of some tetravalent redox systems (acetic acid–water = 1:1, v/v; 25.0°C; $K_2Cr_2O_7$)[40]

System	E'_{mh} mV	E_i mV	$E'_2 - E'_1$ mV	K
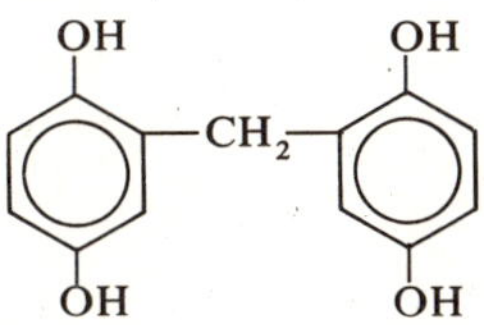	587.2	21.3	39.6	21.9
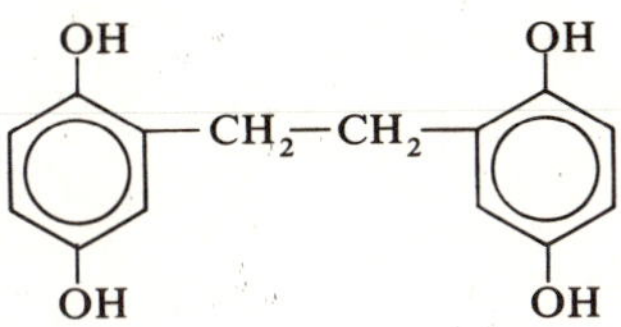	574.3	18.5	32.6	13.2[41]
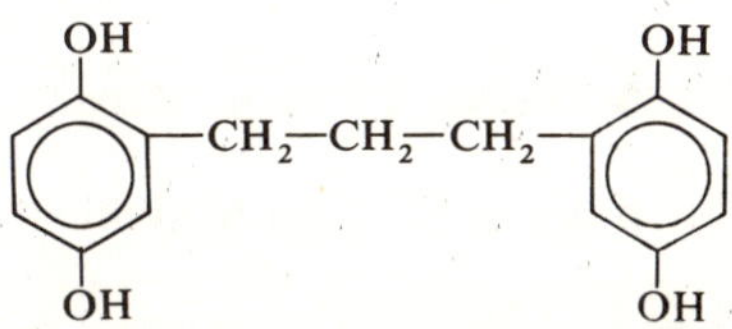	568.2	16.8	(17.8)	(4.0)
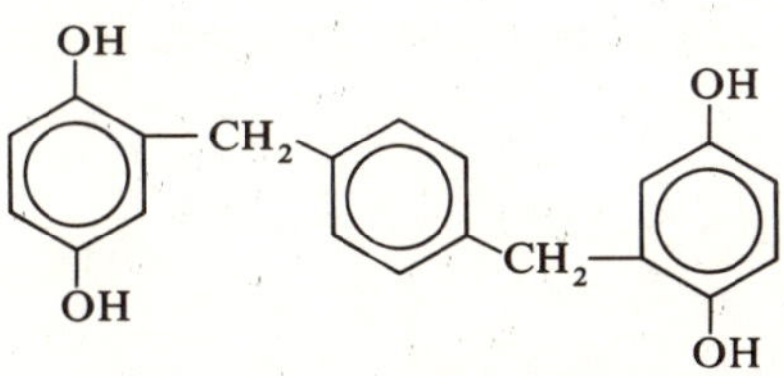	575.3	15.2	21.9	5.5
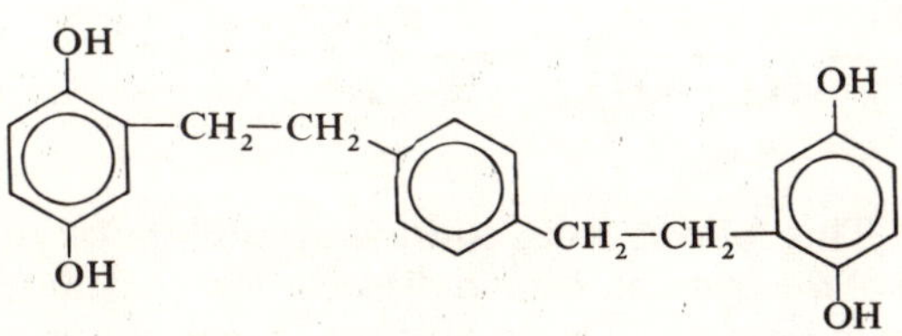	569.7	14.5	18.7	4.3
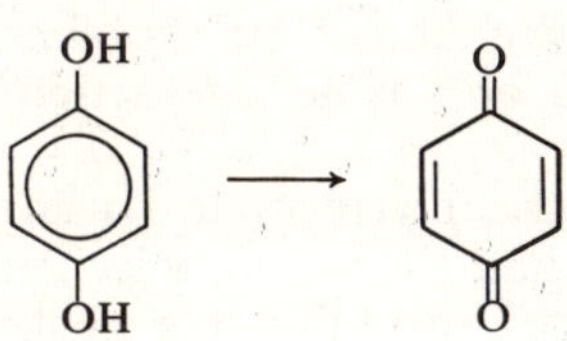	621.3	14.2	—	—

From the Nernst equations at constant pH

$$E = E_1' + \frac{RT}{2F} \ln \frac{[\mathrm{I}]}{[\mathrm{H} \sim \mathrm{H}]} \tag{1}$$

$$E = E_2' + \frac{RT}{2F} \ln \frac{[\mathrm{Q} \sim \mathrm{Q}]}{[\mathrm{I}]} \tag{2}$$

$$E = \frac{E_1' + E_2'}{2} + \frac{RT}{4F} \ln \frac{[\mathrm{Q} \sim \mathrm{Q}]}{[\mathrm{H} \sim \mathrm{H}]} \tag{3}$$

and the redistribution reaction

$$\mathrm{Q} \sim \mathrm{Q} + \mathrm{H} \sim \mathrm{H} \rightleftharpoons 2\mathrm{I}$$

$$\frac{[\mathrm{I}]^2}{[\mathrm{H} \sim \mathrm{H}][\mathrm{Q} \sim \mathrm{Q}]} = K \tag{4}$$

results equation 5.

$$E_2' - E_1' = \frac{RT}{2F} \ln K \tag{5}$$

The difference in the redox potential of the individual divalent steps is connected with the equilibrium constant K.

E_1' and E_2' are the pH-dependent midpoint potentials of both divalent oxidation steps.

$E_m' = (E_1' + E_2')/2$ is the pH-dependent measured midpoint potential of the redox system.

The constant K can be obtained from the titration curves by means of the index potential.

$$E_i = \frac{RT}{2F} \ln \left[\tfrac{1}{2}K^{\frac{1}{2}} + \tfrac{1}{2}(K + 12)^{\frac{1}{2}}\right]$$

The maximal content of the intermediate I is given by

$$\left(\frac{[\mathrm{I}]}{[\mathrm{H} \sim \mathrm{H}] + [\mathrm{I}] + [\mathrm{Q} \sim \mathrm{Q}]}\right)_{\max} = \frac{K^{\frac{1}{2}}}{2 + K^{\frac{1}{2}}}$$

The limiting value for K is 4. This follows from considerations similar to those used for deriving the ratio of the first and second dissociation constants of dibasic acids which should be at least 4[43]. This means that according to a statistical point of view the difference $E_2' - E_1'$ should not be less than 18 mV(20°C). This limiting value will be obtained only if no interactions exist between the redox groups.

We were able to elucidate the constitution of the intermediate I using n.m.r. spectroscopy[40]. The compounds IV and V and their equimolecular mixture show in tetradeuterioacetic acid at 100°C the following peaks (solvent peak 2.04 p.p.m. as reference).

		chemical shift for the methyl group δ(p.p.m.)
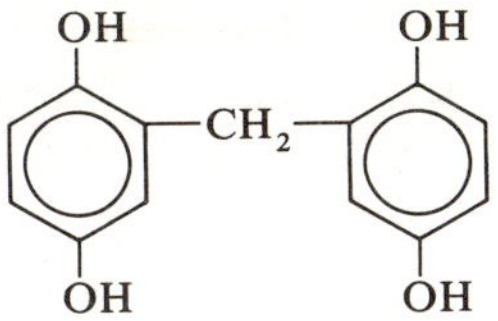	IV	3.83 (s)
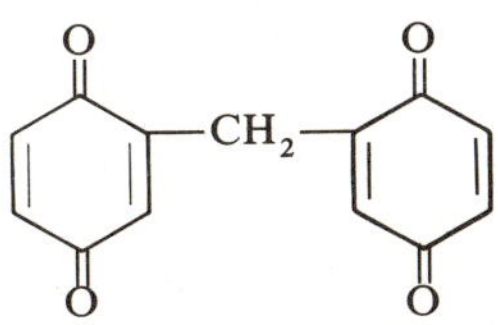	V	3.61 (m)
equimolar mixture of IV and V		3.70 (m) and peaks at 3.83 and 3.61 (according to IV and V having much less intensities)

These n.m.r. spectra show that a new peak is produced in the mixture. This signal is situated nearly half-way between the former peaks and gives evidence for a new compound VI.

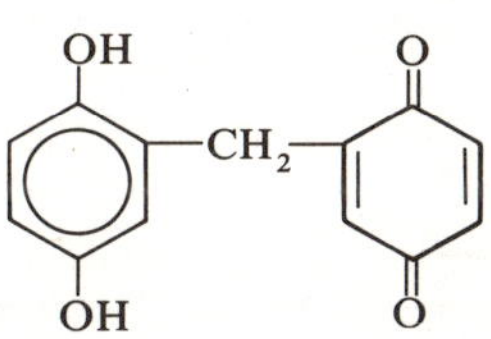

VI

Table 2 shows that the elongation of the connecting chain between the redox systems results in a decrease of the index potential close to the limiting value of 14.1 mV. This value is not completely reached because there are other factors influencing the index potential. In addition to the inductive effect intramolecular complexes (quinhydrones) can also be formed to a certain extent.

Table 3 illustrates the results of our calculations of existing complexes in tetravalent redox systems at 50 per cent oxidation. The explanation of this procedure would be too extensive.

The longer the intramolecular distance the less is the complexing. Compounds with sterically favoured complex formation also show higher amounts of the quinhydrone.

Steric effects arising from hindered free rotation of the connected redox systems have an obvious influence on the midpoint potential. This is shown

Table 3. Species present at 50% oxidation in solutions of some tetravalent systems (acetic acid–water 1:1, v/v; 25.0°C; total concentration *ca.* 5×10^{-4} mol/l.)[40]

System	E_i mV	H ~ H %	H ~ Q %	(H ~ Q) complex %	Q ~ Q %
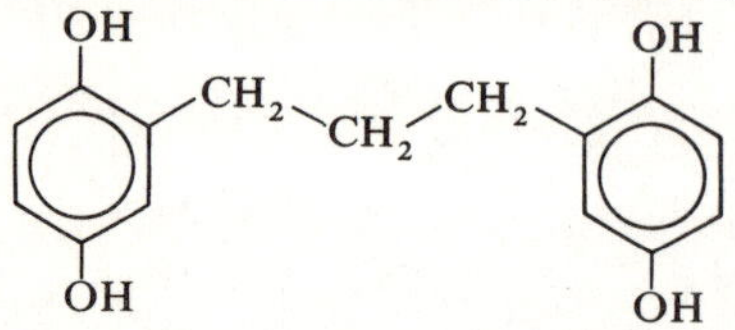	16.8	19.1	38.1	23.1	19.1
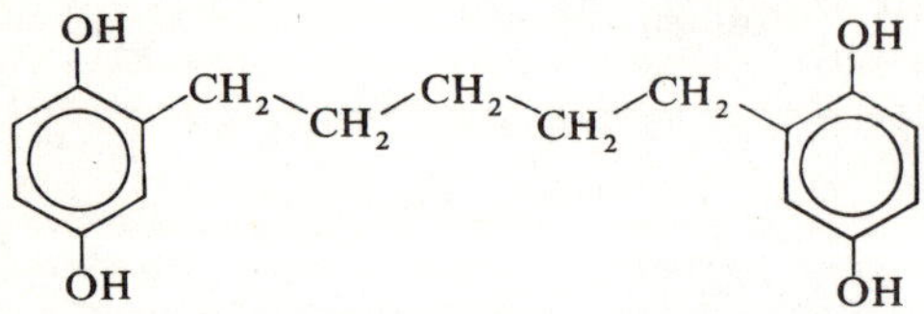	15.0	23.4	46.7	6.6	23.4
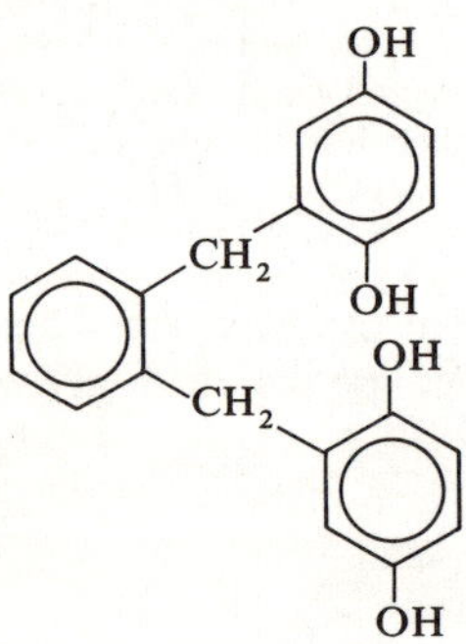	24.7	11.9	30.3	45.9	11.9

Table 4. Potentiometric data of tetravalent naphthoquinones (acetic acid–water = 4:1, v/v: 25°C; potassium dichromate)

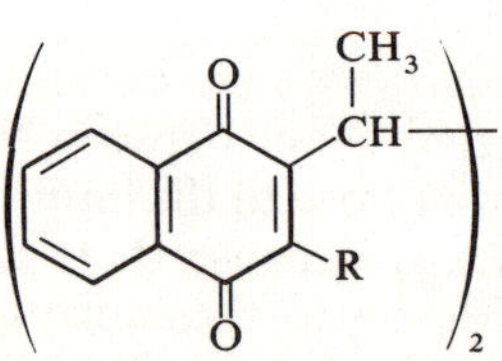

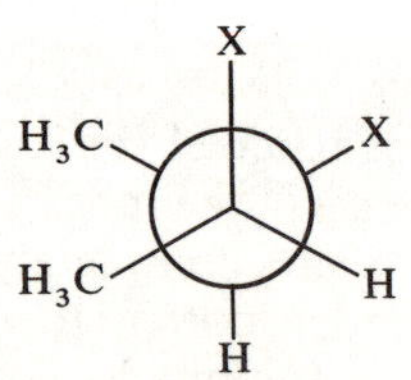

meso

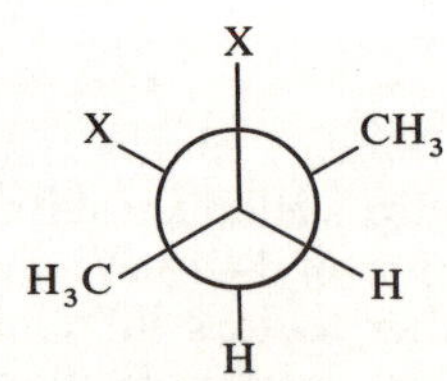

racemic

Constitution	R	E'_{mh} mV	E_i mV	colour at 50% oxidation
meso	H	383.4	17.9	brownish
racemic	H	383.3	21.0	reddish
meso	CH_3	299.6	17.5	brownish
racemic	CH_3	286.3	18.5	reddish
benzoquinone	—	670.2	14.4	—

by some tetravalent naphthoquinone derivatives[44] which can be obtained in meso and racemic forms. A methyl group in the quinone nuclei changes distinctly the midpoint potential, due to hindered rotation. The index potential for racemic forms are always higher because they can exist in a conformation in which complex formation is favoured.(see *Table 4*).

Hexavalent redox systems have been synthesized either with a symmetric structure[45] (see *Table 5* below) or a reduced symmetric structure[42a, 46].

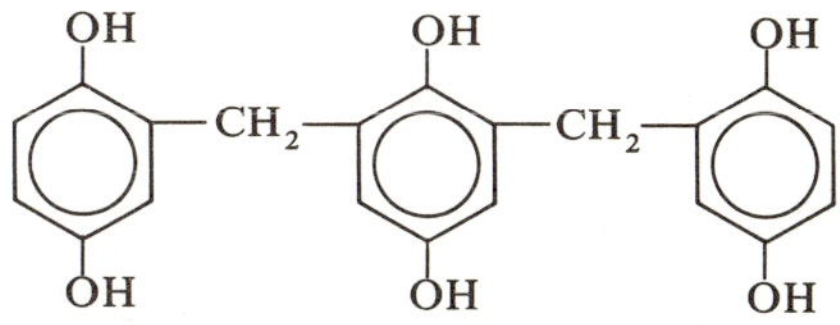

During oxidation the symmetric hexavalent systems produce in each bivalent redox step only one intermediate.
In abbreviated form:

$$(H \sim H \sim H) \overset{E'_1}{\rightleftharpoons} (H \sim Q \sim H) \overset{E'_2}{\rightleftharpoons} (H \sim Q \sim Q) \overset{E'_3}{\rightleftharpoons} (Q \sim Q \sim Q)$$

Compounds with reduced symmetry give two different intermediate compounds in each step. The analytical treatment of the potentiometric titration curves of the symmetric compounds is similar to that of the symmetric tetravalent systems[45, 47]. The results are:

(i) The titration curves are symmetrical to the midpoint.

(ii) The formation constant K for the intermediate is equal by definition for the first and second redox step. This is concluded from the symmetry of the molecule.

(iii) K has a limiting value of 3. In the ideal case ($K = 3$) the titration curve shows an index potential of 14.1 mV (25°C) because there exists no interaction between the redox systems. The differences of the midpoint potentials $E'_2 - E'_1$ and $E'_3 - E'_2$ are equal and have the same limiting value of 14.1 mV (25°C).

(iv) The equilibrium constant K for the intermediate can be determined using the index potentials. The results of the potentiometric titration of some compounds are shown in *Table 5*. They confirm our analytical approach.

The titration curve of hexavalent systems with reduced symmetry[46] can be analysed only by making several assumptions. The redox curve can approximately be considered to result from two tetravalent steps overlapping each other.

$$H \sim H \sim H \overset{E'_1}{\rightleftharpoons} H \sim Q \sim H \overset{E'_2}{\rightleftharpoons} H \sim Q \sim Q$$

$$H \sim Q \sim H \overset{E'_2}{\rightleftharpoons} H \sim Q \sim Q \overset{E'_3}{\rightleftharpoons} Q \sim Q \sim Q$$

The potentiometric titration of polymer redox systems is characterized by a very slow establishment of a stable potential after each addition of titrant. This is attributed to a slow establishment of the redox equilibrium within

Table 5. Potentiometric data of symmetrical hexavalent systems (acetic acid–water 1:1, v/v; 25°C; potassium dichromate)[45]

System		E'_{mh} mV	E_i mV	K	E'_1 mV	E'_2 mV	E'_3 mV
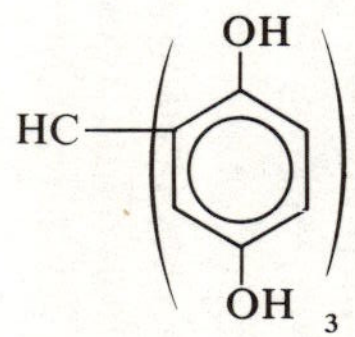		584.4	25.7	13.1	551.3	584.4	617.5
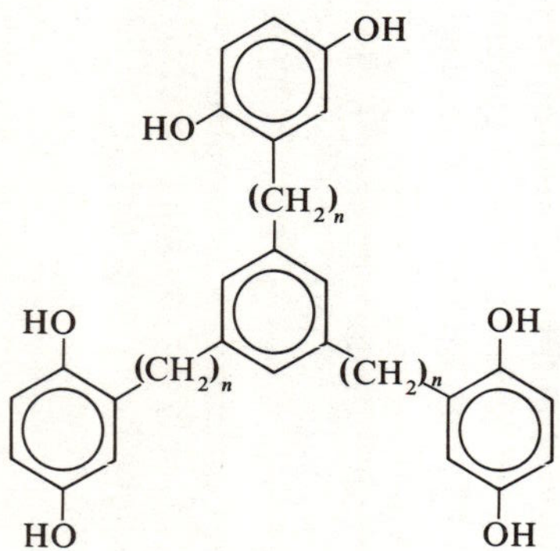	$n=1$	575.8	16.1	4.03	557.9	575.8	593.7
	$n=2$	568.6	15.2	3.54	552.4	568.6	584.8
Benzoquinone		621.3	14.2	—	—	—	—

the polymer and/or between mediators and the oxidant or reductant and the polymer.

The diffusion of the reactants in the redox resin is the rate-determining step. Therefore, the swellability of the resins is an important factor. Excessive swelling leads to mechanically unstable particles. For practical applications a compromise must usually be found by experiments. The titration curves of the redox resins deviate in shape and energy (e.g. position of the midpoint potentials) from those of the corresponding monomer redox systems. This is illustrated by the anthraquinone redox resins[48]. Polymers with a high content of anthraquinone (50 mol %) gave a midpoint potential (at pH = 1.17) of 122 mV and an index potential of 19 mV. Diminishing the redox content in the redox resin to 9 mol %, the titration gave a midpoint (at the same pH 1.18) of 108 mV and an E_i of 14.5 mV. The titration curve of the redox resin with the low content of anthraquinone had the same shape as the normal monomer anthraquinone system. This resin shows a ratio of *ca.* 8 styrenesulphonic groups per quinone unit. In the former resin the ratio is 0.9 : 1. It means that in both resins the redox units are in different environments. It is known e.g. that a change of the titration medium from water to ethanol raises the midpoint potential of benzoquinone by 13 mV. Another reason for the different potentials can be found in the shape of the titration curves of the polymers due to the nonuniformity of the redox systems and their immediate vicinity. The polymer redox systems show the same behaviour as the corresponding soluble monomers if the redox capacity is low and the micro-environment is similar to that in solution. This was confirmed by Cassidy and his co-workers with soluble oxidation–reduction polymers[49, 50].

APPLICATIONS OF REDOX RESINS

Redox resins are insoluble but swellable macromolecular compounds with reversible redox systems, which can exchange electrons and protons with the ambient medium. They can be used as regenerable insoluble oxidizing or reducing agents which can easily be separated by filtration from the substrate solution. That means the solution which has to be oxidized or reduced is not contaminated by the oxidizing or reducing agent. This is the great advantage of these substances compared with the usual soluble oxidants or reductants. There have been many proposals for the application of redox resins. In this paper only some examples of their applications which have been studied in our laboratories can be mentioned[51, 52]. Redox resins in the reduced form (hydroquinone form) have been used for the reduction of ions and for depositing metals. Using a poly(vinyl anthraquinone) resin[51] the following metals could be deposited by a batch process: silver from neutral $AgNO_3$ solutions, copper from 2N sulphuric acid–Cu^{2+} solutions, Te from neutral tellurite solutions, and red Se from 2N sulphuric acid–selenite solutions. The deposited metals could be dissolved without affecting the redox resin. Fe^{3+} could be reduced to Fe^{2+}, UO_2^{2+} to U^{4+}, and VO_2^+ to V^{3+} via VO^{2+}.

Redox resins in the reduced form can remove oxygen from solutions[53, 54]. In the presence of stabilizers they can reduce oxygen to hydrogen peroxide[37, 55–57] and hydrogen peroxide solutions can be produced[51]. We applied the anthraquinone-containing resin in its reduced form in a column for the reducing of *p*-benzoquinone. Further we applied the condensation redox resin prepared from hydroquinone, phenol and formaldehyde in the quinone form for the dehydrogenation of various organic compounds[52, 56]. Cycloheptatriene can be dehydrogenated to the tropylium cation, 1,4-dihydronaphthalene to naphthalene, 9,10-dihydroanthracene to anthracene. 1,4-Dihydrocarbazole and 1,2,3,4-tetrahydrocarbazole could be dehydrogenated to carbazole, while the hexa- and dodeca-hydrocarbazole could not be dehydrogenated. Hydrazobenzene could be dehydrogenated to azobenzene, cystein to cystine, ascorbic acid to dehydroascorbic acid.

The possible applications of the redox resins in biochemistry and pharmacy are of great interest. For instance we were successful in dehydrogenating NADH to NAD^+ with a hydroquinone redox resin in the oxidized form. The redox resin in the quinone form also performed the Strecker degradation of alanine to acetaldehyde and also of other α-amino acids[52]. Redox resins in the reduced form (hydroquinone form) were used successfully as antioxidants. Many other applications have been reported[3].

With this selective and by no means complete review we intended to show that there are many possibilities of synthesizing redox polymers with various redox potentials. Many interesting electrochemical investigations have been carried out. There is a great field for physical chemists to extend the theoretical knowledge. Further applications seem to be promising. In conclusion the author wishes to underline the statement of Cassidy[3]: 'The whole field is ripe for exploitation'.

ACKNOWLEDGEMENTS

I would like to thank Dipl. Chem. E. Ehrenthal and Dr. W. Storck for their helpful assistance in compiling and preparing this manuscript.

REFERENCES

1 H. G. Cassidy and K. A. Kun, *Oxidation–Reduction Polymers* (*Redox Polymers*), Interscience: New York (1965).
2 G. Manecke, *Angew. Makromol. Chem.* **4/5**, 26 (1968).
3 H. G. Cassidy, *J. Polymer Sci., Pt. D* **6**, 1 (1972).
4 I. H. Updegraff and H. G. Cassidy, *J. Amer. Chem. Soc.* **71**, 407 (1949).
5 (a) G. Manecke and G. Bourwieg, *Chem. Ber.* **92**, 2958 (1959).
(b) G. Manecke and G. Bourwieg, *Chem. Ber.* **95**, 1413 (1962).
(c) G. Manecke and G. Bourwieg, *Chem. Ber.* **96**, 2013 (1963).
6 K. A. Kun and H. G. Cassidy, *J. Polymer Sci.* **56**, 83 (1962).
7 G. Manecke and G. Bourwieg. *Makromol. Chem.* **99**, 175 (1966).
8 G. Manecke and W. Storck, *Chem. Ber.* **94**, 300 (1961).
9 G. Manecke, G. Ramlow and W. Storck, *Chem. Ber.* **100**, 836 (1967).
10 G. Manecke and J. Melzer, unpublished.
11 G. Manecke and W. Storck, *Chem. Ber.* **102**, 3584 (1969).
12 G. Manecke and W. Storck, *Chem. Ber.* **94**, 3239 (1961).
13 G. Manecke and W. Storck, *Angew. Chem.* **74**, 903 (1962).
14 G. Manecke, K. Creutzburg and J. Klawitter, *Chem. Ber.* **99**, 2440 (1966).
15 G. Manecke and K. Creutzburg, *Makromol. Chem.* **93**, 271 (1966).
16 E. Merck, Darmstadt.
17 G. Manecke and H. J. Beyer, *Makromol. Chem.* **123**, 223 (1969).

[18] G. Manecke and G. Wehr. *Makromol. Chem.* **136**, 121 (1970).
[19] G. Manecke, H. J. Beyer, G. Wehr and J. Beier, *J. Electroanal. Chem.* **28**, 139 (1970).
[20] G. Manecke and H. J. Beyer, *Makromol. Chem.* **116**, 26 (1968).
[21] G. Manecke, J. Beier and G. Wehr, *Makromol. Chem.* **134**, 231 (1970).
[22] G. Manecke, Chr.-St. Rühl and G. Wehr, *Makromol. Chem.* **154**, 121 (1972).
[23] G. Manecke, H. J. Beyer and R. Mölleken, *J. Electroanal. Chem.* **22**, 152 (1969).
[24] G. Manecke and H. J. Beyer, unpublished.
[25] G. Manecke and R. Mölleken, *Makromol. Chem.* **145**, 53 (1971).
[26] G. Manecke, G. Wetzel and W. Storck, unpublished.
[27] G. Manecke, H.-J. Kretzschmar and W. Hübner, *J. Makromol. Sci. Chem.* **A7**(5), 1181 (1973).
[28] G. Manecke and G. Ramlow, *Chem. Ber.* **101**, 1987 (1968).
[29] G. Manecke, G. Ramlow, W. Storck and W. Hübner, *Chem. Ber.* **100**, 3413 (1967).
[30] G. Manecke and G. Ramlow, *J. Polymer. Sci. Pt C* **22**, 957 (1969).
[31] F. Acosta and G. Manecke, unpublished.
[32] G. Manecke and H. J. Kretzschmar, *Chem. Ber.* **103**, 3862 (1970).
[33] G. Manecke and H. J. Kretzschmar, *Makromol. Chem.* **169**, 15 (1973).
[34] G. Manecke, *Z. Elektrochem., Ber. Bunsenges. Physik. Chem.* **57**, 189 (1953).
[35] G. Manecke, *Z. Elektrochem., Ber. Bunsenges. Physik. Chem.* **58**, 369 (1954).
[36] G. Manecke and Ch. Bahr, *Naturwissenschaften* **44**, 260 (1957).
[37] G. Manecke and Ch. Bahr, *Z. Elektrochem., Ber. Bunsenges. Physik. Chem.* **62**, 311 (1958).
[38] G. Manecke and G. Koßmehl, *Makromol. Chem.* **80**, 22 (1964).
[39] L. Michaelis, *J. Biol. Chem.* **96**, 703 (1932).
[40] W. Storck and G. Manecke, unpublished.
[41] G. Manecke and D. Zerpner, *Makromol. Chem.* **108**, 198 (1967).
[42] (a) G. Manecke and H. J. Förster, *Makromol. Chem.* **52**, 147 (1962).
(b) G. Manecke, H. J. Förster and H.-J. Panoch, *Electrochim. Acta* **13**, 1101 (1968).
[43] E. Q. Adams, *J. Amer. Chem. Soc.* **38**, 1503 (1916).
[44] G. Manecke and W. Storck, *Chem. Ber.* **104**, 1207 (1971).
[45] W. Storck and G. Manecke, unpublished.
[46] G. Manecke and H. J. Panoch, *Makromol. Chem.* **96**, 1 (1966).
[47] J. E. C. Mills, 'Redox polymers', *Dissertation*, University of Toronto (1971).
[48] (a) G. Manecke and W. Storck, *J. Polymer Sci., Pt C* **4**, 1457 (1963).
(b) G. Manecke and W. Storck, *Makromol. Chem.* **75**, 159 (1964).
[49] G. Wegner, N. Nakabayashi, S. Duncan and H. G. Cassidy, *J. Polymer Sci.*, [*A–1*] **6**, 3395 (1968).
[50] N. Nakabayashi, G. Wegner and H. G. Cassidy, *J. Polymer Sci.*, [*A–1*] **7**, 583 (1969).
[51] G. Manecke, G. Koßmehl, G. Hartwich and R. Gawlik, *Angew. Makromol. Chem.* **2**, 86 (1968).
[52] G. Manecke, G. Koßmehl, R. Gawlik and G. Hartwich, *Angew. Makromol. Chem.* **6**, 89 (1969).
[53] G. Manecke, *Angew. Chem.* **67**, 613 (1955).
[54] B. Sansoni, *Dissertation*, München, (1956).
[55] G. Manecke, *Angew. Chem.* **68**, 582 (1956).
[56] G. Manecke, Ch. Bahr and Ch. Reich, *Angew. Chem.* **71**, 646 (1959).
[57] G. Manecke and G. Koßmehl, *Makromol. Chem.* **70**, 112 (1964).

POLYMER-ANALOGOUS REACTIONS OF POLYENES IN POLY(VINYL CHLORIDE)

F. TÜDÖS, T. KELEN, T. T. NAGY and B. TURCSÁNYI

Central Research Institute for Chemistry
Hungarian Academy of Sciences
Budapest, Hungary

ABSTRACT

The primary process in the thermal degradation of PVC in an inert atmosphere and at relatively low temperatures, is the elimination of hydrogen chloride molecules and the formation of polyene sequences of different lengths. The polyenes formed are highly reactive and participate in different secondary reactions.

It has been shown for PVC samples degraded both in dilute solution and in the solid state that the amount of polyenes does not correspond to the degree of dehydrochlorination: proportionality may be observed only at the initial stage of degradation and, later on, the number of double bonds in the system is much lower than the amount of hydrogen chloride molecules split off. With the increase of conversion the increase in the amount of polyenes containing three or more double bonds gradually slows down and, later on, practically ceases. This phenomenon is also quantitatively well interpretable by intramolecular cyclization (backbiting) of polyenes. The reaction leads to cyclohexadiene structures. On completing the kinetic treatment of the process, we determined the relative rate constant of cyclization.

Reactive dienophilic reagents readily react with the polyenes formed in degraded PVC. Based on the changes observed in the u.v. and visible spectra, a kinetic study was made on the Diels–Alder reaction with chloromaleic anhydride. The kinetic treatment, performed on the basis of a presumed mechanism, allowed the determination of the rate constants and their temperature dependence.

Depending on the experimental conditions, the polyenes formed in the course of degradation may participate in other reactions as well. Some of the possible reaction routes are: the change in molecular weight, the formation of low molecular weight aromatic compounds (e.g. benzene) and the considerable proton exchange observed in the reaction of polyenes with hydrochloric acid.

INTRODUCTION

Considering the use and consumption of synthetic materials, polyvinyl chloride stands second in the world. Although it has numerous excellent properties, degradation takes place at a considerable rate at the usual temperature of processing.

Publications dealing with the degradation of PVC (in an inert atmosphere, at 150–250°C) all agree on a dehydrochlorination process and simultaneous formation of conjugated double bond sequences (polyenes) in the polymer chain[1].

These polyenes give characteristic u.v. and visible spectra of degraded PVC. To each band of these complex spectra, we may assign a very intense main absorption band[2] appearing at the longest wavelength in the individual spectrum of each polyene.

There is no essential difference of opinion regarding the polyene lengths assigned to the bands. Several authors suggest that the u.v. and visible spectra of degraded PVC can be quantitatively evaluated and the extinction coefficient of polyenes of different lengths can be determined[3,4].

Most authors agree that following relatively slow initiation (elimination of the first hydrogen chloride molecule) the reaction will 'zip' along the polymer chain to leave conjugated unsaturated sequences. There is, however, a divergence of opinion regarding the mechanism of the reaction, the initiation step, chain termination, the length of the 'zip' and other important problems, which are the subjects of research all over the world today[5].

The further fate of the polyenes so formed is, however, a field hardly touched by investigators, although it is both theoretically and practically a very important terrain. The applicability of PVC is much affected by even a relatively small amount of polyenes formed in the course of processing. Conjugated polyenes may initiate further different reactions, e.g. oxidation. Polymer analogous reactions converting the polyenes into other less sensible groups bear, therefore, considerable significance considering the stabilization of PVC. Another important problem in the processing and application of PVC is discolouration, which may also be reduced or eliminated by appropriate polymer analogous reactions.

In the course of our earlier investigations, it has been found and verified[6], as evidence of polyene reactivity, that polyenes react with hydrogen chloride and thus, evidently, also with the hydrogen chloride formed in the course of degradation. The considerable proton exchange observed in this type of polymer analogous reaction of polyenes supports the possibility of readdition of hydrogen chloride to the degraded polymer.

In this work, we deal with two further polymer analogous reactions: intramolecular cyclization reaction of polyenes and their Diels–Alder reaction with active dienophilic components.

It should be noted that, depending on the experimental conditions, the polyenes formed during degradation take part in other—non-polymer analogous—reactions as well. The changes in molecular weight and the formation of low molecular weight aromatic compounds (e.g. benzene) also point to such reactions.

1. CYCLIZATION OF POLYENES DURING DEGRADATION

In this work we mainly applied the experimental methods described in detail in our previous papers[6]. We used the polymers Ongrovil S 470, a commercial PVC (further on: S 470 PVC), and prepolymerized laboratory PVC (further on: E PVC), both suspension-polymerized Hungarian products.

Our measurements were carried out in a 1,2,4-trichlorobenzene solution of 1 per cent PVC, and on powder or film samples, in the temperature ranges

160–210°C and 160–230°C, respectively. The samples were degraded in glass vessels flushed with purified argon and placed in a silicone oil thermostat. The hydrogen chloride evolved in the reaction was transferred with argon into a conductivity cell; the conductivity was measured and registered. From the calibration, we may obtain the amount of hydrogen chloride evolved, and from this (knowing the weight of the samples) the conversion of hydrogen chloride elimination can be calculated.

The PVC films investigated were prepared from a solution of PVC in previously deperoxidized tetrahydrofuran (THF). In order to remove the solvent, the films were kept *in vacuo* at 80–100°C for several hours. For the purpose of degradation, a 6–7 cm^2 of the film was glued on a U-form glass frame with a few drops of tetrahydrofuran.

The degradation flask holding the sample was first flushed with an argon stream for about half an hour and then placed in a thermostat preheated to the temperature of degradation. U.V. and visible spectrophotomeric investigations were performed on the films by interrupting the degradation. The concentration of polyenes was calculated by using the extinction coefficients determined by Popov and Smirnov for polyenes formed in the degradation of poly(vinyl alcohol) films[4]. It should be noted here that the assignment of the polyene lengths to the individual bands as well as the extinction coefficient values are less reliable for diene and triene and for polyenes containing more than 10–12 conjugated double bonds, than for those containing 5–10 double bonds.

The degraded powder samples were dissolved in peroxide-free tetrahydrofuran, for spectrophotometry. The extinction coefficients of polyenes for THF solution were determined by dissolving in THF the degraded PVC film after spectrophotometry, taking into account the extinction coefficients of the film.

Spectrophotometry of the degraded PVC solution was carried out by taking samples without interruption of the degradation process. In order to prevent polymer precipitation the solution was diluted with THF (1:1) prior to photometry. The extinction coefficients were equivalent to those of pure THF solution. Owing to the self-absorption of TCB, the spectra could be evaluated only for polyenes containing more than four conjugated double bonds.

Our studies on the formation of benzene were carried out with powder samples. The benzene formed in the course of degradation was passed with an argon stream into 0.1–0.2 ml isooctyl acetate in a trap cooled with dry ice–acetone, and the product was determined by gas chromatography.

In our earlier publications dealing mainly with degradation of PVC dissolved in diisooctyl adipate[6] we showed that, generally, in the initial phase of degradation (up to some tenth per cent conversion) the length distribution of polyenes is approximately a geometrical distribution. Similar results were found in experiments carried out on PVC film and powder samples.

Studying the reaction at somewhat higher—though still low—conversions (of some few per cent), we ascertained that the concentration of polyenes is, in fact, proportional to the extent of dehydrochlorination only in the initial phase of the reaction.

This fact is well demonstrated in *Figures 1* and *2* where the conversion dependence of the concentrations of polyenes containing 5, 7 and 9 conjugated double bonds (p_5, p_7 and p_9 in the percentage of monomeric units) and that of the concentration-ratio p_5/p_{10} is shown (S 470 PVC film sample, 210°C).

In the course of degradation, the increase in concentration gradually slows down, first for longer then for shorter polyenes, and the rates become stationary at a certain degree of conversion. Since these changes are faster in the case of long polyenes, it may also be established that polyene distribution changes with the time, and the average length of polyenes decreases with increasing conversion—as illustrated by the increase in the value of p_5/p_{10}.

The amount and length distribution of polyenes change with the conversion as outlined above, irrespective of the sample used for degradation (film, powder or solution), a fact very important in the interpretation of the phenomena.

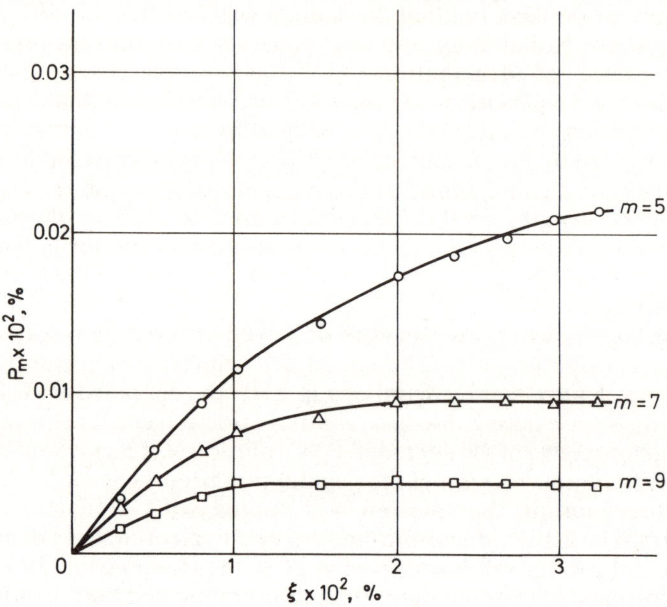

Figure 1. Relative amount of polyenes containing 5, 7 or 9 double bonds, plotted against dehydrochlorination conversion (S 470 PVC film, 18 µm, 210°C).

Some authors have earlier made qualitative observations[7] according to which the amount of polyenes formed in the course of degradation is not proportional to the conversion of dehydrochlorination. However, no quantitative conclusions have been drawn, that is, the phenomena have not been interpreted.

Obviously, the decrease in the amount of polyenes (related to the conversion) is not caused by a radical change of the mechanism of degradation. The formation rate of polyenes is proportional to the rate of dehydrochlorination even in the later phases of degradation. This points, therefore, to a chemical reaction consuming the polyenes formed. The fact that the consumption of polyenes in dilute solution takes place at a higher rate than in the solid phase (see later), suggests a first order reaction. On the other hand, under the conditions of degradation there is no such agent which might cause polyene con-

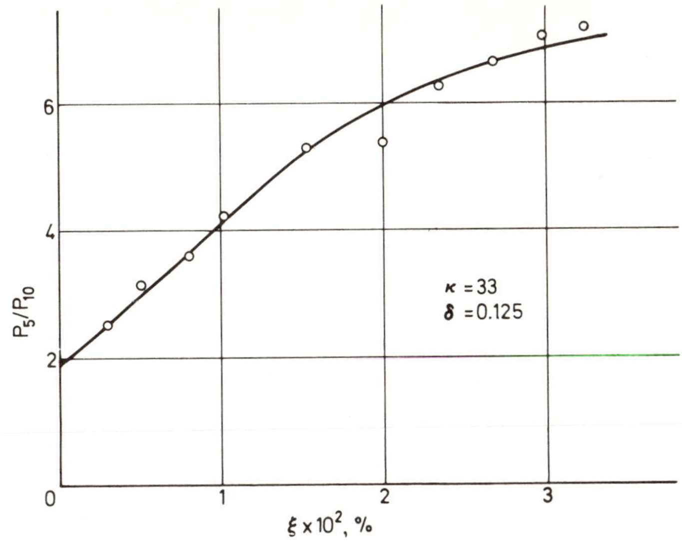

Figure 2. Concentration ratio (p_5/p_{10}) of polyenes containing 5 and 10 double bonds, plotted against the conversion of dehydrochlorination. The curve drawn designates the calculated function (S 470 PVC film, 18 μm, 210°C).

sumption in its reaction with polyenes: the degradation was carried out in a carefully purified intense argon stream, which assures oxygen-free conditions and removes the volatile products of degradation. Neither can the experimental data be explained by assuming chain scission at the unsaturated sites since this would involve considerable decrease in the molecular weight, which conflicts with the experimental results.

A satisfactory explanation of the observations is the consideration of an intramolecular cyclization reaction (backbiting-like isomerization) of polyenes, in the course of which the conjugated sequences become shorter or split into two shorter parts separated by a ring structure: the number of π bonds decreases by one to bring about ring closure with the formation of a new σ bond (*Figure 3*).

As is known, compared to rings of other structures, six-membered rings are preferred for steric and energetic reasons. Also in the intramolecular cyclization of polyenes, i.e. the formation of mainly six-membered rings, cyclohexadiene groups is the most probable reaction path. Light absorption of the conjugated diene within the ring structure is not observable in the spectrum studied, so formation of a ring is manifested in the disappearance of three double bonds in the photometric studies.

Such cyclization of polyenes is supported by the fact that—according to other authors[8] and our own results—at 180–200°C benzene is also formed, although compared to hydrogen chloride in relatively small amount (*Figure 4*).

The formation of benzene proceeds at first at a low rate and then continuously accelerates in the initial phase of degradation. This suggests that the formation of benzene from polyenes takes place through some intermediate (probably through a cyclohexadiene structure). A closer kinetic study of the formation of benzene is now in progress.

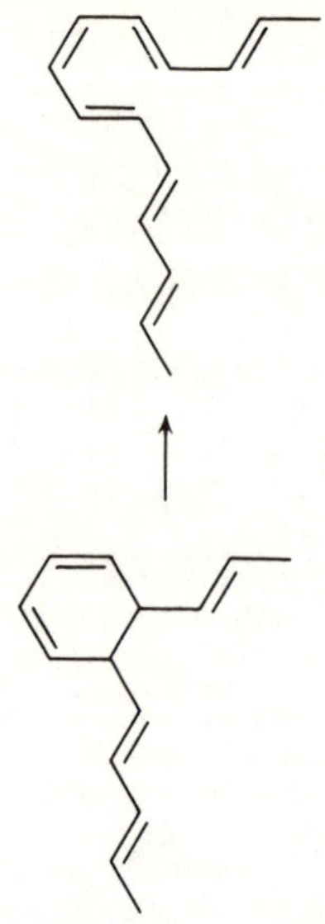

Figure 3. Scheme of intramolecular cyclization of polyenes.

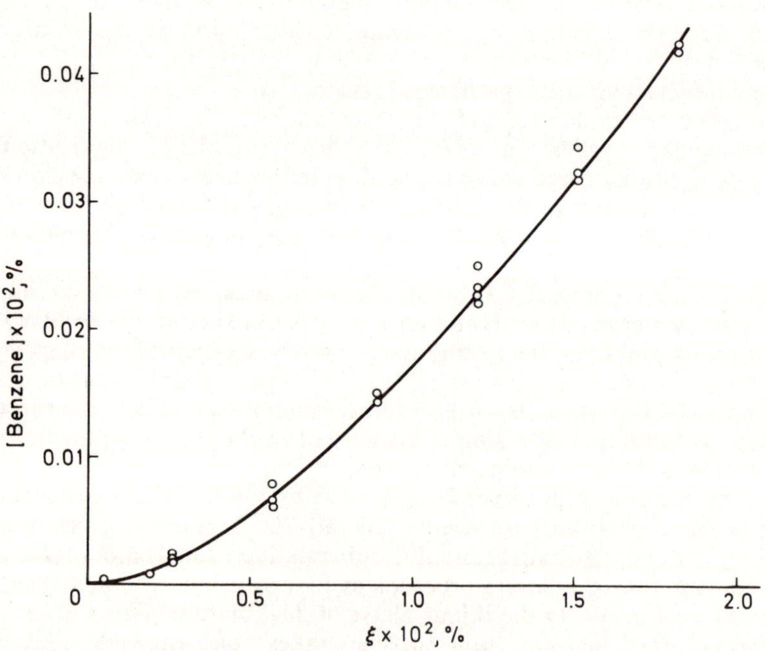

Figure 4. Relative amount of benzene, plotted against dehydrochlorination conversion (S 470 PVC, powder sample, 180°C).

The cyclization isomerization of unsaturated hydrocarbons of low molecular weight (conjugated trienes and tetraenes) has been described by several authors[9]. The presence of cyclohexadiene in degraded PVC has so far not been directly shown, owing to experimental difficulties. There is, however, some reference in the literature to its possible formation[10].

Kinetic treatment of the cyclization model

The kinetics of the formation of polyenes in the course of PVC degradation, as well as their length distribution, has been discussed in our earlier publications[6] and so the subject will only be briefly outlined here.

The formation of polyenes takes place in a special zip reaction whose initiation step is the primary elimination proceeding randomly or activated by defects. With the consumption of defects the initiation rate decreases and approaches the rate of pure random initiation.

Chain propagation proceeds in a series of activated elimination steps at a rate higher by orders of magnitude than the initiation rate. There are different assumptions for the interpretation of chain termination, which we shall revert to later on. Chain termination will not be considered here as a concrete reaction step but as an approximation described by a probability parameter: hydrochloric acid elimination from a monomeric unit results in a product (double bond) activating further decomposition with $(1-\delta)$ probability, or in an 'inactive' product with δ probability.

Since chain propagation is very fast compared to initiation, the concentration of chain propagating intermediates becomes stationary (and rather low) from the very beginning of the degradation and the length distribution of polyenes formed in the zip reaction is determined by the δ probability parameter. Based on the above outlined mechanism the primary formation rate of polyenes each containing m double bonds (v_m^+) is:

$$v_m^+ = \delta^2(1-\delta)^{m-1}\frac{d\xi}{dt} \tag{1}$$

where ξ denotes the conversion of hydrochloric acid elimination.

Accordingly, the amount of polyenes of m length at the onset of the reaction is proportional to the conversion (as long as the effect of secondary consumption and formation reactions is negligible) and the length distribution of polyenes is a geometrical distribution:

$$p_m = \delta^2(1-\delta)^{m-1}\xi \tag{2}$$

where p_m stands for the concentration (related to monomeric units) of m length polyenes.

As an approximation, we assume that cyclization occurs with identical probability on an arbitrary double bond triad of polyenes containing at least three conjugated double bonds. In our model only six-membered cyclohexadiene rings have been considered.

Intramolecular cyclization proceeds in several consecutive steps (isomerization processes and ring closure), the rate determining process of which is probably the *cis–trans* isomerization of the double bond. Designating the cyclohexadiene group with C, we may assume the overall reaction to be:

$$P_m \xrightarrow{k_c} P_i + P_{m-i-3} + C \qquad m \geqslant 3; i \geqslant 0 \tag{3}$$

where k_c is the unimolecular rate constant of the rate determining step of cyclization, and i and $m - i - 3$ stand for the lengths of the new sequences formed.

These new polyenes may then take part in further cyclization provided $i \geqslant 3$, or $m - i - 3 \geqslant 3$. These reactions are also considered to have the rate constant k_c; in other words, the newly formed polyene sequences are not considered to be different from the first-formed polyenes of identical length. (This assumption is essential for the mathematical formulation of the model and is not in contradiction with the experimental results.)

Since the formation of one ring consumes an arbitrary double bond triad of the polyene, m length polyenes may cyclize in $m - 2$ ways. The rate of decrease w_m^- of such polyenes, owing to the cyclization reaction, is as follows:

$$w_m^- = -(m-2)k_c p_m \tag{4}$$

In the course of cyclization reactions, however, the secondary formation of m length polyenes is also possible, namely from all P_j polyenes containing at least three more double bonds than m ($j \geqslant m + 3$). This may take place (if $j \neq 2m + 3$) at either one or the other end of a P_j polyene. If, however, $j = 2m + 3$, the only possibility is the simultaneous formation of two polyenes of m length.

According to the above, in the course of cyclization m long polyenes are formed from j long polyenes at $w_{m,j}^+$ rate:

$$w_{m,j}^+ = 2k_c p_j \qquad j \geqslant m + 3 \tag{5}$$

The overall secondary formation rate (w_m^+) of P_m may be obtained by summing the above equation from $j = m + 3$ to $j = \infty$:

$$w_m^+ = \sum_{j=m+3}^{\infty} w_{m,j}^+ = 2k_c \sum_{j=m+3}^{\infty} p_j = 2k_c p(m+3:\infty) \tag{6}$$

with introduction of the following sum

$$\sum_{j=k}^{n} p_j \equiv p(k;n) \tag{7}$$

for the total concentrations of polyenes longer than $k - 1$ but shorter than $n + 1$.

The complete time dependence of p_m polyene concentration may therefore be written as

$$\frac{dp_m}{dt} = v_m^+ + w_m^+ + w_m^- \tag{8}$$

Substituting equations 1, 4 and 6, we have:

$$\frac{dp_m}{dt} = \delta^2(1-\delta)^{m-1}\frac{d\xi}{dt} + 2k_c p(m+3;\infty) - k_c(m-2)p_m \tag{9}$$

or in another form:

$$\frac{dp_m}{dt} = \delta^2(1-\delta)^{m-1}\frac{d\xi}{dt} + 2k_c p(m;\infty) - k_c(mp_m + 2p_{m+1} + 2p_{m+2}) \qquad (10)$$

For the amount $p(m;\infty)$ in equation 10, another differential equation may be written, since

$$\frac{dp(m;\infty)}{dt} = \frac{d}{dt}\sum_{j=m}^{\infty} p_j = \sum_{j=m}^{\infty}\frac{dp_j}{dt} \qquad (11)$$

After summing, we obtain

$$\frac{dp(m;\infty)}{dt} = \delta(1-\delta)^{m-1}\frac{d\xi}{dt} + k_c[4p_m + 2p_{m+1} + \xi(m;\infty) - 2(m+1)\,p(m:\infty)] \qquad (12)$$

where we introduced an amount similar to $p(k;n)$

$$\xi_p(k;n) \equiv \sum_{j=k}^{\infty} jp_j \qquad (13)$$

for the total amount of double bonds in polyenes longer than $k-1$ but shorter than $n+1$. According to this definition we may derive the differential equation of $\xi_p(m;\infty)$

$$\frac{d\xi_p(m;\infty)}{dt} = \frac{d}{dt}\sum_{j=m}^{\infty} jp_j = \sum_{j=m}^{\infty}\frac{jdp_j}{dt} \qquad (14)$$

i.e., after summing:

$$\frac{d\xi_p(m;\infty)}{dt} = [1+(m-1)\delta](1-\delta)^{m-1}\frac{d\xi}{dt} + k_c\{-3\xi_p(m;\infty) + [6-(m-1)m]\,p(m;\infty) + 2(2m-3)\,p_m + 2(m-1)\,p_{m+1}\} \qquad (15)$$

The system of differential equations 10, 12 and 15 is, in this form, not yet suitable for comparison with the experimental data since the equations contain also the polyene concentrations p_{m+1} and p_{m+2}. If new differential equations were set up for these amounts, they would require the concentration of further longer polyenes (p_{m+3}; p_{m+4}, etc), which would lead to a system of an infinite number of differential equations. In order to avoid this, we substituted p_{m+1} and p_{m+2} with approximate expressions. This can be done, all the more so since the terms containing these values in the equation are substantially smaller than the others. The approximations were chosen so as to obtain correct values if the length distribution of polyenes is geometrical, a condition well satisfied at the initial stage. The corresponding approximations are:

$$p_{m+1} = p_m\left[1 - \frac{p_m}{p(m;\infty)}\right] \qquad (16)$$

$$p_{m+2} = p_m \left[1 - \frac{p_m}{p(m;\infty)}\right]^2 \tag{17}$$

The knowledge of $d\xi/dt$ is also necessary for the calculations. This has been determined parallel to our photometric measurements or in separate experiments by measuring the extent of dehydrochlorination. The conversion curve approached a straight line, in other words, the rate of dehydrochlorination in each experiment (in the low conversion range studied, $0 \leqslant \xi < 0.05$) was considered constant, and characterized by the overall rate (k_{HCl}) of random elimination. With this approximation, i.e. by neglecting the activated initiation, at the initial stage of degradation (where the effect of the secondary cyclization process is, in any case, not significant) cyclization has received undue consideration.

For better comparison, the measurements were evaluated in the function of conversion. Introducing the relative rate constant

$$\kappa \equiv k_c \bigg/ \frac{d\xi}{dt} \approx k_c/k_{HCl} \tag{18}$$

and applying equations 16 and 17, the system of differential equations 10, 12 and 15 can be numerically integrated and applied for the evaluation of experimental results.

Comparison of experiments and calculations

Owing to the above-mentioned technical difficulties, we were able to obtain the most reliable data for the concentration of polyenes containing 5–10 conjugated double bonds. Therefore, for comparison of the calculated and experimental results we used the following two quantities:

(a) the concentration ratio of polyenes containing 5 and 10 double bonds, p_5/p_{10}, which is indirectly characteristic of polyene distribution,

(b) the total amount of double bonds in polyenes longer than four but shorter than 11, $\xi_p(5; 10)$, which is indirectly characteristic of the total amount of double bonds present in the system.

The probability parameter δ can be determined by the extrapolation of the ratio p_5/p_{10} to zero conversion. According to equation 2, this ratio is:

$$\frac{p_5}{p_{10}} = \frac{1}{(1-\delta)^5} \tag{19}$$

The other parameter, the relative rate constant κ, can be determined by a systematic method of trial-and-error, using the p_5/p_{10}–ξ diagram. Thus, both parameters can be determined without applying the experimental results on $\xi_p(5, 10)$–ξ.

As can be seen in *Figures 2* and *5*, the conversion dependence of the ratio p_5/p_{10}, characteristic of the average length of polyenes, is fairly well described by the model for film samples of both polymers under investigation. Similarly, there is good agreement between the calculated and measured p_5/p_{10} values for PVC samples degraded in powder form (*Figure 6*); here and in *Figures 2* and *5* the calculated curve has been plotted.

With the knowledge of κ and δ, obtained from the diagrams p_5/p_{10}–ξ, and on the basis of the model, we calculated the functions $\xi_p(5, 10)$–ξ (curves A in *Figures 7, 8* and *9*).

The concentrations of the individual polyenes have also been calculated, as a function of dehydrochlorination conversion. As an example, we present the calculated p_m values for polyenes of $m = 5$ and $m = 10$, plotted in *Figure 10* with curves A.

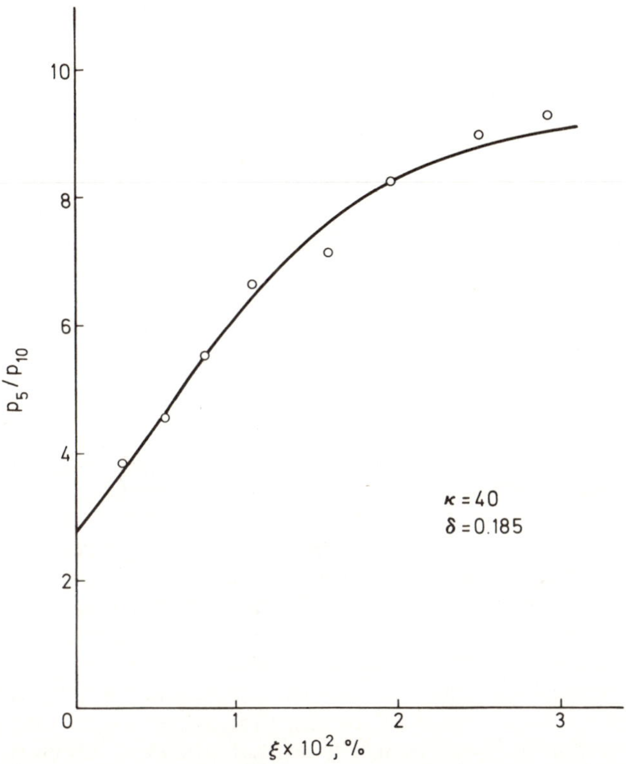

Figure 5. Concentration ratio (p_5/p_{10}) of polyenes containing 5 and 10 double bonds, plotted against the conversion of dehydrochlorination. The curve drawn designates the calculated function (E PVC film, 10 μm, 210°C).

The calculated amount of polyenes, as well as the calculated amount of $\xi_p(5; 10)$ shows a trend qualitatively similar with that of the measured amounts. The agreement is, however, not satisfactory: the calculated value in the total range under investigation is higher than the measured. But this deviation is not in contradiction with the above described cyclization model: on the contrary, it supports it. Thus, in the above calculations the chain termination has not been given as a concrete reaction step and was only considered with the probability parameter δ. A more detailed analysis shows that among the assumed chain termination steps, some may result in the formation of fewer conjugated double bonds than the number of hydrogen chloride molecules eliminated.

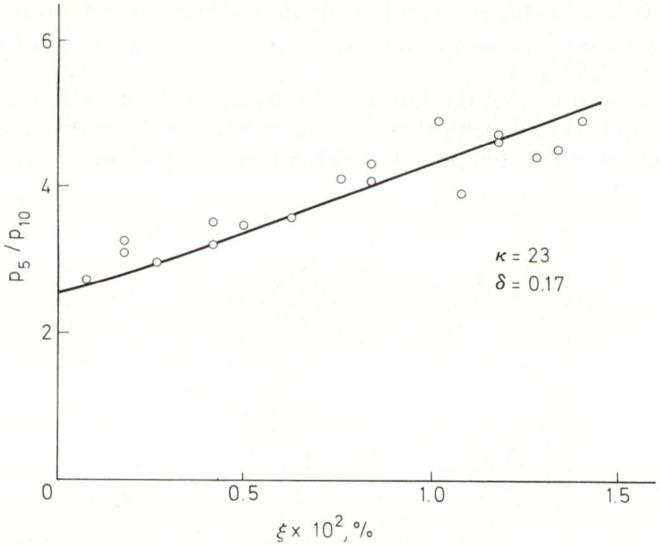

Figure 6. Concentration ratio (p_5/p_{10}) of polyenes containing 5 and 10 double bonds, plotted against the conversion of dehydrochlorination. The curve drawn designates the calculated function (S 470 PVC, powder sample, 210°C).

Such termination occurs also in the α,γ-elimination (cyclopropane formation) earlier proposed by us[6] and shown in *Figure 11*.

Chain termination by formation of a cyclohexadiene ring has a similar effect (*Figure 12*). This may occur if there are three adjacent C—C bonds in *cisoid–cis–cisoid* conformation at the end of the chain still propagating in the course of elimination. Then owing to the favourable steric position, instead of further chain propagation, cyclization may occur immediately after the last elimination step, implying chain termination.

There is a marked difference between this cyclization process and the secondary cyclization discussed earlier. According to our measurements, the rate of secondary cyclization is several orders of magnitude smaller than that of chain propagation and cannot compete with it, owing to slow *cis–trans* isomerization. Chain termination by cyclization will take place only if dehydrochlorination results in a chain conformation favourable for cyclization proceeding at a higher rate than chain propagation.

If the chain termination characterized by δ involved exclusively such cyclizations, the primary formation rate of polyenes would be considerably lower than given in equation 1:

$$v_m^+ = \frac{\delta^2(1-\delta)^m}{1+2\delta}\frac{\mathrm{d}\xi}{\mathrm{d}t} \tag{20}$$

Considering this primary formation rate of polyenes, the length distribution of polyenes (and the ratio p_5/p_{10}) agrees with that resulting from the model, where no concrete chain termination was formulated and no reduction of the polyene length was taken into account. There are considerable deviations in

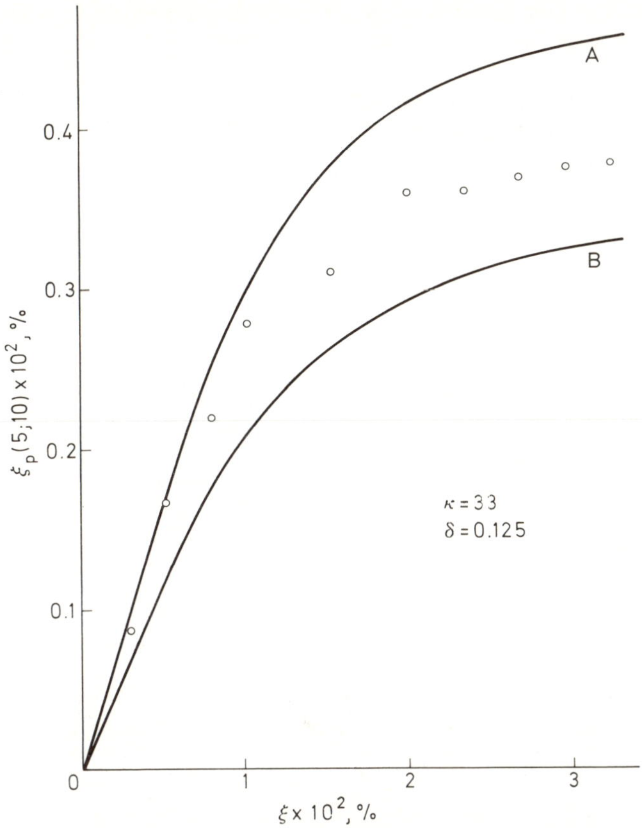

Figure 7. Total amount of double bonds in polyenes of 5–10 length in the function of dehydrochlorination conversion. Curves A and B are the calculated functions (S 470 PVC film, 18 μm, 210°C).

the amount of polyenes and in the sum $\xi_p(5;10)$. The B curves of *Figures 7–10* were plotted using equation 20, in other words, as if chain termination would proceed exclusively with cyclization, in the way demonstrated in *Figure 12.* This calculation resulted in polyene concentrations lower than the experimental values. The polyene concentrations measured are between the values calculated by the two different chain termination models. It may be assumed, therefore, that the average consumption of termination is less than three double bonds.

Detailed examination of the thermal degradation of PVC in trichlorobenzene solution is still under way. From experiments carried out so far, we may state that polyene distribution is shifted towards shorter polyenes with increasing dehydrohalogenation conversion. The conversion dependence of the characteristic p_5/p_{10} ratio is depicted in *Figure 13*.

The change in the p_5/p_{10} ratio is greater than in the case of samples degraded in the solid phase. The length distribution of polyenes at the beginning of the reaction cannot be approached with a geometric distribution as shown by the fact that the p_5/p_{10} ratio extrapolated to 0 is less than 1; therefore the experiments cannot be evaluated by the above differential equations.

It may be seen from the conversion dependence of $\xi_p(5:10)$ shown in *Figure 14* that in dilute solution it becomes stationary at a lower conversion than in the solid phase. The value of κ, the relative rate constant, may be estimated on the basis of the diagram $\xi_p(5:10)$–ξ; in dilute TCB solution,

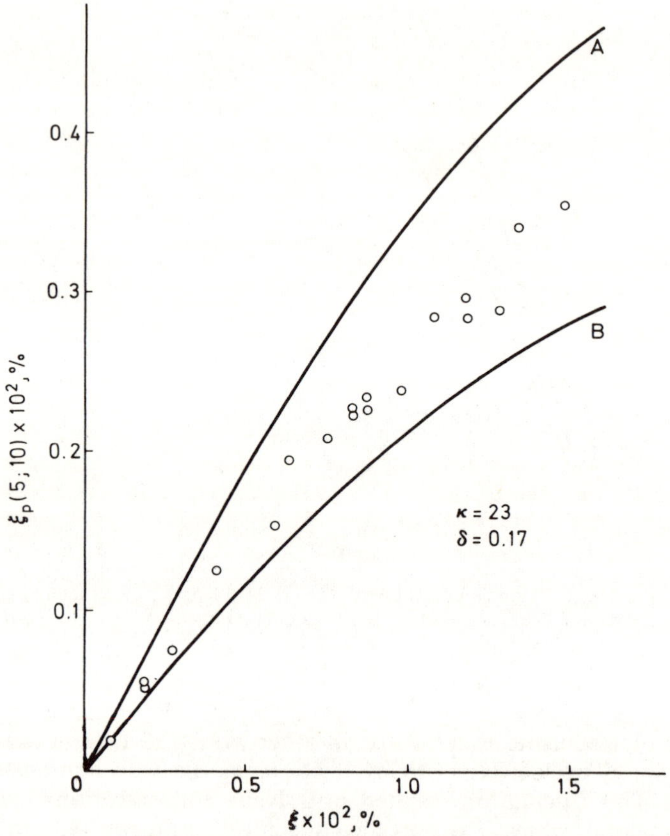

Figure 8. Total amount of double bonds in polyenes of 5–10 length in the function of dehydrochlorination conversion. Curves A and B are the calculated functions (S 470 PVC, powder sample, 210°C).

under the conditions given in *Figure 13*, κ = 55–60. This value is much higher than those obtained for the same polymer (S 470 PVC) at the same temperature (210°C), for samples of powder (κ = 23), or for film (κ = 33). On the other hand, the degradation of the film samples of the other polymer (E PVC) gives a relative rate constant (κ = 40) fairly similar to that obtained with the film S 470.

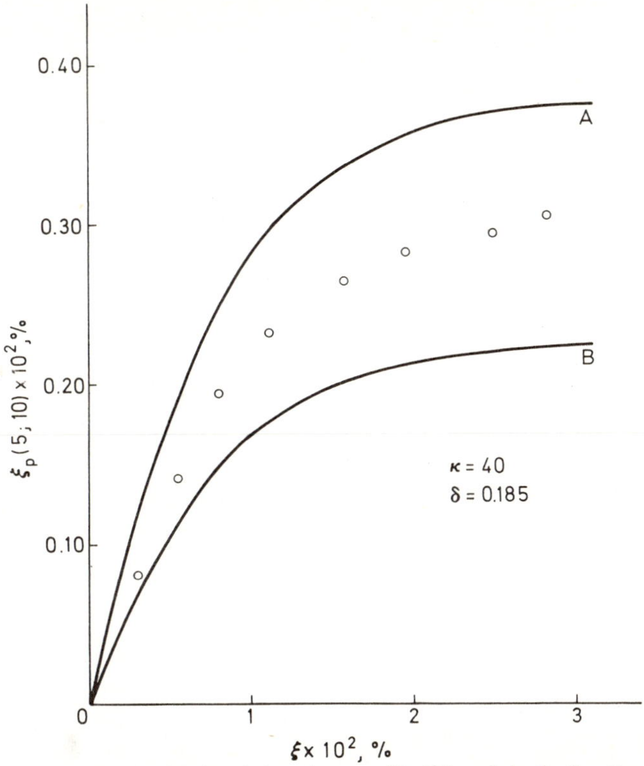

Figure 9. Total amount of double bonds in polyenes of 5–10 length in the function of dehydrochlorination conversion. Curves A and B are the calculated functions (E PVC film, 10 μm, 210°C).

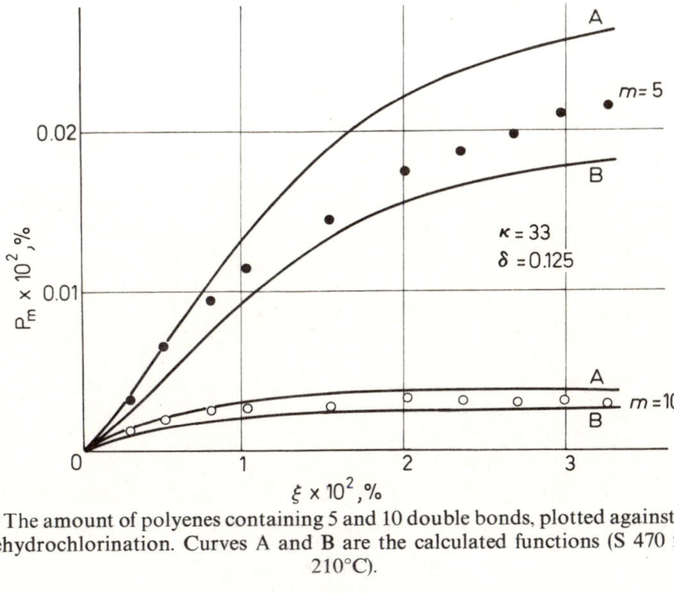

Figure 10. The amount of polyenes containing 5 and 10 double bonds, plotted against the conversion of dehydrochlorination. Curves A and B are the calculated functions (S 470 film, 18 μm, 210°C).

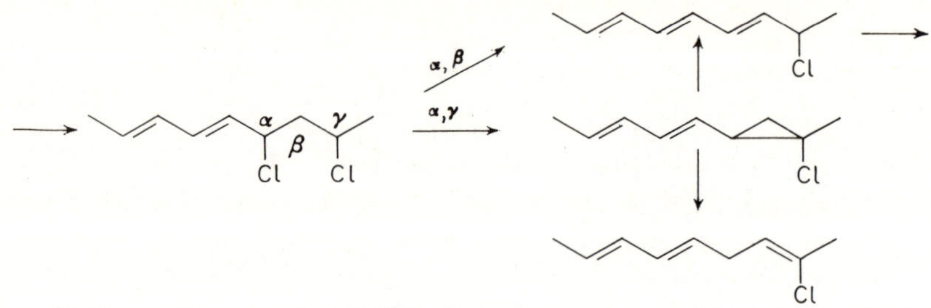

Figure 11. Propagation of activated HCl elimination, and a possible chain termination via cyclopropane ring structure.

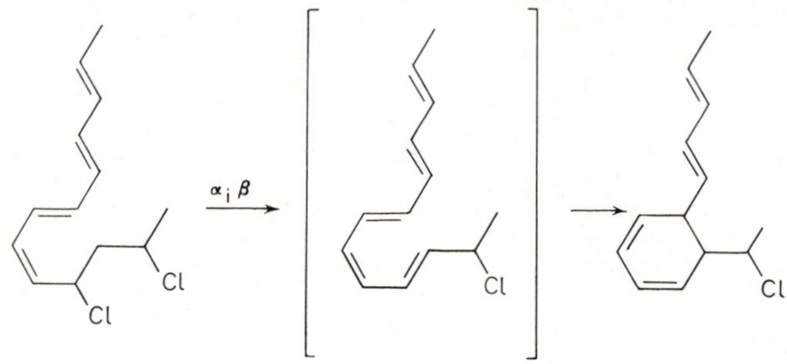

Figure 12. Chain termination due to the cyclization reaction of the propagating chain.

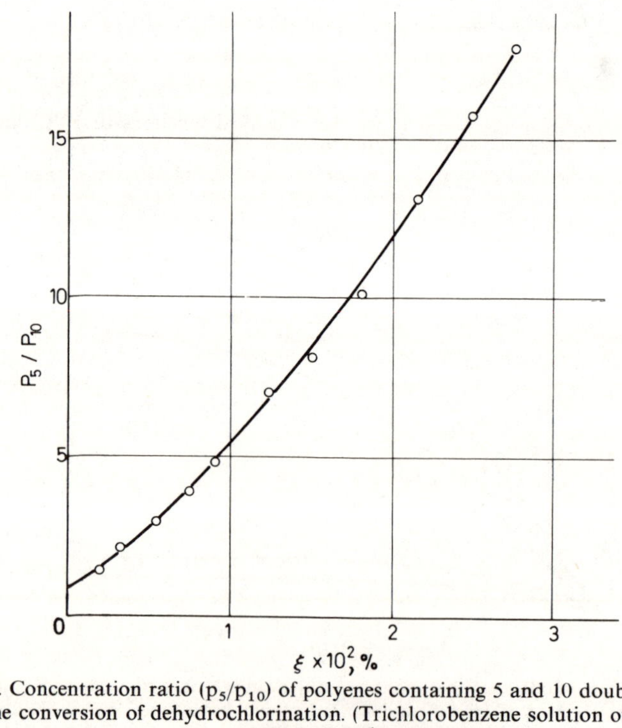

Figure 13. Concentration ratio (p_5/p_{10}) of polyenes containing 5 and 10 double bonds, plotted against the conversion of dehydrochlorination. (Trichlorobenzene solution of 1 % S 470 PVC, 210°C, 3 l.h^{-1} argon).

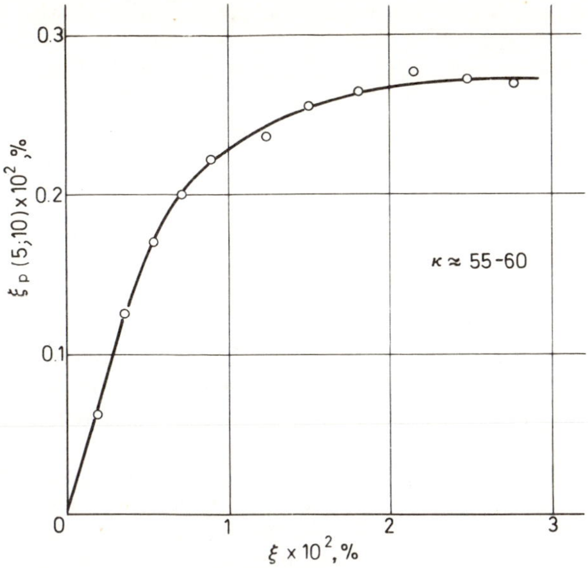

Figure 14. Total amount of double bonds in polyenes of 5–10 length, plotted against the conversion of dehydrochlorination. (Trichlorobenzene solution of 1% S 470 PVC, 210°C, 3 $l.h^{-1}$ argon).

The difference between the solid and dissolved samples is remarkable and can be interpreted by the isomerization–cyclization model outlined above, chain mobility in dilute solution being obviously much greater than in the solid phase.

In addition to polymer analogous secondary processes, in the course of thermal degradation, other types of reactions also proceed involving a change in the molecular weight of the polymer. During the thermal degradation of PVC powder samples, osmometric measurements show a slow increase in the number average molecular weight (*Figure 15*). Such an increase has also been observed by other authors[11].

The rise in molecular weight is evidently caused by bimolecular reactions of polyenes. The Diels–Alder type addition between two polyenes or the reaction of polyenes with polyene radicals always present[6] during thermal degradation, may serve as examples for such reactions. No rise in molecular weight could be observed in samples degraded in dilute solution.

2. THE DIELS–ALDER REACTIONS OF POLYENES

As previously outlined, polyenes primarily formed in PVC degradation may undergo considerable transformation, owing to intramolecular cyclization reactions (backbiting). These processes apparently have a favourable effect since the amount of double bonds as well as the visually observable discolouration of the polymer is smaller than that corresponding to the extent of dehydrochlorination. It may, however, give rise to processes leading to further degradation. It should be noted here that the scission of polymer chain will most probably occur at cyclohexadiene structures (formation of benzene). Cyclohexadiene structures may also play an important role in the course of oxidative degradation; as is well known, upon excitation, this structure easily reacts with oxygen with the formation of ascaridole type structures[12]:

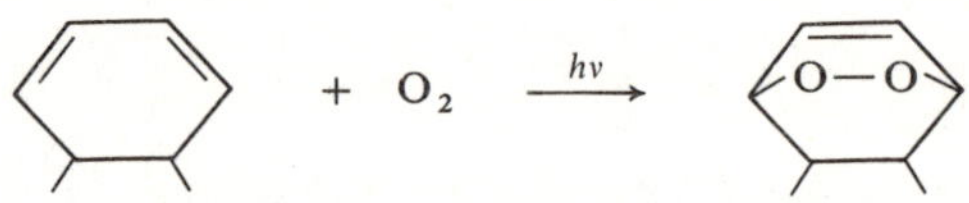

(21)

Thus, owing to thermal degradation, PVC contains reactive polyenes and unsaturated ring structures formed in primary and secondary processes. Therefore all information on their reactivity may provide helpful data for a closer knowledge of the system. Such data may be obtained, e.g., by studying the addition reactions of polyenes.

It is a well established fact that polyenes readily react with different ionic and radical reagents. These processes cannot, however, be considered quantitative as the intermediates formed are relatively stable with long lifetimes. Furthermore, radical addition may give rise to secondary reactions of polyene radicals and ionic processes may lead to equilibrium at an early stage. Our studies on hydrogen chloride incorporation[6] and the fact that polyenes are capable of halogen addition to only a small proportion of the total amount of double bonds[13], support the above statements. These observations justified the study of the Diels–Alder reactions of polyenes formed in PVC. The fact that certain important additives of PVC, containing dienophilic sections, have a stabilizing effect[14] also justified a closer study of these processes. Since diene reactions decrease the extent of discolouration on the one hand, and both hinder the formation of and transform cyclohexadiene structures on the other, the Diels–Alder reactions of polyenes may have considerable practical importance.

In our investigations a solution of S 470 PVC (~ 1 wt %) was previously degraded over a period of 300 minutes in an argon stream at 180°C. Purified 1,2,4-trichlorobenzene was the solvent used and the volume of the solution was 40 ml. After degradation, the reaction vessel was placed in a thermostat preheated to the temperature of the Diels–Alder reaction (100, 120, 140, 160°C), under a steady stream of argon gas.

Owing to its great reactivity and easy measurement, we have chosen chloromaleic anhydride (CMA) as the reagent. The commercial product (Fluka) was distilled.

The samples taken from the reaction mixture were diluted with peroxide free tetrahydrofuran of identical volume and then measured by spectrophotometry.

Above 28000 cm^{-1} strong absorption can be observed in the reaction mixture, which may probably be assigned to the charge-transfer complex between the solvent molecules and the reagent. This excluded the possibility of the determination of shorter polyenes. Owing to the considerable error in the spectrophotometry of long polyenes we measured the time dependence of the amount of $m = 6–11$ polyenes during the reaction.

Experimental results; kinetic treatment of the process

Immediately after addition of the reagent, the absorption of polyenes starts to decrease; the extinctions generally remain measurable over a reaction period of 2–3 hours (*Figure 16*).

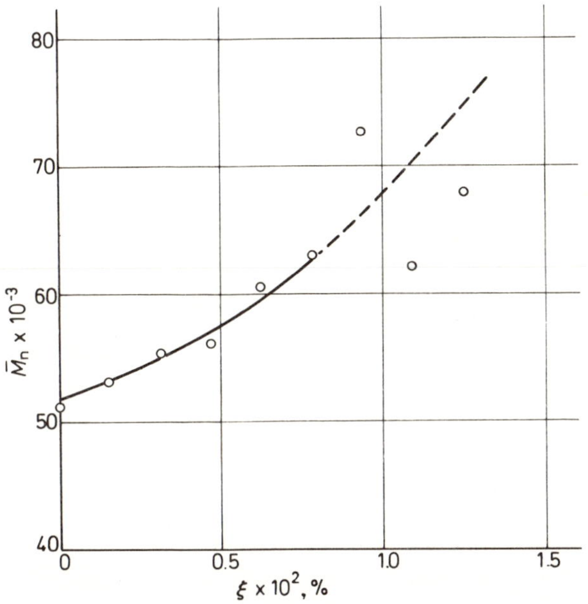

Figure 15. Change in the number average molecular weight of PVC, plotted against the conversion of dehydrochlorination. (S 470 PVC, powder sample, 180°C; above 1% conversion, owing to the decrease in solubility of the samples, the error of determination is greater than at lower conversions).

As is known, the Diels–Alder reaction is for both components a first order process[15]. In our experiments we used CMA in great excess compared to the amount of polyenes, so the consumption reaction could be considered as pseudo-unimolecular. Applying different CMA concentrations, we established that the initial rate is proportional to the concentration of the reagent (*Figure 17*).

Plotting the logarithm of extinctions against reaction time, we obtained characteristic curves (*Figure 18*).

It may be ascertained that the decrease in the consumption rate of polyenes is faster than it would be in the case of a simple first order reaction; $\ln (E_0/E)$ is a linear function of time only at the beginning of the reaction. It is notable, however, that the change of relative extinctions for different polyenes proceeds practically in the same way; this means that the reaction rate is independent of polyene length. Therefore, in the following, we considered the average values of results obtained for $m = 6$–11 polyenes.

The conclusion may be drawn, that the reaction can be regarded as a simple addition only at the initial stage and the complexity of the process becomes evident in the course of reaction. Similar curves may also be

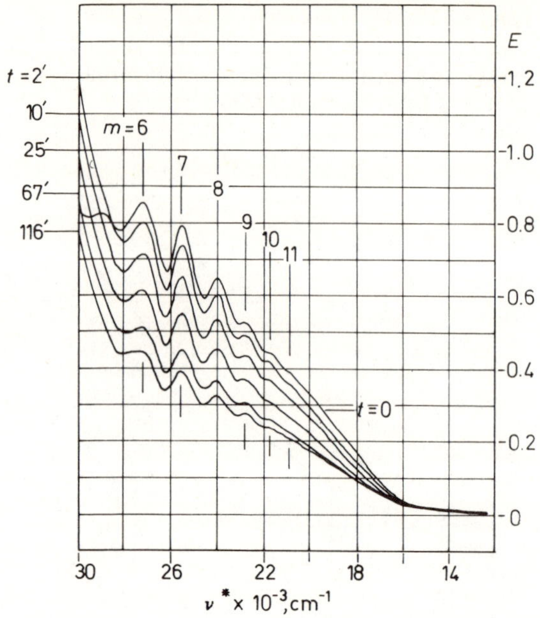

Figure 16. Decrease in the extinction of polyenes, owing to the effect of the Diels–Alder reaction. The spectra are taken at $t = 0$, prior to CMA addition; and at $t = 2$, 10, 25, 67 and 116 min. the corresponding reaction times. 140°C, [CMA] = 0.5% = 5.81×10^{-2} mol l.$^{-1}$ The wavenumbers belonging to the $m = 6$–11 polyenes are designated.

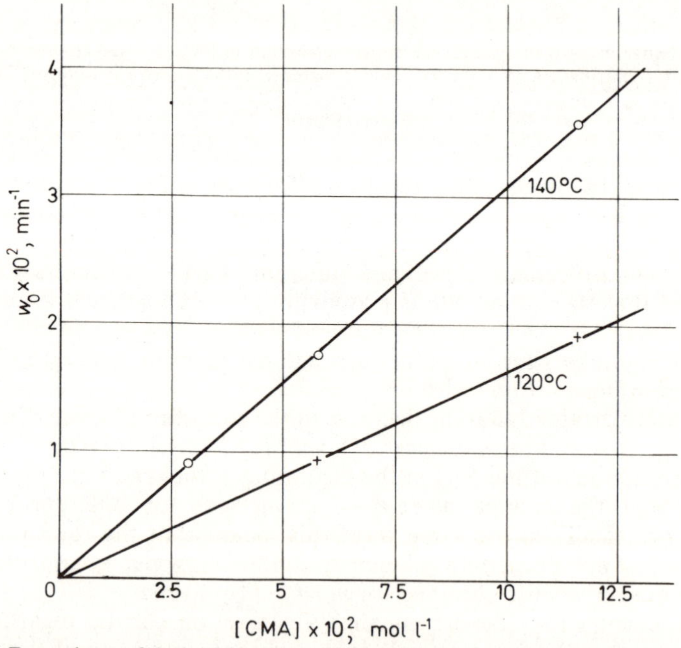

Figure 17. Dependence of the initial rate (w_0) of the Diels–Alder reaction on the concentration of CMA.

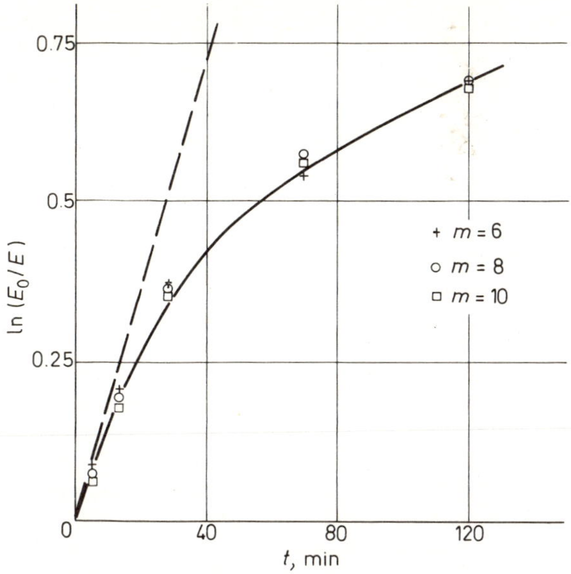

Figure 18. Change in conversion (x) of the Diels–Alder reaction in the function of reaction time. Logarithmic presentation, $\ln(E_0/E) \equiv \ln(1 - x)$. 140°C; [CMA] = 5.81×10^{-2} mol l^{-1}. The continuous curve shows the calculated function and the dotted line represents the first order irreversible reaction.

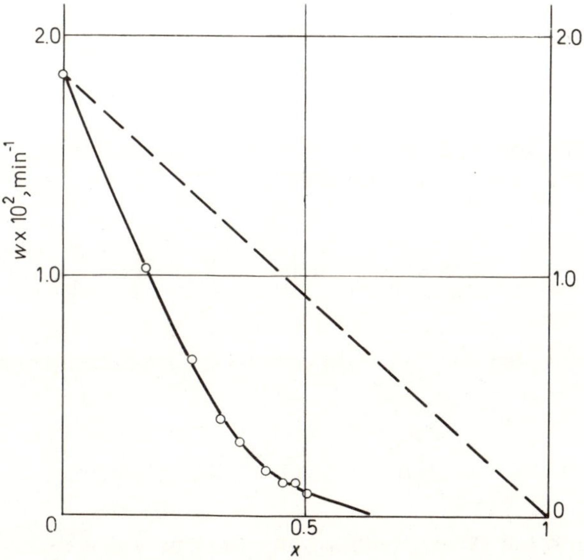

Figure 19. Dependence of the Diels–Alder reaction rate ($w = dx/dt$) from conversion (x). 140°C [CMA] = 5.81×10^{-2} mol $l.^{-1}$ The continuous curve shows the calculated function and the dotted line represents the first order irreversible reaction.

obtained for reversible first order reactions and, indeed, the reversibility of Diels–Alder reactions is well known[15]. In order to clarify the problem we have plotted the rates against conversion. The relationship obtained for experiments carried out at 140°C (0.5 per cent CMA) is given in *Figure 19.*

It may be established that the process is neither an irreversible pseudo-unimolecular addition (dotted line) nor a simple reversible reaction. In the latter case the rate would linearly decrease from the initial value w_0 up to the final conversion x_∞.

The bend found in the rate–conversion relationship as well as the fact that the time dependence of the conversion for polyenes of different length is identical, points to a complex reversible addition. We tried to describe the process with the following simplified scheme:

Let us assume that addition always takes place on the two final double bonds of the intact polyene sequence, and that the adduct always dissociates in the position adjacent to the intact polyene sequence. Thus, in these processes the polyene length always changes by two. Owing to steric and polar effects, it may be assumed that once the polyene has reacted with the rate constant $k'_a(k'_a = k_a[\text{CMA}])$, it will take part in the further addition only at a lower rate (rate constant: $k'_b = \alpha k'_a$, $\alpha < 1$). The rate of dissociation of the adduct (rate constant k_d) is, on the other hand, independent of the number of previous additions.

The spectrophotometric method applied does not make any distinction between an intact polyene and an intact sequence of a partly reacted polyene: the extinction measured depends only on the length of the intact polyene sequence. If the concentration of polyenes, which contained originally $m + 2i$ conjugated double bonds, and participated in i number of additions is designated with ${}_i\text{p}_{m+2i}$, the complete extinction caused by m length polyenes may be calculated as follows:

$$E_m = \varepsilon_m \cdot c_m = \varepsilon_m \sum_{i=0}^{\infty} {}_i\text{p}_{m+2i} \tag{22}$$

where ε_m is the corresponding extinction coefficient, and c_m stands for the total concentration of m long polyenes. The conversion of the process can be defined as:

$$x = 1 - \frac{c_m}{c_m(O)} = 1 - \frac{E_m}{E_m(O)} = 1 - \frac{E}{E_0} \tag{23}$$

since, as was pointed out before, the conversion is independent of the polyene length:

$$\frac{E_6}{E_6(O)} = \frac{E_7}{E_7(O)} = \cdots = \frac{E}{E_0} \tag{24}$$

According to the experimental results, the length distribution of polyenes in the previously degraded PVC used in the study of the Diels–Alder reaction may be well approximated by a geometrical distribution: therefore the following relationship is valid for any m and i value ($m \geqslant 6$, $0 \leqslant i < \infty$):

$$p_{m+2i}(O) = \beta^i p_m(O) \tag{25}$$

where $\beta = (1 - \delta)^2$ is the parameter of geometrical distribution.

With the application of the above assumptions and relationships, we discuss first the addition process considered irreversible. In the course of subsequent unimolecular additions, an originally k long polyene is transformed as follows:

$$P_k \xrightarrow{k'_a} {}_1P_k \xrightarrow{\alpha k'_a} {}_2P_k \xrightarrow{\alpha k'_a} \cdots \xrightarrow{\alpha k'_a} {}_iP_k \xrightarrow{\alpha k'_a} \cdots \tag{26}$$

The usual treatment of this consecutive set of reactions gives the following result for the concentration of ${}_iP_k$ polyenes formed after the ith addition step:

$${}_iP_k = P_k(O)\frac{(-\alpha)^{i-1}}{(1-\alpha)^i}\left\{\exp(-\alpha k'_a t)\sum_{j=0}^{i-1}\frac{[-(1-\alpha)k'_a t]^j}{j!} - \exp(-k'_a t)\right\} \tag{27}$$

if $i \geqslant 1$, and the consumption of the initial polyene can be described by the relationship

$$P_k = P_k(O)\exp(-k'_a t) \tag{28}$$

For summing according to equation 22, we need the concentration of $k = m + 2i$ long polyenes, which may be obtained from equations 25, 27 and 28:

$${}_iP_{m+2i} = P_m(O)\left(\frac{\beta}{1-\alpha}\right)^i(-\alpha)^{i-1}\left\{\exp(-\alpha k'_a t)\sum_{j=0}^{i-1}\frac{[-(1-\alpha)k'_a t]^j}{j!} - \exp(-k'_a t)\right\} \tag{29}$$

if $i \geqslant 1$, and the originally m-length polyene is consumed according to the expression

$$P_m = P_m(O)\exp(-k'_a t) \tag{30}$$

After substituting equations 29 and 30 into equation 22 and summing, we obtain the conversion defined in equation 23:

$$x = 1 + \frac{k'_1 - w_0}{k'_2 - k'_1}\exp(-k'_2 t) - \frac{k'_2 - w_0}{k'_2 - k'_1}\exp(-k'_1 t) \tag{31}$$

where

$$w_0 = (1 - \beta)k'_a \tag{32}$$

$$k'_1 = \alpha(1 - \beta)k'_a = (1 - \beta)k'_b \tag{33}$$

$$k'_2 = k'_a \tag{34}$$

Exact treatment of the reversible reaction

$$P_k \underset{k_a}{\overset{\alpha k'_a}{\rightleftharpoons}} {}_1P_k \underset{k_a}{\overset{\alpha k'_a}{\rightleftharpoons}} {}_2P_k \underset{k_a}{\overset{\alpha k'_a}{\rightleftharpoons}} \cdots \underset{k_a}{\overset{\alpha k'_a}{\rightleftharpoons}} {}_iP_k \underset{k_a}{\overset{\alpha k'_a}{\rightleftharpoons}} \cdots \tag{35}$$

is even more complicated. Instead, we have applied a method of approximation analogous with the procedure used for the exact treatment of the simple reversible system

$$\mathrm{A} \underset{\overleftarrow{k}}{\overset{k}{\rightleftharpoons}} \mathrm{B} \tag{36}$$

Kinetic treatment of the equilibrium reaction 36 may also be performed in the following way: instead of applying the usual differential equation

$$\frac{\mathrm{d}x}{\mathrm{d}t} = k(1 - x) - \overleftarrow{k}x \tag{37}$$

(where x is the conversion; $x = B/A_0$), we may describe the reaction rate as:

$$\frac{\mathrm{d}x}{\mathrm{d}t} = k'(x_\infty - x) \tag{38}$$

where the constants k' and x_∞ can be expressed with k and $\overleftarrow{k}$, on the basis of the equilibrium state.

The approximation performed analogously for the reversible consecutive set of reactions 35 and treatment of the results according to equations 22 and 23 gave for the conversion:

$$x = x_\infty + \frac{k_1 x_\infty - w_0}{k_2 - k_1} \exp(-k_2 t) - \frac{k_2 x_\infty - w_0}{k_2 - k_1} \exp(-k_1 t) \tag{39}$$

where

$$x_\infty = \frac{\alpha k_a'^2}{\alpha k_a'^2 + \alpha k_a' k_d + k_d^2} = \frac{k_a' k_b'}{k_a' k_b' + k_b' k_d + k_d^2} \tag{40}$$

$$w_0 = (1 - \beta) k_a' \tag{41}$$

$$k_1 = \alpha(1 - \beta) k_a' \frac{\alpha\beta k_a'^2}{\alpha\beta k_a' - k_d^2 x_\infty} = (1 - \beta) k_b' \frac{\beta k_a' k_b'}{\beta k_a' k_b' - k_d^2 x_\infty} \tag{42}$$

$$k_2 = k_a' \frac{(1 - \beta) k_a'}{(1 - \beta) k_a' - k_d x_\infty} \tag{43}$$

Comparison of the relationships 31–34 and 39–43 showed that in the case $k_d = 0$ (the adduct does not dissociate), the latter convert into the former. In other words, the relationships 39–41 also give a correct description of the irreversible reaction.

The relationship 39 obtained for the reversible process and knowledge of the parameter β, allow the evaluation of experiments. In the course of the previous degradation performed as outlined above, a distribution with parameter $\beta = 0.498$ can be observed. Since $\beta = (1 - \delta)^2_{,}$ this value corresponds to $\delta = 0.294$, i.e. $\bar{n} = 3.4$. It should be noted, that polyene distribution remained practically constant during the Diels–Alder reaction, since the time dependence of the conversion was identical for polyenes of different length. The results obtained for the numerical values of constants k_a, k_b and k_d and their temperature dependence are shown in *Figure 20*.

For the rate constant of the first addition (k_a) we obtained an activation energy of 10.4 kcal mol^{-1}, for that of the further additions (k_b): 12.9 kcal mol^{-1} and that of the adduct dissociation (k_d): 4.0 kcal mol^{-1}.

The applicability of relationship 39 obtained for the conversion is demonstrated by the calculated curves drawn in *Figure 17* and *19*; the calculations were performed with the following values:

$t = 140°C$ $\qquad$ $k_a = 0.636\ l\ mol^{-1}\ min^{-1}$

$[CMA] = 0.0581\ mol\ l^{-1}$ $\qquad$ $k_b = 0.151\ l\ mol^{-1}\ min^{-1}$

$k_d = 9.06 \times 10^{-3}\ min^{-1}$

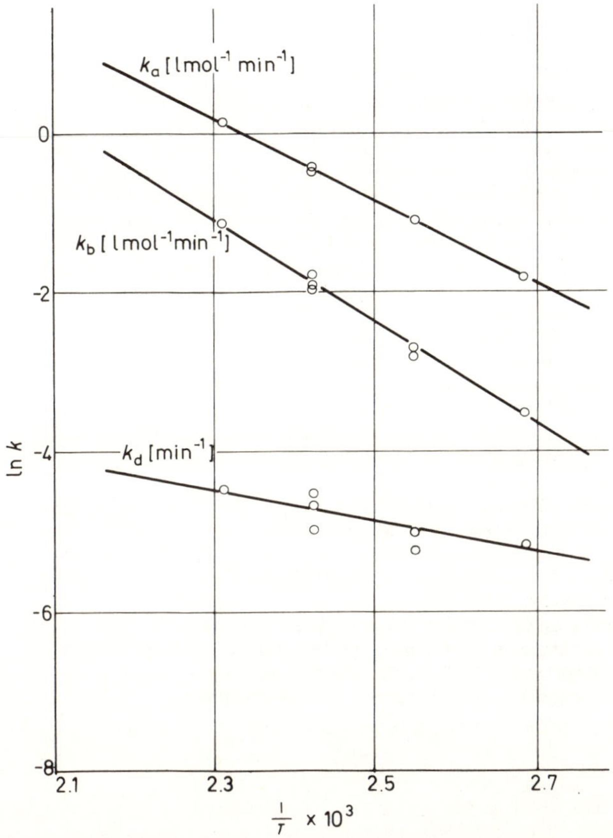

Figure 20. Temperature dependence of rate constants of the polyene–CMA reaction; k_a is the bimolecular rate constant of the first addition, k_b is that of the further reaction and k_d is the rate constant of unimolecular dissociation in the adduct.

The excellent agreement shows that the assumptions applied in the derivation of the conversion are correct enough and the reaction proceeds—at least essentially—according to the mechanism assumed. The rate constant k_a and its activation energy obtained for the first addition process, directly derived from the well reproducible initial rate, are, by all means, reliable. In the temperature range of thermal degradation (in spite of the relatively low activation energy) the k_a value is considerably higher compared to the overall rate of dehydrochlorination (at 210°C $k_a \approx 4 \, l \, mol^{-1} \, min^{-1}$, and the rate of elimination is $w_{HCl} \approx 10^{-4} \, min^{-1}$). In the case of sufficient CMA concentration, the rate of Diels–Alder addition may be expected to compete even with the very high rate component ('zip') of the degradation process; simultaneous study of the addition and degradation may give detailed information on the degradation mechanism.

REFERENCES

[1] D. Braun and W. Quarg. *Angew. Makromol. Chem.* **29/30**, 163 (1973); I. K. Varma and S. S. Grover. *Angew. Makromol. Chem.* **7**, 29 (1969).

[2] G. Palma and M. Carenza. *J. Appl. Pol. Sci.* **16**, 2485 (1972); D. Braun and M. Thallmaier. *Makromol. Chem.* **99**, 59 (1966).

[3] G. C. Marks, J. L. Benton and C. M. Thomas. *S.C.I. Monograph No 26*. 204 (1967); D, Braun, M. Thallmaier and D. Mepp. *Angew. Makromol. Chem.* **2**, 71 (1968); D. Braun and M. Thallmaier. *Makromol. Chem.* **108**, 241 (1967).

[4] K. R. Popov and V. L. Smirnov. *Optika I Spektroskopiya* **14**, 787 (1963).

[5] D. Braun. *Gummi Asbest. Kunststoffe* **24**, 1116 (1971); J. N. Hay. *J. Pol. Sci. A-1* **8**, 1201 (1970).

[6] T. Kelen, G. Bálint, G. Galambos and F. Tüdös. *Europ. Polym. J.* **5**, 597 (1969); T. Kelen, G. Galambos, F. Tüdös and G. Bálint. *Europ. Polym. J.* **5**, 617 (1969); T. Kelen, G. Galambos, F. Tüdös and G. Bálint. *Europ. Polym. J.* **5**, 629 (1969); T. Kelen, F. Tüdös, G. Galambos and G. Bálint. *Europ. Polym. J.* **6**, 127 (1970); T. Kelen. G. Bálint, G. Galambos and F. Tüdös. *J. Pol. Sci.* **C 33**, 211 (1971); F. Tüdös and T. Kelen. *Macromolecular Chemistry–8* (Ed. K. Saarela) p. 393. Butterworths, London (1973).

[7] T. Morikawa. *Kagaku Kogyo* **41**, 571 (1967); W. I. Bengough and I. K. Varma. *Europ. Polym. J.* **2**, 61 (1966).

[8] M. B. Nejman and P. A. Papko. *Vysokomol. Soed.* **A 10**, 841 (1968); B. B. Troitskii, L. S. Troitskaya, V. M. Myakov and A. F. Lepaev. *Preprint IV.*–34. International Symposium on Macromolecules. Helsinki, 1972.

[9] R. R. Huisgen, A. Dahmen and H. Huber. *J. Amer. Chem. Soc.* **89**, 7130 (1967); R. B. Woodward and R. Hoffmann. *Angew. Chem.* **81**, 797 (1969).

[10] T. Morikawa. *Kagaku Kogyo* **39**, 325 (1965).

[11] W. C. Geddes. *Europ. Polym. J.* **3**, 747 (1967); K. Kurzweil and P. Kratochvil. *Coll. Czech. Chem. Commun.* **34**, 1429 (1969).

[12] G. O. Schenck and K. Ziegler. *Naturwissenschaften* **32**, 157 (1944).

[13] *Methoden der Organischen Chemie* (Houben–Weyl), *V./l d* 199–203 (1972): *V./4* 80 (1960).

[14] L. S. Troitskaya and B. B. Troitskii. *Plasticheskye Massy*, **9**, 19 (1968).

[15] J. Sauer. *Angew. Chem.* **78**, 233 (1966); *ibid.* **79**, 76 (1967).

ADDITION COMPOUNDS AND COMPLEXES WITH POLYMERS AND MODELS

ROLF C. SCHULZ

Institute of Macromolecular Chemistry, Technische Hochschule Darmstadt, D 61 Darmstadt, FRG

with contributions by
D. FLEISCHER, A. HENGLEIN, H. M. BÖSSLER, J. TRISNADI and H. TANAKA

ABSTRACT

Most chemical transformations of polymers consist of either the formation or the splitting of covalent bonds. There are other reactions with polymers, however, which depend upon the reversible formation of adducts or complexes. Several examples of these types of addition compounds or complexes with polymers are described. Lanthanide complexes with homopolymers and with copolymers of ring acetal (trioxane, dioxolane, trioxocane, etc.) bring about signal shifts and splitting in ^{1}H- and ^{13}C-n.m.r. spectra. These make it possible to obtain important information about microstructure and sequences.

Iodine produces coloured addition compounds with various polymers. Newer results are presented concerning the kinetics of the iodine–amylose reaction, which were examined with the assistance of pulse radiolysis.

Iodine–amylose shows strong induced Cotton effects in o.r.d. and c.d. These depend upon the helical structure of the inclusion compound.

Polysaccharides form complexes with Schweizer reagent or with other heavy metals as well. The structure of the Cu–amylose complexes is deduced from the changes in the optical rotation and from the induced Cotton effects.

Even adducts between amylose and several azo dyes show strong induced Cotton effects. The composition of these adducts, the equilibrium constants and dissymmetry factors have been determined.

Finally various types of EDA complexes from polymer donors and/or acceptors are depicted. Our own research is concerned with polymer and oligomer esters with spirocyclic fluorene or tetranitrofluorene groups.

INTRODUCTION

There is a large number of chemical reactions with polymers. They are mainly carried out for the purpose of modifying physical and technical properties. But these reactions can also serve in explaining the composition and structure of homopolymers or the sequence of copolymers. In much of the research done in this area functional groups are introduced in order that these reactive polymers may be utilized as exchange resins, as supports for enzymes or solid phase syntheses, or for the preparation of graft copolymers.

By far the greater number of these reactions on macromolecules include either solvolyses, eliminations or rearrangements as well as nucleophilic or electrophilic substitutions. They consist accordingly in the formation and/or splitting of covalent bonds. Most of these reactions are irreversible. They can be described through the following equation:

$$P_n(X)_a + R \rightarrow P_n(X)_a(Y)_b \qquad (1)$$

In this equation $P_n(X)_a$ signifies a polymer molecule with a functional group X which is transformed to group Y by means of the reagent R. The conversion is frequently incomplete and a small amount of the original group remains unchanged in the reaction product.

In addition to these reactions there are several transformations with polymers which involve the formation of complexes or adducts. No covalent bonds are formed in these, but instead they contain heteropolar bonds, chelates, H-bonds, c.t.-complexes, etc. These reactions are generally reversible and lead to equilibria. They can be described through the following schematic equation:

$$P_n(B)_a + C \overset{K}{\rightleftharpoons} P_n(B)_{a-x}(B \cdot C)_x \qquad (2)$$

This means that a complex-forming, low molecular weight compound C together with a certain number of functional groups B in polymer P_n form a complex (B·C) in a reversible reaction. Here too, it is possible that a fixed number of groups B remains uncomplexed.

In low molecular weight complexes the equilibrium constant is dependent upon the chemical nature of the components B and C, and upon the solvent and temperature. In related polymer complexes it has been demonstrated that the structure of the polymer, the tacticity and the secondary structure as well as the degree of polymerization also have a great influence upon the equilibrium constant. Sometimes the complex formation takes place only with polymers above a fixed degree of polymerization while oligomers or low molecular weight model compounds are not capable of complex formation.

Frequently the properties of the polymers are considerably altered through complex formation, even when the equilibrium constant is very small. The number of complexed groups is determined spectroscopically On the other hand complex formation can also be used for analytic purposes. Several of these polymer complexes have a technical significance, e.g. in electrophotographic techniques. Of course reversible complex and adduct formation with macromolecules also plays an important role in many biochemical processes and enzyme reactions. It is also the basis of the very recent and important affinity chromatography. Several reactions of the type of equation 2 will be described in the following report. The examples are taken from the literature or from our own research, and they apply to synthetic as well as to naturally occurring polymers.

LANTHANIDE COMPLEXES WITH POLYACETALS

It had been known for some time that paramagnetic transition metal ions had a very strong influence on signals of complex-bonded ligands in the ^{1}H n.m.r. spectrum. But it was only through the systematic research done by

Hinkley[1] that the practical importance of these shifts was generally known. Since then, the so-called 'shift reagents' have become important in n.m.r. spectroscopy[2]. Various complexes of europium or praseodymium are the most frequently used (formulae I and II).

$Eu(DPM)_3$:

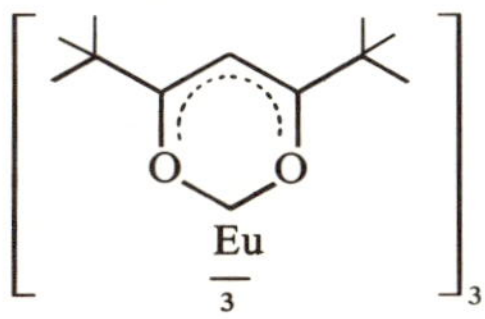

I

$Eu(fod)_3$:

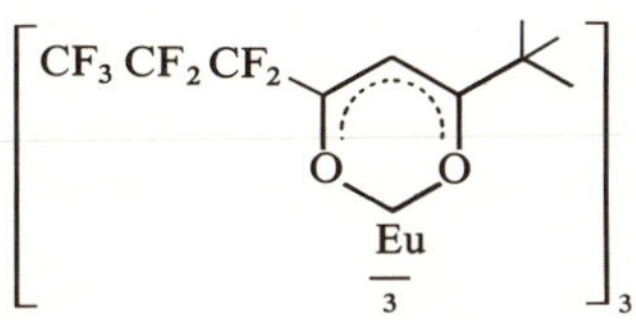

II

In these chelates, two coordination positions are still unoccupied. Therefore, they may add two molecules bearing nonbonded electron pairs (for example, amines, alcohols or ethers). Within these coordination complexes, the protons of the ligands are not only exposed to the external field but also to the magnetic field of the paramagnetic central ions (eg. Eu^{3+}. Pr^{3+}). The intensity of this additional field depends on the distance and the angle between the ligand proton and the central ion as well as on the concentration of the complex. Consequently, normally identical protons may show different shifts in the n.m.r.-spectrum according to their geometrical position within the coordination complex, which results in a very differentiated splitting of the signals and thus in a gain of information.

As already mentioned, the isotrope shift δ_{exp} also depends among other things on the concentration of the lanthanide chelate [L]. By measuring δ_{exp} with varying chelate concentrations, the equilibrium constant K for the complex formation as well as the shift for a lanthanide (L) to substrate (S) ratio of 1:1 may be determined under specific conditions[2, 3]. This ratio is called the ΔEu-value. This value is very useful for the assignment of the signals, being characteristic of each proton within a specific complex. Although shift reagents have nearly always been used for low-molecular compounds up to now, they have also been found to be very valuable for elucidating polymer structures[4].

In the following, it will be demonstrated by some examples that by means of complex formation between europium chelates and polyacetals it is possible to identify sequences up to 7 base units using n.m.r. spectroscopy, and to determine them quantitatively.

In the n.m.r. spectra of 1,3-dioxolane (III) and polydioxolane (IV), two singlets appear respectively which are to be assigned to the protons of the formaldehyde or the ethylene oxide units. When adding $Eu(DPM)_3$, both signals are shifted by certain amounts. The shift intensities of the ethylene and methylene protons, expressed by ΔEu-value[5] are as shown in the

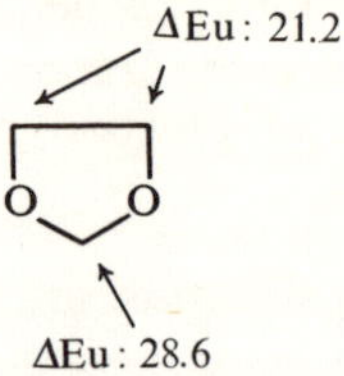

III

ΔEu: 29.7 (on the two CH_2 groups of CH_2—CH_2); 16.4 (on the O—CH_2—O group)

$$\text{+}CH_2\text{—}CH_2\text{—}O\text{—}CH_2\text{—}O\text{+}_n$$

IV

formulae III and IV. In the ^1H n.m.r. spectrum of a trioxane–dioxolane copolymer (*Figure 1*) four peaks appear[5]. The three singlets between $\delta = 4.74$ and $\delta = 4.89$ p.p.m. originate in the protons of the formaldehyde unit, three different arrangements of which can be found in these copolymers. More explicitly, it is neighboured either by two other methylene units, or by one methylene unit and one ethylene unit, or by two ethylene units. The signal (E) with $\delta = 3.73$ p.p.m. is to be assigned to the ethylene protons of the dioxolane unit. When $Eu(DPM)_3$ is added to the polymer solution, this singlet is split up into three lines (*Figure 2*).

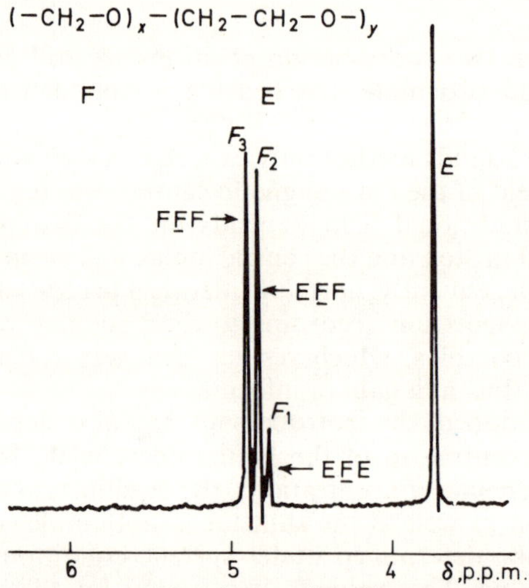

Figure 1. 60 MHz ^1H n m.r. spectrum of a trioxane–dioxolane copolymer in $CDCl_3$ (62.5 mol % trioxane) F_1: $\mathrm{E\underline{F}E}$; F_2: $\mathrm{E\underline{F}F}$ and or $\mathrm{F\underline{F}E}$; F_3: $\mathrm{F\underline{F}F}$.

Again, this splitting can only originate in the varying neighbourhood of the ethylene units. As each ethylene unit, however, can only be placed between two CH_2O units and, consequently, only one triad with EFE is possible, this splitting must be caused by more distant units. Therefore, heptads must be taken into consideration. By means of the ΔEu-values and the copolymerization statistics, the signals could be assigned to the following heptads:

E_1: $\mathrm{FEF\underline{E}FEF}$; E_2: $\mathrm{FEF\underline{E}FFF}$; $(\mathrm{FEF\underline{E}FFE})$; E_3: $\mathrm{FFF\underline{E}FFF}$; $(\mathrm{EFF\underline{E}FFF})$; $(\mathrm{EFF\underline{E}FFE})$

Recently, Fleischer[6] was able to demonstrate that shift reagents also led to

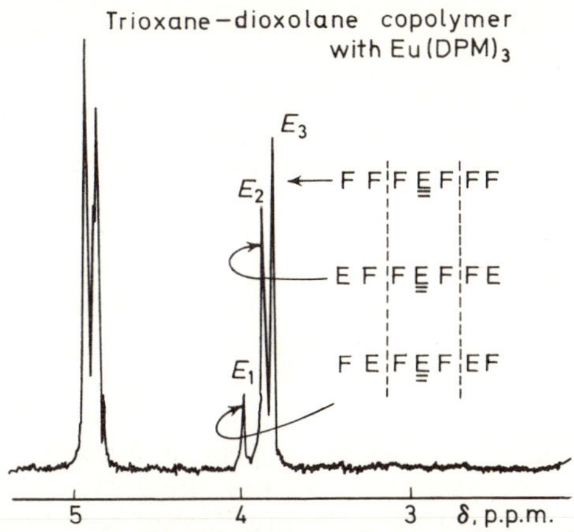

Figure 2. The same spectrum as in *Figure 1* upon addition of $Eu(DPM)_3$
E_1 : FE | FE̳F | EF and or FE | FE̳F | FF
E_2 : EF | FE̳F | FE and or EF | FE̳F | FF
E_3 : FF | FE̳F | FF

further splittings of the resonance signals in the ^{13}C c.m.r. spectra of polyacetals. The ^{13}C spectrum of the above-mentioned trioxane–dioxolane copolymer shows instead of four signals as in the proton spectrum, eight signals which can be assigned to pentads of methylene- and ethylene-oxide units. On addition of $Eu(fod)_3$, the eight signals are split up into 10–13 signals. These can again be assigned to sequences of seven basic units. The assignments following from the proton spectra are not only confirmed but also supplemented, the splitting being so extensive in the methylene signal region as to make heptads perceptible. Moreover, the proportions of these sequences may be determined quantitatively, thus permitting an extensive elucidation of these polymer and copolymer structures[6].

By means of shift reagents, the overall structure, the quantity of monomer diads (DOL–DOL; DOL–TOC and TOC–TOC) as well as the sequences of methylene and ethylene units of trioxocane(TOC)-dioxolane (DOL) copolymers were determined in a similar way[7].

IODINE COMPLEXES WITH SOME POLYMERS

There are some synthetic and naturally occurring polymers which produce intense colour reactions with iodine either in aqueous solution or in the solid state. These reactions are reversible, and the iodine can be easily removed. These reactions therefore are of the type depicted in equation 2. in which the kind of complex or adduct formation depends upon the respective polymers. Iodine adducts are formed for example with polyamides[8]. polyethyleneoxide[9], polyvinylalcohol[10], polyvinyl carbazole[11] and some polysaccharides.

In 1927, the blue polyvinylalcohol–iodine complex was discovered simultaneously by Staudinger *et al.*[12] and by Herrmann and Haehnel[13]. This complex is of practical importance as an antiseptic in medicine and in the production of polarizing sheets[14]. It is also used for analytical purposes in the determination of polyvinylalcohol[15]. Following a proposal made by Zwick[10], the polyvinylalcohol–iodine complex is presumably composed of helical segments of the polymer, in the hollow cavities of which are included extended chains of associated iodine molecules. This structure is stabilized through the addition of boric acid. The rate of complex formation, the structure of the complex, the wavelength of the absorption maxima and its extinction coefficient are dependent upon various parameters. The influence of the tacticity, molecular weight and branching of polyvinyl alcohol, and of the content of residual acetyl groups upon the complex formation with iodine, has been studied in numerous recent papers[16]. The solutions are deep blue, green, red-orange or yellow, depending upon the conditions. The absorption maxima appear at 600–620 nm, 650–680 nm, 480–500 nm and approximately 350 nm. Recently, experiments on the resonance Raman spectra of the polyvinylalcohol–iodine complex have also been published[17, 18]. However, this method mostly furnishes information about the structure of the enclosed iodine atom but does not say very much about the structure of the polymers.

The deep blue colour which appears when mixing a dilute aqueous solution of starch and iodine has been known for over a hundred years[19]. In spite of many experiments, the cause of this colour reaction remained uncertain for a long period of time. In 1939 Freudenberg[20] suggested that a helical inclusion compound could be the structure for the colour complex. In 1944 this hypothesis was verified by Rundle *et al.*[21] with the help of x-ray diffraction. The accepted model for the solid state is a complex composed of a helix with six glucose units per turn. A chain of 10–15 iodine atoms is packed in the tube-shaped interior. On the average, 6–8 glucose units are associated with two iodine atoms.

It follows from various experiments that the amylose–iodine complex in a dilute aqueous solution has a very similar structure, and consists of helical segments with included iodine chains. It is to be emphasized that the iodine complex is formed only above a certain degree of polymerization (DP > 12) (i.e. oligomers produce no colour reaction with iodine). Furthermore the absorption maximum of the complex increases with the DP. With the DP $>$ 100, a maximum at approximately $\lambda = 640$ nm is eventually attained[22]. The colour reaction is thermoreversible. Above 70 °C the colour of the solution fades and on cooling the blue colour appears again. For further details, the original papers or reviews should be referred to[19, 23–25]. In the following only two problems will be dealt with, viz. the kinetics of complex formation and the optical activity of the complexes.

It was of course known previously that the reaction between iodine and amylose proceeds very rapidly, but it was not until 1970 that quantitative kinetic measurements were published for the first time by Hiromi *et al.*[26] as well as by Thompson and Hamori[27]. They made use of the stopped-flow method. The dead time of the instruments was only 0.5 ms and with an amylose concentration of 0.025 per cent and an iodine concentration of 10^{-5} molar, the complex formation was already as much as 50 per cent completed within the dead time. For that reason Hamori *et al.* used much smaller amylose and iodine concentrations in order to slow down the reactions.

Henglein *et al.*[28] used pulsed radiolysis for these measurements. With this method the time resolution is from 10 to 100 times better, i.e. from 10^{-4}–10^{-5} s. Aqueous solutions of amylose containing 10^{-2} mol KI were irradiated with a short pulse of 1–30 μs of high energy electrons. Within this time a definite amount of molecular iodine is formed from the initially produced OH-radicals and the iodine ions. The iodine reacts with the amylose forming the coloured complex. The increase of extinction with time at certain wavelengths is indicated on an oscillograph with a time scale of from 0–0.5 ms. The results are discussed with reference to the mechanism suggested by Thompson and Hamori[27].

As a result[28] an equilibrium should very quickly be attained between free iodine and the empty helices (H_0) whereby, in a first stage, unstable complexes with 1–3 molecules of iodine per helix are produced (H_1, H_2, H_3) (see scheme 1). Additional iodine is taken up into these H_3 complexes, until iodine chains of approximately 7 iodine molecules (H_7) are formed. When more than 4 iodine molecules are included, the reverse reaction is so small as to be unobservable.

The diffusion of the iodine molecules into the empty helices follows with a rate constant k' of approximately 10^9 l mol^{-1} s^{-1} which corresponds to a diffusion controlled reaction (*Scheme 1*). At low iodine concentrations

Thompson–Hamori mechanism for iodine–amylose reaction

$$H_0 + I_2 \underset{k_1}{\overset{k'}{\rightleftharpoons}} H_1 \qquad H_3 + I_2 \underset{k_r}{\overset{k'}{\rightleftharpoons}} H_4$$

$$H_1 + I_2 \underset{k_2}{\overset{k'}{\rightleftharpoons}} H_2 \qquad H_4 + I_2 \underset{k_r}{\overset{k'}{\rightleftharpoons}} H_5$$

$$H_2 + I_2 \underset{k_3}{\overset{k'}{\rightleftharpoons}} H_3 \qquad H_{z-1} + I_2 \underset{k_r}{\overset{k'}{\rightleftharpoons}} H_z$$

(H_n: helical segment with n iodine molecules)

Kinetics of I_2 uptake into amylose helices (H_z) (A. Henglein *et al.*)

$$\text{at } [I_2]: \sim 10^{-6} \text{ mol l}^{-1} \qquad \frac{d[H_z]}{dt} \sim k' [I_2]^4$$

$$\text{at } [I_2]: \sim 10^{-4} \text{ mol l}^{-1} \qquad \frac{d[H_z]}{dt} \sim k' [I_2]$$

$$k' \sim 10^9 \text{ l mol}^{-1} \text{ s}^{-1}$$

amylose: 0.025 %; $[C_6H_{10}O_5]$: $\sim 10^{-5}$ mol l^{-1}

Scheme 1. Mechanism and kinetics of amylose–iodine reaction

($\sim 10^{-6}$ mol l^{-1}) the initial rate of the reaction is proportional to between the fourth and fifth power of $[I_2]$. At high iodine concentrations ($\sim 10^{-4}$ mol l^{-1}) the reaction is of first order with respect to $[I_2]$.

Our further experiments deal with the optical activity of the amylose–iodine complexes. Like all other polysaccharides amylose is optically active, since each base unit contains five asymmetric carbon atoms. As the wavelengths decrease, the rotation values steadily increase[29]. This results in a so-called 'plain' rotatory dispersion curve, since amylose contains no chromophoric groups which absorb between 800 and 200 nm. A chiral secondary structure, such as a helix, contributes essentially to the rotation value as well, but in this case it cannot be separated from the basic rotation. In the circular dichrogram no Cotton effects appear and therefore no conclusions can be drawn from the o.r.d. and c.d. measurements about the secondary structure of the amylose. If, nevertheless, the helix model holds true for the amylose–iodine complex, then the chiral helix should asymmetrically induce the electron transitions of the enclosed chain of iodine atoms. Therefore the normally optically inactive iodine chromophore must become optically active in the inclusion compound and must exhibit a so-called 'induced Cotton effect'. This prediction, as a matter of fact, proves to be correct[30, 22]. The rotation values at 600 nm are approximately three times greater for the amylose–iodine complex than for pure amylose. And in this region too a maximum in circular dichroism appears which is lacking in pure amylose.

We have studied in detail the chiroptical properties of iodine–amylose and moreover have examined the dependence of the induced Cotton effects upon the iodine concentration, upon the molecular weight of the amylose, upon the solution and upon other parameters. The following are the most important results: (i) the optical activity is greatest in a proportion of 1 I_2 to 7 or 8 glucose base units; (ii) the wavelength for the long-wave Cotton-effect increases with the polymerization degree of the amylose; (iii) the circular dichroism ($\Delta\varepsilon$) is dependent upon the degree of polymerization and has a sharp maximum at a DP of 70–90; (iv) the induced Cotton effects appear only when an amylose–iodine complex is formed.

These results can only be explained by accepting the hypothesis of helical iodine inclusion compounds. They further confirm a model of the so-called interrupted helix much discussed at the present time; it is composed of many long inflexible helical sections with short intervals of randomly coiled segments between them[19].

METAL COMPLEXES WITH POLYSACCHARIDES

It is well known that many biopolymers, particularly enzymes, form very stable heavy metal complexes and they are biologically effective only in the presence of these metals. Many synthetic polymers likewise form various kinds of metal complexes. In the following, several complexes with polysaccharides will be very briefly covered.

In 1857, Schweizer[31] observed that cellulose dissolves in an ammoniacal copper tetramine solution. A copper–cellulose complex is thereby formed, which contains one copper atom for each glucose base unit. The complex is destroyed by acidification and the cellulose is precipitated. At a later date still other cellulose solvents were discovered, consisting of aqueous solutions of basic cadmium or iron complexes[32]. These soluble cellulose heavy metal complexes play an important role in cellulose chemistry, for example, in the production of synthetic fibres and in determining molecular weight[32].

It is very surprising that amylose does not dissolve in the Schweizer reagent. Instead an insoluble complex is formed, about whose structure nothing was known up to now. It can be demonstrated that the amylose–copper complex does not precipitate when mixing a very dilute solution of Schweizer reagent and amylose (amylose concentration about 10^{-3} mol glucose units l^{-1}). In examining the composition and structure of the complex, we again made use of the rotatory dispersion and the circular dichroism. Here too appear considerable deviations from the normal o.r.d. curve of amylose as well as several induced Cotton effects[33].

It must therefore be a chiral complex. If the difference of the optical rotation between amylose and the amylose complex is plotted as a function of the molar fraction of copper (according to JOB) then maximum deviations are found at a mole ratio for glucose:Cu of 1:1. The overall composition is therefore the same as for the copper complex of cellulose. An inclusion compound, such as that formed with iodine, is very unlikely with this mole ratio. The measurements of circular dichroism have a big advantage in that the Cotton effects can also give some evidence about the configuration of the complex. According to investigations made by Bukhari *et al.*[34] on copper complexes of sugar derivatives, a short-wave negative c.d. peak and a long-wave positive peak occur when the complex is formed with hydroxyl groups in a bis-equatorial position (*Scheme 2*). As a matter of fact with the

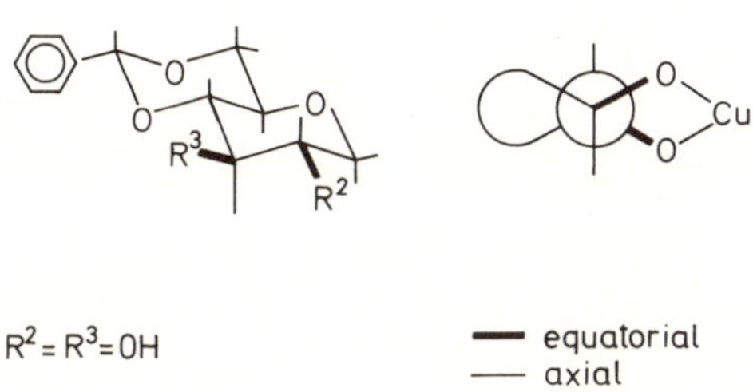

Scheme 2. Chiral copper chelate with bis-equatorial OH groups in sugar derivatives

copper–amylose complex we find a weak positive Cotton effect above 600 nm and a strong negative Cotton effect at 280 nm. In the C1-conformation of the glucose units in amylose only OH groups on the C_2 and C_3 atoms are in the required bis-equatorial position. With this it clearly follows from the circular dichroism measurements that a chiral copper complex is formed from the Schweizer reagent and amylose with both secondary OH groups of the same base unit.

We have further examined complexes of amylose with cobalt amines[33]. These too show abnormal optical rotatory dispersion and several overlapping Cotton effects. In this case, however, it was not possible to determine the composition, since presumably several different complexes are formed and the equilibria between the various chelates are reached only very slowly.

However, in these cases the chiroptical properties can be traced back to the individual asymmetrical base units and no polymer effect or influence of secondary structure can be observed.

AMYLOSE COMPLEXES WITH AZO COMPOUNDS

Some dyes show great affinity to cellulose which bonds them very firmly. They are called direct dyes. Although this dye bonding is very important for cotton dyeing, we have as yet no exact knowledge about the bonding forces[35, 36]. On the other hand, we know that certain dyes form salt-type compounds with polypeptides[37] or nucleic acids[38]. The chiroptical properties of these polymer–dye adducts furnish information about the secondary structure and helix–coil transformations of the polymers. Consequently, we tried to apply these methods also to polysaccharide–dye adducts[39]. Of course, only soluble polymers are suitable, so we chose amylose first. The aim was to get information about the amylose structure and the composition of the dye adducts. Because of the size of the dye molecules it was to be expected that the adducts could not be inclusion compounds of the iodine–amylose type.

For the interaction between a helical macromolecule and dye molecules four models are to be considered, shown in *Scheme 3*. Model A signifies that

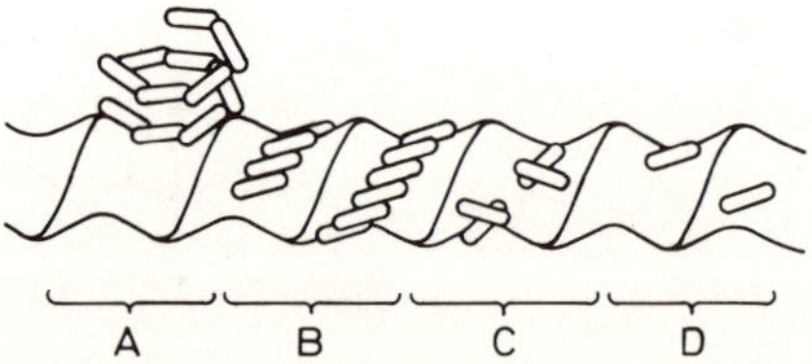

Scheme 3. Different models for dye–polymer interaction

the polymer helix induces the formation of chiral dye associates but does not react itself with the dye[37, 40]. Models B to D represent some alternatives of sterically ordered polymer–dye adducts[41]. In these cases, the chromophoric groups are fixed in specific positions and are asymmetrically induced due to their chiral arrangement. Therefore, extrema of optical activity and Cotton effects should appear within the absorption range of the dyes, although the dye solutions are optically inactive. When adding an amylose solution (concentration $\sim 2.3 \times 10^{-3}$ mol $C_6H_{10}O_5$ l^{-1}) to an aqueous Congo red solution (concentration $\sim 1.2 \times 10^{-4}$ mol l^{-1}), the absorption spectrum changes only very little, but in the circular dichrogram several strong Cotton effects appear and the optical rotation differs wideiy from that of pure amylose[42]. Consequently, a chiral dye adduct is formed from amylose and Congo red. When plotting the circular dichroism $\Delta\varepsilon$ with constant amylose concentration as a function of the dye concentration, a typical saturation curve is obtained which may be described by a Langmuir isotherm[39]. Hence the equilibrium constant of the amylose–Congo red addition may be determined, which is $K \sim 3 \times 10^4$ l mol^{-1}. Moreover, a ratio of 15 glucose units to 1 Congo red molecule is reached at saturation.

The complex structure may also be determined from the $\Delta\varepsilon$ values according to the continuous variation method (JOB-method), and again a molar ratio of glucose:Congo red = 15:1 is obtained by this independent method.

The dissymmetry factor $g_\lambda = \Delta\varepsilon/\varepsilon$ is $g_{515} = 2.6 \times 10^{-4}$ at the strongest induced Cotton effect ($\lambda = 515$ nm) and thus much smaller than with the amylose–heavy metal complex ($g = 10^{-3}$–10^{-2}). Therefore, dye bonding cannot consist in the formation of an inherently dissymmetrical chelate complex.

It should be pointed out that this dye bonding is only possible with polysaccharides or cyclodextrines[43]. Linear oligosaccharides[36] or model compounds, as for example α-methylglucoside or 4,6-ethylidene-α-methylglucoside, do not show any reaction with Congo red, either in the absorption spectrum or in the circular dichrogram. Consequently, there are specific bonding forces typical of polymers which, as well as sterical conditions, are necessary. In fact, these experiments do not reliably demonstrate the helix structure of the amylose in aqueous solution, but it is made probable that there are at least short ordered sections to which dye molecules may regularly add. According to our results, we prefer for these dye adducts model D where non-aggregated, single Congo red molecules are associated with ordered segments of the polymer.

Apparently, the dye molecules must also have a specific size and capacity to be able to add regularly to amylose. We, therefore examined 44 further dyes of different structure[44]. There were only 13 which formed chiral adducts with amylose showing induced Cotton effects. Their common characteristics are that each contains at least one azo group and is a direct dye for cotton. Ten of the dyes may be described by the general formula V:

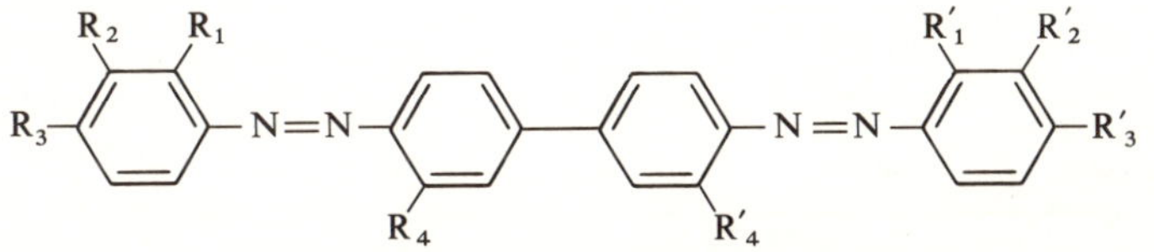

V

The substituents, of course, are of importance for adduct formation We were able to find rules for the kind and position of the necessary groups[44]. At any rate, the length of the dye molecules as well as the presence of at least two groups interacting with the amylose are probably important. In fact, dyes bearing all necessary functional substituents but having only half the length of the general formula do not form chiral adducts. Some examples of the Cotton effects between 400 and 600 nm are shown in *Figures 3* and *4*. The long-wave extrema always coincide with the longest-wave absorption bands. There are also several overlapped, positive and negative Cotton effects at 200 and 400 nm, but these cannot be clearly assigned to specific absorption bands.

It is emphasized that amylose and dye solutions alone do not show the slightest Cotton effect. The respective recording curves coincide with the baseline. For three other dye–amylose adducts (such as Benzoazurine G, Benzo Fast red F and Thiazine red R), the equilibrium constants *K*. the

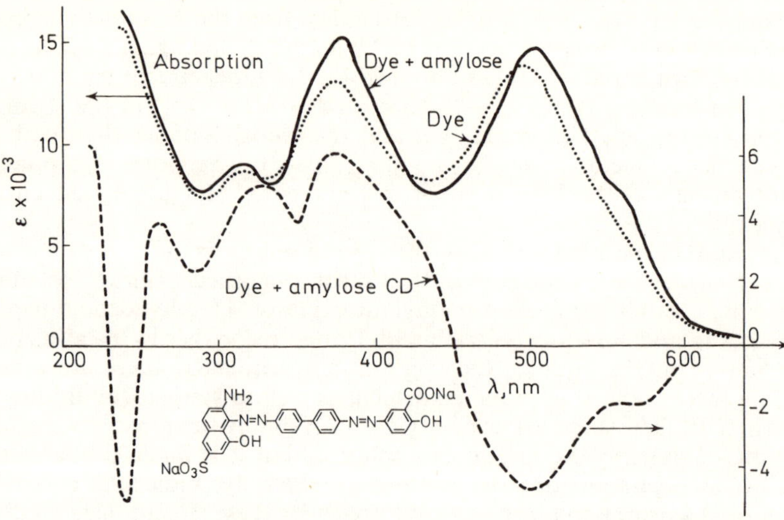

Figure 3. Absorption spectrum of Benzo Fast red F (. . . .) and dye adduct with amylose (———); c.d. spectrum of dye–amylose adduct (– – –) in water (C_{dye} = 1.2×10^{-4} mol l^{-1}; glucose : dye = 20 : 1)

Table 1. Data for the amylose adducts with some azo dyes

Name of dye	Number in Colour Index	Equilibrium constant $K \times 10^{-4}$, $l\ mol^{-1}$	Dissymmetry factor $g \times 10^4$ (for wavelength λ)	Ratio $C_6H_{10}O_5$: dye for saturation
Congo red	22 120	~ 3	2.6 (515 nm)	~ 15 : 1
Benzo azurine G	24 140	3.9	1.5 (310 nm)	~ 42 : 1
Benzo Fast red F	22 310	5.6	3.1 (500 nm)	~ 12 : 1
Thiazine red R	14 780	6.0	0.3 (510 nm)	~ 7 : 1

dissymmetry factors *g*, and the molar ratios at saturation were determined according to the above-mentioned methods[44] (see *Table 1*). In all cases, *K* is of the order of $10^4\ l\ mol^{-1}$ similar to *K* for Congo red. The *g* values are very similar, too. Consequently, these are probably quite similar complexes. However, very different values are obtained for the molar ratios of glucose units to dye.

DONOR–ACCEPTOR COMPLEXES WITH MACROMOLECULES

Donor–acceptor complexes are formed by combining an electron donating compound (D) with an electron accepting compound (A) according to the following equation[45]:

$$D + A \overset{K}{\rightleftharpoons} [D \ldots A] \leftrightarrow [D^+A^-] \qquad (3)$$

Complexes of this type, in which both components are low molecular weight

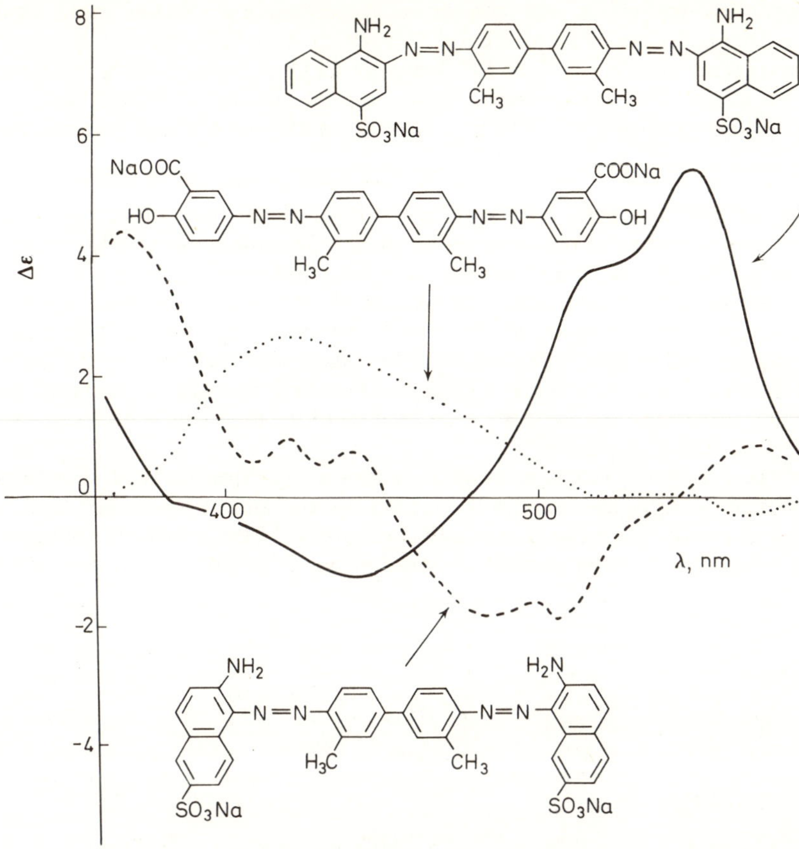

Figure 4. c.d. spectra of amylose adducts with benzopurpurin 4B (———), benzopurpurin 5B extra (---) or chrysamine (...) in water. (C_{mylose}: 2.4×10^{-3} base mol l^{-1} C_{dye}: 1.1×10^{-4} mol l^{-1})

compounds, have already been amply investigated. In recent years it has been recognized that such EDA complexes are also important in polymer science, for example, as initiators for polymerizations[46] or for producing alternating copolymers[47].

The following explanations are limited to such EDA complexes, in which at least one of the components is a macromolecule. It is possible to distinguish four different types of complexes.

$$P_n(D)_x + A \overset{K'}{\rightleftharpoons} P_n(A \ldots D)_x \qquad \text{type I} \qquad (4)$$

$$P_n(A)_x + D \overset{K'}{\rightleftharpoons} P_n(A \ldots D)_x \qquad \text{type II} \qquad (5)$$

$$P_n(A)_x + P_m(D)_x \overset{K'}{\rightleftharpoons} P_n(A \ldots D)_x P_m \qquad \text{type III} \qquad (6)$$

$$P_n(A)_x(D)_y \overset{K'}{\rightleftharpoons} P_n(A \ldots D)_z \qquad \text{type IV} \qquad (7)$$

$P_n(A)_x$ and $P_n(D)_x$ symbolize polymers with a large number (x) of donor or acceptor groups which are connected by covalences to the main chain. There are also other examples known in which A and D groups are found in the same macromolecule (type IV). Polymer EDA complexes have interesting properties and can, for example, be utilized as organic photoconductors in electrophotography[48, 49]. For this reason, many papers and patents have recently appeared in this area[50–53, 58].

It has been shown that the c.t. bands in the absorption spectrum of polymer complexes are generally very similar to those of the low molecular weight model compounds. On the other hand, the equilibrium constants for complex formation in polymers and models are very different. In many cases, the constants are dependent upon the degree of polymerization, and upon the quantity and distribution of the complex-forming groups in the macromolecule as well as upon the tacticity. The reasons for such 'polymer effects'[55] have not yet been completely clarified. Of course it must be accepted that complex formation also causes a change in molecular shape and can eventually lead to crosslinking[56, 57].

So far, the most widely examined complexes belong to type I, in which the donor groups are pendantly joined to the macromolecules and are complexed with low molecular weight acceptors. These polymers are in most cases obtained through radical or ionic polymerizations. Several examples

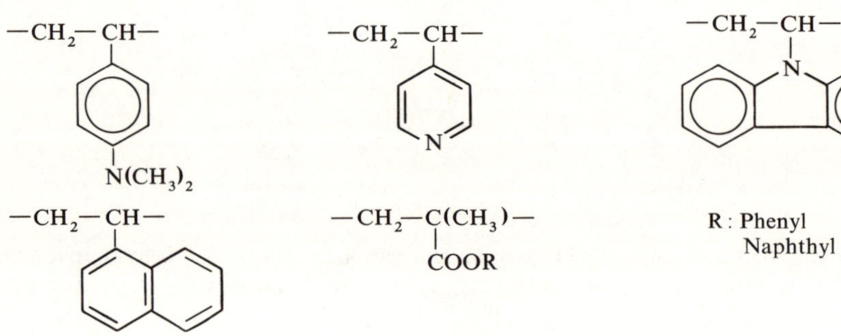

Scheme 4. Vinyl polymers with pendant donor groups (P(D))

of such polymer donors are presented in *Scheme 4*. Commonly used acceptors are shown in *Scheme 5*. There are also a number of donor polymers which are produced by means of polycondensation, for instance starting from the diols VI[51].

$$HO-CH_2-CH_2-\underset{\displaystyle R}{\underset{|}{N}}-CH_2-CH_2-OH \qquad \text{VI}$$

R : phenyl, *m*-tolyl, *p*-anisyl, 2-fluorenyl

Of special interest is the product VII obtained through the polycondensation of monobromopyrene and formaldehyde, since it has many properties advantageous for use as electrophotographical material[49].

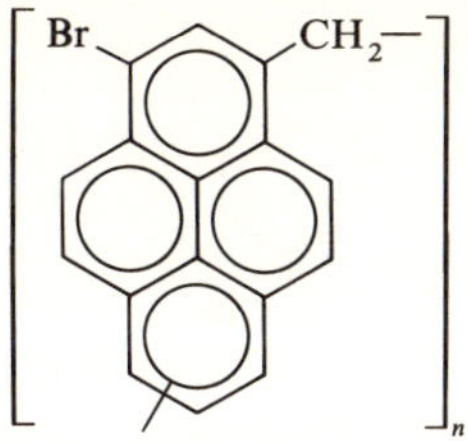

$n = 6\text{–}10$

VII

In the case of bromopyrene resin, acceptors with especially high electron affinity were employed in order to shift the absorption bands in the long-wave area and to increase the light sensitivity (for instance tetranitrofluorenone or dicyanomethylene fluorene, see *Scheme 5*).

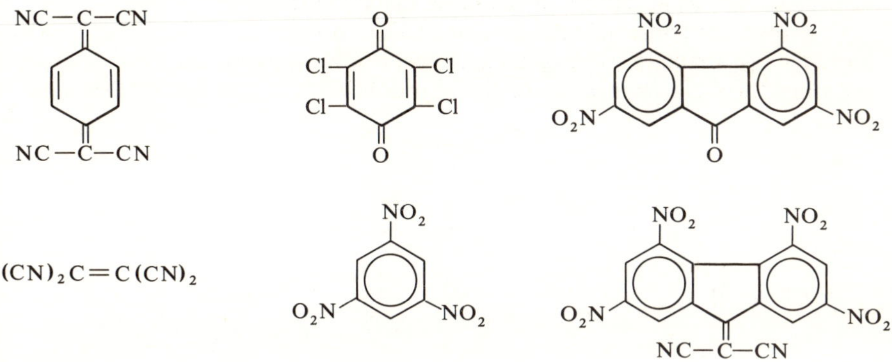

Scheme 5. Examples of low molecular acceptors

Polymer EDA complexes of type II have up to now been only slightly investigated. Two examples of polymers with acceptor groups are VIII[59] and IX[51].

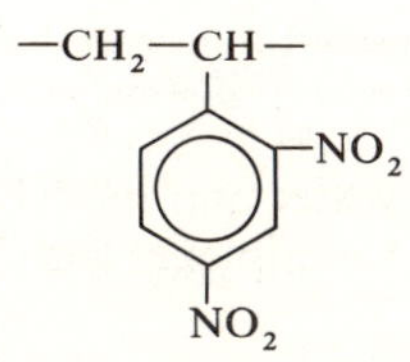

VIII

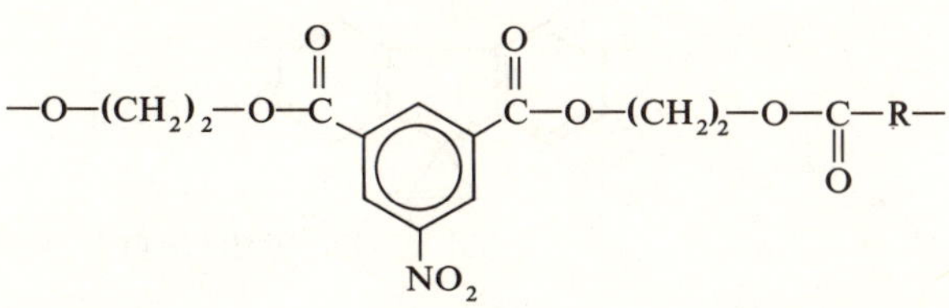

IX

By combining polymer donors with polymer acceptors, type III complexes are obtained[51].

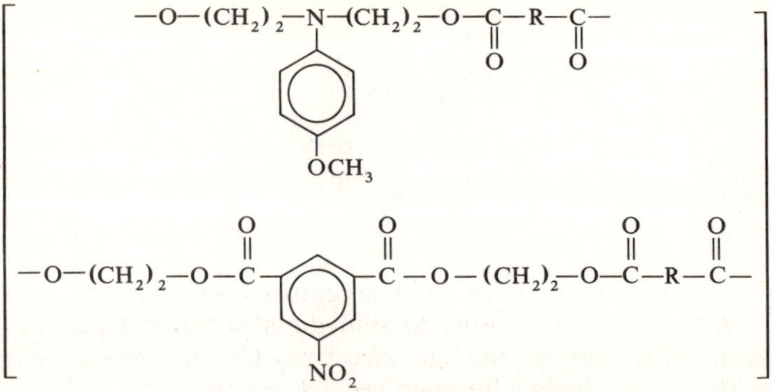

X

The most difficult to prepare are polymer EDA complexes which contain donors as well as acceptors in the same macromolecule[51, 60].

Our own experiments with polymer EDA complexes are based upon the spirocyclopropanefluorene group. Appropriate starting material can be obtained through the reaction of diazofluorene with double bonds (*Scheme 6*).

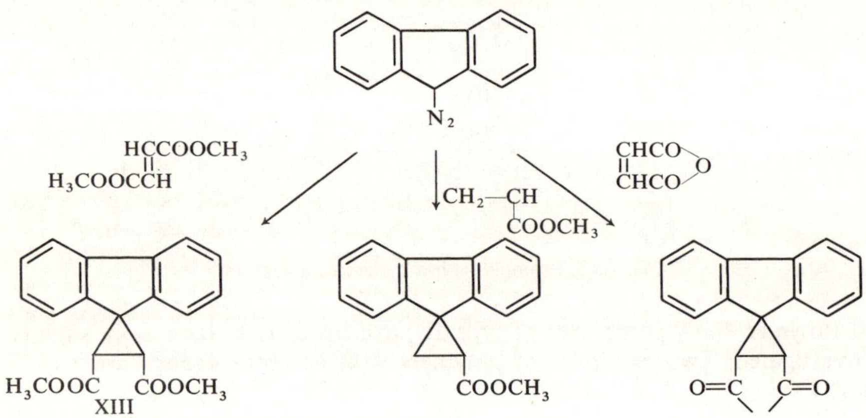

XIII

Scheme 6. Synthesis of spirocyclopropane-1,9′-fluorene derivatives; starting materials for donor polyesters (P(D)) and models

From this, various donor polyester and model compounds can be prepared (XI) which form coloured EDA complexes with TCNE or TCNQ as acceptors in dichloromethane.

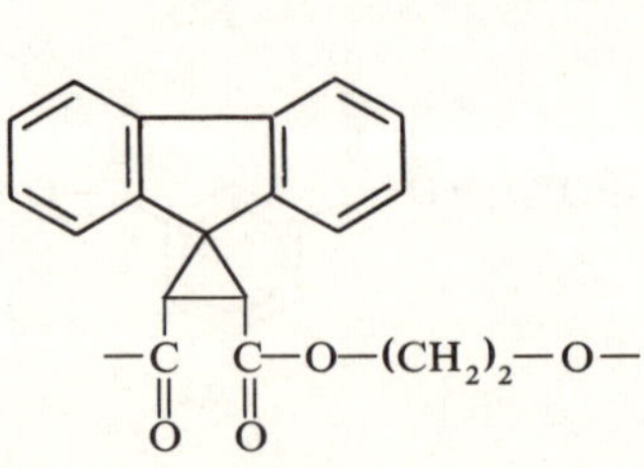

XI

It can be demonstrated that the equilibrium constant is dependent upon the degree of polymerization, and it is much higher for polyesters than it is for the monomeric analogues. It is possible that sandwich-like EDA complexes are present because of the rigidity of the spirocyclic base units. This work has already been published in detail elsewhere[61].

Acceptor polymers of analogous structure are obtained through the introduction of nitro groups in the spirocyclopropanefluorene system[62] according to *Scheme 7.*

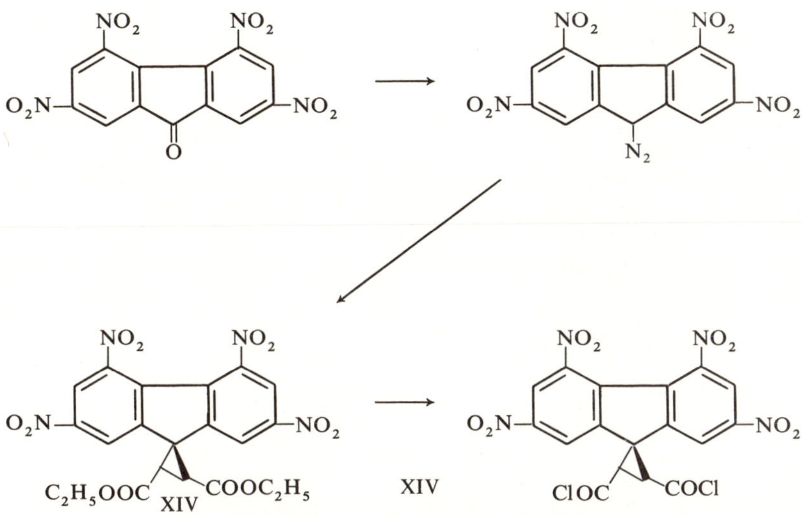

Scheme 7. Synthesis of spirocyclopropane-1,9′-tetranitrofluorene derivatives; starting materials for acceptor polyesters (P(A)) and models

The starting material for this is tetranitrofluorenone. It is transformed into the diazo compound, which is reacted with diethyl fumarate. From this, the dicarbonyl chloride is prepared. A polyester is formed from this through polycondensation in solution with ethyleneglycol.

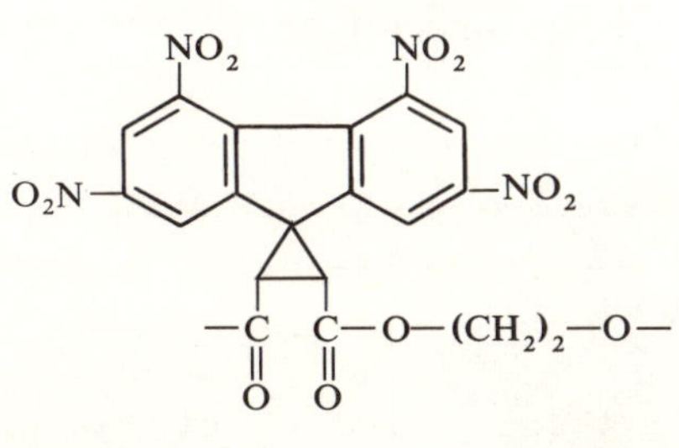

XII

With fluorene as a donor, a type II EDA complex is formed. The absorption spectrum is very similar to that in the complex between the monomer spiroester and fluorene. In both cases, a c.t. band at ~ 400 nm appears. The structure of both complexes may for that reason be the same. Nevertheless, the values for $K\varepsilon$ or for the equilibrium constant K, as determined by Benesi–Hildebrand–Scott, are here larger in the polymer complex than in the monomer model complex.

Table 2. Complex formation between fluorene (donor D) and acceptor polymer (P(A) XII) or acceptor model (A) (XIII)

	P (A) XII + D			A + D		
λ nm	$K\varepsilon$ $mol^{-2} l^2 cm^{-1}$	ε $mol^{-1} l cm^{-1}$	K $mol^{-1} l$	$K\varepsilon$ $mol^{-2} l^2 cm^{-1}$	ε $mol^{-1} l cm^{-1}$	K $mol^{-1} l$
410	603	600	1	462	1210	0.4
430	487	510	1	388	910	0.4
450	400	490	0.8	325	725	0.5

The acceptor polymer XII can also be complexed with the analogue donor polymer XI mentioned above[62]. This results in a type III complex. Although the spectrum indicates no distinct maxima, the broad absorption between 410 and 500 nm can certainly be attributed to complex formation since the individual components do not absorb in this area. A combination of the corresponding monomer acceptor XIV and donor XIII show a similar

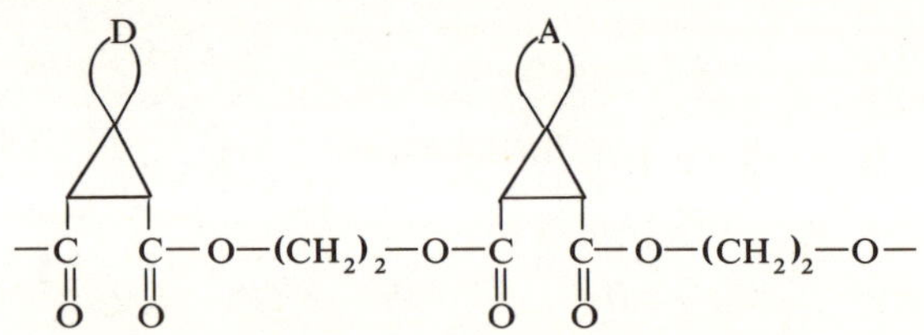

XIVa

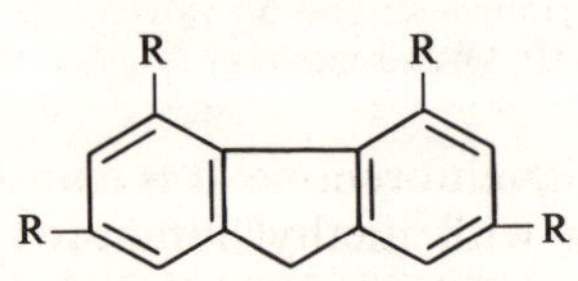

D : R = H XIII

A : R = NO_2 XIV

spectrum. But even in this case, a constant is found for the polymer EDA complex which is clearly higher than for the corresponding monomer complex (see *Table 3*).

Table 3. Complex formation between P (D) XI and P (A) XII compared with model complexes

	P (A) XII + P (D) XI			D XIII + A XIV		
λ nm	$K\varepsilon$ $mol^{-2} l^2 cm^{-1}$	ε $mol^{-1} l cm^{-1}$	K $mol^{-1} l$	$K\varepsilon$ $mol^{-2} l^2 cm^{-1}$	ε $mol^{-1} l cm^{-1}$	K $mol^{-1} l$
410	2270	1436	1.6	400	700	0.6
450	1390	870	1.6	210	400	0.5

Starting from the dichlorides of the fluorene compound and of the tetranitrofluorene compound, a copolyester can be produced which contains donor as well as acceptor groups in the molecule and therefore belongs to type IV[62] (formula XIVa).

Copolyesters with D:A ratios of approximately 4:1 and 1.4:1 have been produced. During the synthesis, the reaction mixture attains an intense green-brown colour. The absorption spectrum of the polyester is distinctly different from the mixture of the monomer esters and also from the above-mentioned mixture of the donor and acceptor polyesters.

At 600–650 nm there appears a strong absorption band. The optical density of this band is strictly proportional to the concentration of acceptor groups in the solution. From this it can be concluded that these bands actually derive from intramolecular and not from intermolecular interactions between the donor and acceptor groups. By comparison with model compounds it can be estimated that in spite of intense colouration only about ten per cent of the groups available are complexed. The rigid fixation of the complexed groups by means of the spiran structure certainly plays a role in this particular type of polyester.

In conclusion, we have also produced polymers which carry a spirofluorene ring as a freely mobile side group on a polyvinylalcohol backbone, formula XV[62].

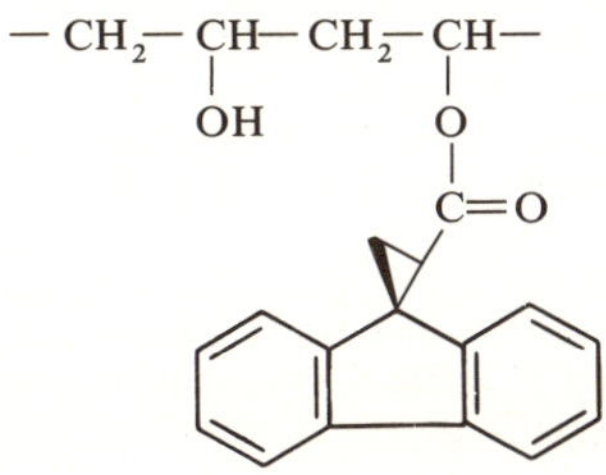

XV

The conversation lies between 35 and 60 mol per cent, depending upon the conditions. The modified polyvinylalcohol forms EDA complexes with TCNE and TCNQ. The c.t. bands show up at 555 nm or at 565 nm and agree with the low molecular weight model compounds. Presumably, the remaining OH groups of the PVA also play a role in the complex formation with polyesters, since the ξ_{k1} values, calculated by Liptay[63], are not constant. Therefore only approximate values for $K\varepsilon$ according to Benesi–Hildebrand–Scott can be calculated. Nevertheless it follows that in this case no polymer effect appears and the $K\varepsilon$ values for polymer complexes and for models can be considered the same within the limits of experimental error.

CONCLUSIONS

This short survey, with the help of several chosen examples, is meant to show that synthetic and naturally occurring macromolecules are capable of forming various types of adducts, complexes or chelates. Although mostly labile compounds are dealt with, many have properties of interest for scientific as well as practical reasons. In many cases the reactions take place only with macromolecules and are not possible with low molecular weight compounds or oligomers. Frequently a very specific secondary structure is also a prerequisite for complex formation. It is to be pointed out that adduct formation with biopolymers certainly plays an important role in biological processes as well. Perhaps the experiments on synthetic polymers and model systems can contribute toward a better understanding of biochemical processes.

ACKNOWLEDGEMENTS

I want to express my sincerest thanks to all my co-workers for their active and skilful cooperation. We all gratefully acknowledge the financial support of Fonds der Chemischen Industrie and of Deutsche Forschungsgemeinschaft through the programme of the Sonderforschungsbereich 41.

REFERENCES

[1] C. C. Hinckley, *J. Am. Chem. Soc.* **91**, 5160 (1969); J. K. M. Sanders and D. H. Williams, *Chem. Commun.* 422 (1970); P. V. Demarco, Th. K. Elzey, R. B. Lewis and E. Wenkert, *J. Am. Chem. Soc.* **92**, 5734 (1970), *ibid.* **92**, 5737 (1970).
[2] R. V. Ammon and R. D. Fischer, *Angew. Chem.* **84**, 737 (1972); *Angew. Chem. Internat. Edit.* **11**, 675 (1972); L. Pohl, *Kontakte (Merck)* 17 (1972).
[3] R. E. Rondeau and R. E. Sievers, *J. Am. Chem. Soc.* **93**, 1522 (1971); J. K. M. Sanders and D. H. Williams, *Tetrahedron Lett.* 2813 (1971); I. Armitage *et al. Chem. Commun.* 1281 (1971): D. R. Kelsey, *J. Am. Chem. Soc.* **94**, 1764 (1972); B. L. Shapiro and M. D. Johnston, *J. Am. Chem. Soc.* **94**, 8185 (1972).
[4] F. F. L. Ho, *J. Polymer Sci.* **B9**, 491 (1971); A. R. Katritzky and A. Smith, *Tetrahedron Lett.* 1756 (1971); J. E. Guilett, I. R. Peat and W. F. Reynolds *Tetrahedron Lett.* 3493 (1971): R. Anderson, A. H. Haines and B. P. Stark, *Angew. Makromol. Chem.* **27**, 151 (1972); J. F. Chandler and B. P. Stark, *Angew. Makromol. Chem.* **27**, 159 (1972).
[5] D. Fleischer and R. C. Schulz, *Makromol. Chem.* **152**, 311 (1972).
[6] See the contributions of D. Fleischer in the review of W. Bremser, *Chemiker. Ztg* **97**, 259 (1973).
[7] D. Fleischer and R. C. Schulz, *Makromol. Chem.* **162**, 103 (1972).
[8] I. Matsubara and J. H. Magill, *Polymer*, **7**, 199 (1966).
[9] R. Springer and R. Lang, *Wehrmed.* **5**, 146 (1967).
[10] M. M. Zwick, *J. appl. Polymer Sci.* **9**, 2393 (1965).
[11] A. M. Hermann and A. Rembaum, *J. Polymer Sci.* **C17**, 107 (1967).
[12] H. Staudinger, K. Frey and W. Starck, *Chem. Ber.* **60**, 1782 (1927).
[13] W. O. Herrmann and W. Haehnel, *Chem. Ber.* **60**, 1658 (1927).
[14] J. G. Pritchard, *Poly(vinyl alcohol)*. Polymer Monographs (Ed. H. Morawetz) Vol. 3, McDonald Techn. and Scientific, London (1970).
[15] J. G. Pritchard and D. A. Akintola; *Talanta*, **19**, 877 (1972).
[16] K. Kikukawa, S. Nozakura and S. Murahashi, *Polymer Journal* **3**, 52 (1972); Y. Kazuo, M. Shuji and G. Yukichi, *Kubunshi Kagaku* **26** (288) 265 (1969); *Chemical Abstracts* **71**, 22411c (1969); S. Hayashi and K. Takizawa, *Kogyo Kagaku Zasshi* **71**, 2051 (1968); S. Hayashi, M. Takayama and Ch. Kawamura, *Kogyo Kagaku Zasshi*, **73**, 412 (1970).
[17] F. Inagaki, I. Harada, T. Shimanouchi and M. Tasumi, *Bull. Chem. Soc. Japan* **45**, 3384 (1972).
[18] M. E. Heyde, L. Rimai, R. G. Kilponen and D. Gill, *J. Am. Chem. Soc.* **94**, 5222 (1972).
[19] For a review see: J. Szejtli and S. Augustat, *Stärke*, **18**, 38 (1966); J. Szejtli, *Periodica polytechnica (Budapest)* 7, 259 (1963); J. Szejtli, M. Richter and S. Augustat, *Biopolymers* **5**, 5 (1967).
[20] K. Freudenberg, E. Schaaf, G. Dumpert and T. Ploetz, *Naturwiss.* **27**, 850 (1939).
[21] R. E. Rundle, J. F. Foster and R. R. Baldwin, *J. Am. Chem. Soc.* **66**, 2116 (1944).
[22] B. Pfannemüller, H. Mayerhöfer and R. C. Schulz, *Makromol. Chem.* **121**, 147 (1969); *Biopolymers* **10**, 243 (1971).
[23] P. K. Rawlings and F. W. Schneider, *J. Chem. Phys.* **52**, 946 (1970).
[24] W. Banks and C. T. Greenwood, *Stärke-starch* **23**, 300 (1971).

[25] D. A. Brant and P. L. Dubin, Polymer Preprints **14**, 169 (1973); C. L. Cronan and F. Schneider. *J. Phys. Chem.* **73**, 3990 (1969); F. W. Schneider, C. L. Cronan and S. K. Podder, *J. Phys. Chem.* **72**, 4563 (1968).
[26] K. Hiromi, T. Shibaoka and S. Ono, *J. Biochem.* **68**, 205 (1970).
[27] J. C. Thompson and E. Hamori, *Biopolymers* **8**, 689 (1969); *J. Phys. Chem.* **75**, 272 (1971).
[28] M. Grätzel, A. Henglein, M. Scheffler, H. M Bössler and R. C. Schulz, *Ber. Bunsen-Gesellschaft* **76**, 74 (1972).
[29] V. S. R. Rao and J. F. Foster, *Biopolymers* **1**, 527 (1963); R. M. Purvinas and H. F. Zobel. *Carbohydrate Research* **10**, 129 (1969).
[30] R. Wolf and R. C. Schulz, *Tetrahedron Lett.* 1799 (1967); *J Macromol. Sci.* **Á2**, 821 (1968): R. C. Schulz, R. Wolf and H. Mayerhöfer, *Kolloid-Z.* **227**, 65 (1968).
[31] E. Schweizer, *J. Prakt. Chem.* **72**, 109 (1857).
[32] J. Schurz in *Chemie der Pflanzenzellwand* (Ed. E. Treiber) p. 349. Springer Verlag Berlin (1957); G. Jayme, in *Cellulose and Cellulose Derivatives* (Ed. N. M. Bikales and L. Segal) High Polymers, Vol. 5, Part IV, p. 381. Wiley-Interscience (1971).
[33] H. M. Bössler and R. C. Schulz, *Makromol. Chem.* **135**, 87 (1970).
[34] S. T. Bukhari, R. D. Guthrie, A S. Scott and A. D. Wrixon, *Chem. Commun.* 1580 (1968).
[35] H. Gerstner, *Verhalten direktziehender Farbstoffe*, p. 276. Akademie Verlag, Berlin (1957): S. R. Cockett and K. A. Hilton, *Dyeing of cellulosic fibres*, p. 211. Leonard Hill Books, London (1961); H. Bach, E. Pfeil, W. Phillipar and M Reich, *Angew. Chem.* **75**, 407 (1963).
[36] B. Carroll and J. W. Van Dyk, *Science (Washington)* **116**, 168 (1952).
[37] L. Stryer and E. R. Blout, *J. Am. Chem. Soc.* **83**, 1411 (1961).
[38] D. M. Neville jr. and D. F. Bradley, *Biochim. Biophys. Acta* **50**, 397 (1961).
[39] H. M. Bössler and R. C. Schulz, *Angew. Makromol Chem.* **29/30**, 431 (1973).
[40] R. E. Ballard, A. J. McCafferey and S. F. Mason, *Biopolymers* **4**, 97 (1966); S. F. Mason. *Proc. Chem. Soc. (London)* 119 (1964).
[41] K. Yamaoka, *Biochim. Biophys. Acta* **169**, 553 (1968); M. Zama and S. Ichimura, *Biopolymers* **9**, 53 (1970); D. G. Dalgleish, H. Fujita and A. R. Peacocke, *Biopolymers* **8**, 633 (1969).
[42] H. M. Bössler and R. C. Schulz, *Kolloid-Z.* **239**, 578 (1970).
[43] F. Cramer, W. Sänger and H. Ch. Spatz, *J. Am. Chem. Soc.* **89**, 14 (1967); K. Sensse and F. Cramer, *Chem. Ber.* **102**, 509 (1969).
[44] J. A. Trisnadi, H. M. Bössler and R. C. Schulz, *Kolloid-Z.* in press (1974).
[45] R. Foster, *Organic Charge Transfer Complexes*, Academic Press, London (1969).
[46] C. E. H. Bawn, A. Ledwith and A. Parry, *Chem. Commun.* 490 (1965); S. Tazuke and T Yamane, *J. Polymer Sci.* **B9**, 327, 331 (1971).
[47] N. G. Gaylord, B. K. Patnaik and N. Patnaik, *J. Macromol. Sci. Chem.* **Á6**, 81 (1972); J. Furukawa, H. Amano and R. Hirai, *J. Polymer Sci.* **A-1**, 10, 681 (1972).
[48] R. M. Schaffert, *IBM J. Res. Develop.* 75 (1971).
[49] B. Löhr, R. Arneth and D. Winkelmann, in *Current Problems in Electrophotography*, W. de Gruyter, p. 219. Berlin (1972).
[50] W. Klöpffer, H. Tacheci and W. Willicks, *Makromol. Chem.* **126**, 147 (1969); W. Klöpffer and W. Willicks, *Makromol. Chem.* **115**, 156 (1968).
[51] Th. Sulzberg and R. J. Cotter, *J. Polymer Sci.* **A-1**, 8, 2747 (1970); *Macromolecules*, **1**, 554 (1968); *ibid.* **2**, 146, 150 (1969).
[52] M. L. Hallensleben, *European Polymer J.* **7**, 1197 (1971).
[53] D. Braun and H. J. Sterzel, *Ber. Bunsen-Ges. Physik, Chem.* **76**, 551 (1972).
[54] H. Boudevska, *Makromol. Chem.* **143**, 11 (1971); H. Boudevska and V. Popminchev, *Makromol. Chem.* **143**, 19 (1971).
[55] S. Iwatsuki and K. Arai, *J. Polymer Sci. Chem. Ed.* **10**, 3329 (1972).
[56] S. Fakirov, H. Boudevska and N. Aneva, *Makromol. Chem.* **155**, 259 (1972).
[57] M. L. Hallensleben and A. Kuppel, *Makromol. Chem.* **143**, 87 (1971).
[58] V. Neitchev, H. Boudevska and G. Detchev, *European Polymer J.* **9**, 183 (1973).
[59] G. Smets, V. Balogh and Y. Castille, *J. Polymer Sci.* **C4**, 1467 (1964).
[60] A. Inami, K. Morimoto and Y. Hayashi, *Bull. Chem. Soc. Japan* **37**, 842 (1964); N. Yang and Y. Gaoni, *J. Am Chem. Soc.* **86**, 5022 (1964).
[61] A. M. Braun, H. G. Cassidy, R. C. Schulz and H. Tanaka, *Makromol Chem.* **146**, 195 (1971).
[62] H. Tanaka, unpublished results.
[63] W. Liptay, *Z. Elektrochem.* **65**, 375 (1961).

METALLOPROTEINS, METALLO-ENZYMES AND HETEROGENEOUS CATALYSTS

R. J. P. WILLIAMS

Inorganic Chemistry Laboratory, Oxford, UK

ABSTRACT

An inspection of the active site region of metallo-enzymes leads to two conclusions. (i) Individual attacking groups usually have uncommon physical properties. (ii) The active site region is cooperatively linked with a large part of the enzyme. Such an interaction is common to heterogeneous catalysts but is absent in homogeneous catalysts. It is concluded that the energetics of catalysis requires an analysis of the nature of cooperativity in microdomains of well organized material and some general remarks as to the problems which arise are described. Methods of examining large well-organized molecules (microdomains) are described and some suggestions are made as to how man-made 'enzymes' could be produced.

There is a very large number of metalloproteins and metallo-enzymes which are now well characterized. In this article it will be assumed that the reader is familiar with much of the work before 1970[1]. The article will be restricted to those metallo-systems which contain iron, cobalt, copper, zinc, or molybdenum. No reference will be made to either magnesium or calcium enzymes. The assumption will be made that metalloproteins, which have no enzyme activity, differ from metallo-enzymes in that, while the metal sites of metallo-proteins have relatively conventional physical properties, the sites of metallo-enzymes have more unusual properties. Elsewhere we have developed the idea that the metallo-enzymes are unusual because the metal ions in them do not have conventional ground state geometries[2,3]. The geometry may be unusual by virtue of bond length, bond angle, or both properties. These energized ground states have evolved to meet the functional catalytic requirements of biological systems, i.e. so as to carry out a certain number of reactions at high speed. As will be shown below it is most illuminating to compare enzymic catalysts with heterogeneous catalysts evolved by man.

As well as examining one or two new examples of enzymes which illustrate the above themes I shall also attempt to develop a better understanding of the energetics of the catalysis due to enzymes. This requires a very detailed inspection of an enzyme in solution, both in the presence and absence of the substrate. In a final section of speculative review I shall consider if it is possible to produce artificial polymers with a catalytic power equal to that of enzymes.

NEW EXAMPLES OF METALLO-ENZYME SITES

Table 1 gives a summary of some of the older data on metal ions at enzyme active sites[1, 2]. There are many features of this table which remain unexplained. For example the curious character of the cytochrome P-450 iron as seen in absorption spectra has not been related to any feature of the haem molecule with any certainty. In what follows I draw attention to more successful new work, taking proteins one at a time.

Table 1. Some properties of metallo-enzymes

Enzyme	Metal	Physical property
Cytochrome P-450	Fe(II)	Unusual absorption spectrum
Myoglobin	Fe(II)	Five-coordinate
Plastocyanin	Cu(II)	Unusual absorption spectrum
Caeruloplasmin	Cu(?)	'Diamagnetic copper'
Laccase	Cu	High redox potential
Carboxypeptidase	Zn	Very distorted geometry
	Co(II)	Anomalous absorption spectrum
Carbonic anhydrase	Zn	Distorted geometry
	Co(II)	Anomalous absorption spectrum

Iron-containing enzymes

Rubredoxin

The most complete structure determination on a metalloprotein involved in catalytic action is the work of Jensen and co-workers on rubredoxin[4]. The curious pattern of bond lengths and bond angles is shown in *Figure 1*. There is no clear explanation for this geometry. It must be remembered that the crystal structure data are for the oxidized state, Fe(III), of the protein. It may well be that the function of the curious structure can only be discovered once it is known how the protein structure changes with oxidation state of the metal. In fact although it is known that this protein is involved in electron transfer little is known of the system in which it is functional *in vivo*.

Nitrogenase

The iron–sulphur proteins of nitrogenase are now known to be different from all other such proteins[5]. Particularly their e.p.r. spectra is of a new kind which may be due to the Fe(III), spin $= \frac{3}{2}$, state. In any event the iron must be in a site of unusual geometry for such e.p.r. signals have not been observed before. Mössbauer data indicate that the iron sites are under the influence of the binding of adenosine triphosphate (ATP)[6]. Now the function of ATP-binding is to permit its hydrolysis and the release of energy. Thus in this case the iron in the protein is in an unusual ground state energized in a 'hot-spot'. It is as if the iron had been lifted to a steady-state configuration by input of appropriate *local* thermal energy.

Ribonucleotide reductase

The evidence for iron dimers in the ribonucleotide reductase of animals is

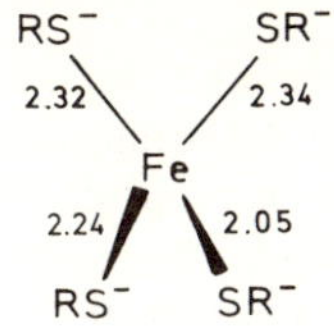

Figure 1. The coordination sphere of iron(III) in rubredoxin

now impressive. However, a more startling finding is that this enzyme contains a stable radical—probably of organic origin[7]. There is another ribonucleotide reductase from *E. coli* which requires coenzyme B_{12} (Co)[8]. Curiously the first intermediate in the reaction path involves the production of an organic radical in this protein too. Paths involving radicals in proteins are not of obvious significance but radicals have been observed trapped in several heterogeneous solid-state catalysts (see below).

Heme-enzymes

Apart from the well-documented cases of the adjustment of spin-state energies by protein geometries as in cytochrome P-450, peroxidase, and hemoglobin there have been several interesting advances in the study of cytochrome oxidase[9]. Thus certain of the e.p.r. signals during the electron transfer process have very high rhombicity; it is clear that high-spin Fe(III) cytochrome states are intermediates; the ground state of some of the heme and some of the copper (blue copper?) does not contribute to the e.p.r. signals and perhaps has a 'zero' magnetic moment. A possible explanation of the low magnetic moment would be a complex of the type

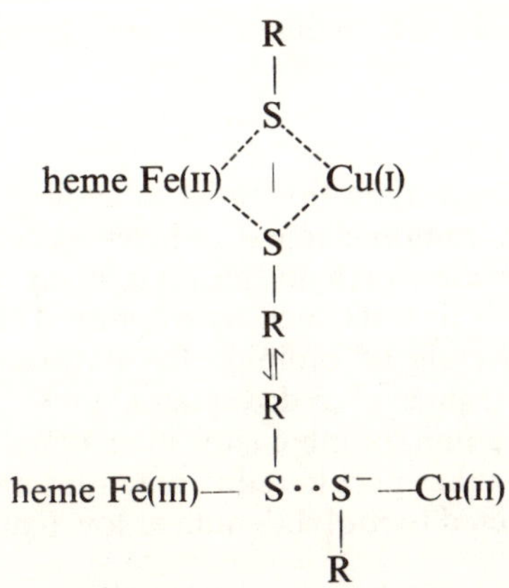

Here oxidation states are difficult to find for certain. A possible one-electron intermediate would be the state heme Fe(III) .... RS—SR Cu(I).

The juxtaposition of organic (—S—S—) and inorganic (Fe) (Cu) reactive centres is one way of generating extremely effective catalysts. Another way of making catalysts is to put together metals of rather different types. (It is, however, very difficult for biological systems to make such catalysts, e.g. in Li–Cu, Al–Ti compounds.)

Problems of heme spectra

The recent work of Hager[10] on the halo-peroxidases throws light on the problem of the nature of the P-450 reactions. Hager found that he could reproduce the Soret absorption band shift from 420 nm to 450 nm as seen in the microsomal P-450 oxidases. Now P-450 and the halo-peroxidases have different ligands coordinated to the iron, probably cysteine and histidine respectively. The fact that they can give almost identical spectra on binding to a strong sixth ligand, carbon monoxide, and that this spectrum differs from that of all other heme proteins in which the coordination of the iron is often fully known, suggests that the iron has lost the fifth ligand. It could well be a feature of these two proteins, unlike oxygen carriers, that the fifth ligand can be lost while the heme is retained in the protein provided that the sixth ligand is very strong[11]. Oxygen is a ligand not too different from carbon monoxide. As pointed out by Dunford and Williams[12] the dependence of the reactivity of peroxidases on a pK of around 0.0 in some peroxidatic activities could arise from a required assistance from protonation of the fifth group histidine in the reaction intermediate. Overall reactivity in oxygenation and peroxidation reactions rests partly, if our conjectures are correct, on the strength of the iron–fifth-ligand bond. The control is a property of the protein as a whole. Elsewhere we have stressed how conformational fluctuations of a fairly large part of molecules such as cytochrome-*c* could be an essential feature of electron transfer[13]. In purified cytochrome aa_3 the addition of *one* mole of CO, O_2, or CN^- alters the properties of *both* heme groups to a much more striking degree than the cooperativity seen in hemoglobin oxygen uptake[14]. This discussion allows us to stress the role of cooperativity between large parts of a protein both in activating a single reactive centre and in making a communication net between active centres. I shall return again to this theme of cooperative energetics.

Molybdenum-containing enzymes

Molybdenum has special catalytic functions in biology, e.g. in nitrogen fixation. Molybdenum is a common element in heterogeneous catalysts and some features of its chemistry are worth noting. (a) It forms high coordination number complexes even with ligands such as sulphur. This is not the case for the lighter transition metals of biology. (b) It forms oxo-complexes Mo═O. (c) It has a wide range of redox states, III–VI. Molybdenum is known (from e.p.r. data) to enter rhombic sites in enzymes and is probably bound to sulphur ligands. It takes part in redox reactions and a possible one, which would be rather restricted to molybdenum at low oxidation–reduction potentials, is

$$\mathrm{Mo^{VI}{=}O + RH \rightarrow Mo^{V}{-}OH + R\cdot}$$

Iron (oxidation state greater than three) brings about such reactions at a potential around + 0.80 V but molybdenum can do so at around 0.0 V. Thus molybdenum could be a special hydrogen atom transfer catalyst.

Molybdenum, like iron in nitrogenase, is activated by ATP. Could the reactions be as follows?

$$Mo^{VI}\underset{O}{\overset{O}{\Diamond}}Mo^{VI} + 6e + 4H^+ \rightleftarrows Mo^{III}\ldots Mo^{III} + 2H_2O$$

$$2H_2O + 2ATP \rightleftarrows 2ADP + 2P$$

$$Mo^{III}\ldots Mo^{III} + N_2 \rightleftarrows Mo^{III}\cdot N_2\cdot Mo^{III}$$

$$Mo^{III}\cdot N_2\cdot Mo^{III} + 2H_2O + 2H^+ \rightleftarrows Mo^{VI}\underset{O}{\overset{O}{\Diamond}}Mo^{VI} + + 2NH_3$$

The lower oxidation state of molybdenum is then stabilized by removal of water (by ATP) and this change is also seen in the iron proteins.

Copper-containing enzymes

Superoxide dismutase

When Vallee and Williams[2] discussed this copper protein of the erythrocyte membrane they thought they were dealing with a metalloprotein which had no enzyme activity. The protein was classified with conventional metal complexes for the literature data at that time indicated that the e.p.r. and absorption spectra were those of a conventional tetragonal complex. Since that time it has been found that the protein is a very common *enzyme* for dismutation of superoxide and it therefore seemingly represented a case of a non-entatic metal site. However, more detailed study of physical properties have since shown that the copper(II) has a rhombic symmetry, that its absorption spectrum is abnormal for the three nitrogen ligands known to be bound, and that its redox potential is high. Rotilio and his co-workers[16] have subsequently discovered a model copper compound of known crystal structure which has very similar e.p.r. and visible absorption spectra to those of the Cu(II) dismutase. The structure is shown in *Figure 2*. This

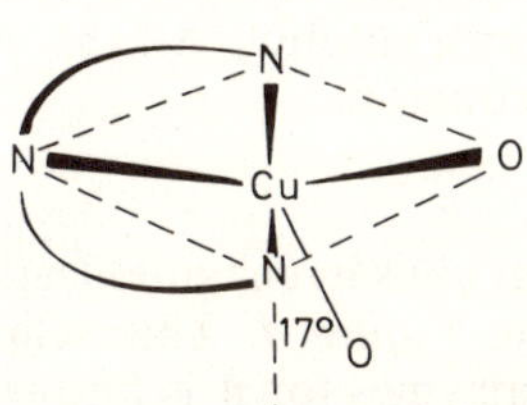

Figure 2. A schematic version of the coordination sphere of copper(II) in superoxide dismutase

structure may well be like that in the dismutase and it is obviously unusual. The structure does not look very probable for the Cu(I) site of the reduced enzyme, and perhaps there is a change of geometry on reduction. It should be possible to follow the local changes in geometry of this enzyme during redox cycles using the methods described later in this paper[17].

An additional problem presented by the dismutase is the function of the water in the fourth (and fifth?) coordination positions. It is known that water exchanges rapidly from around Cu(II) $(H_2O)_6$, rate of exchange $\sim 10^{10}\,s^{-1}$. In distorted five coordinate complexes the rate of water exchange may be much slower[18], $\sim 10\,s^{-1}$. The rate of reaction of the dismutase is much faster than this so in this case it could well be that the electron-transfer step

$$Cu(II) + O_2\cdot^- \rightarrow Cu(I) + O_2$$

is an outer-sphere reaction, not requiring replacement of the water of coordination. Can water be curiously activated in enzyme sites? The question arises again in the discussion of zinc enzymes.

Caeruloplasmin

There are three types of copper atom in caeruloplasmin. The first, type II copper, could be said to resemble the copper of superoxide dismutase. The second, the deep blue copper, (type I) is like the blue copper of other enzymes and may well be copper bound to a thiolate. The third type of copper is of an unknown kind associated with absorption at around 330 nm. A possible structure for this copper has been suggested and is shown in *Figure 3*[19]. Whether this is correct or not, an additional peculiarity of the

Reduced enzyme	Oxidized enzyme	Anion bound oxidized enzyme
Cu(I) . . . Cu(I) RS^- RS^-	Cu(I) . . . Cu(I) RS—SR	Cu(II)X^- Cu(II)X^- RS^- RS^-
no absorbance bands above 300 nm	330 nm band	420 nm band

Figure 3. A proposed relationship between the states of the 'diamagnetic' copper in caeruloplasmin and laccase

copper sites arises from the appearance of an absorption band at around 350–400 nm in both inhibited forms of the enzyme and in some of the reaction intermediates.

It will be noticed that a common feature of iron, molybdenum and copper centres is the association with sulphur. All the inorganic sulphides of these elements are good semiconductors.

Different Cu(II) *sites*

Our knowledge of copper sites in enzymes leads us to postulate a possible relationship between them, *Figure 4*. This scheme could be extended to include heme-containing enzymes for it is known that half the copper(II) in cytochrome oxidase cannot be detected by e.p.r., see above. Perhaps there is a similar series of non-heme non-ferredoxin iron sites, in which case oxidases of three different origins could have much in common.

Selectivity of metal-binding

It can be asked, why do sites such as that in the superoxide dismutase contain copper and not zinc? As we shall see later some of the copper sites suggested here, three histidines, are exactly the same as the sites for zinc in carbonic anhydrase and probably class II aldolase, while the inter-

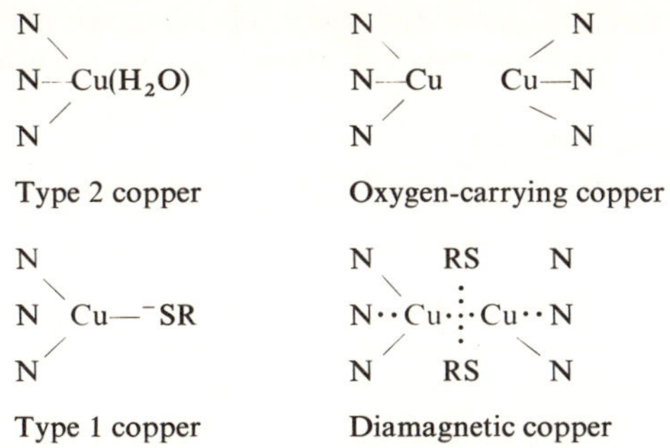

Figure 4. A possible relationship between the coordination spheres of different types of copper found in various enzymes

action of copper with nitrogen and sulphur ligands in, say, caeruloplasmin differs little in chemical composition from that which we believe to exist in alcohol dehydrogenase, for zinc. As was pointed out in 1953 the difference must lie in the geometry which the protein subtends[20]. If we compare the small molecule $N(CH_2CH_2NH_2)_3$ which fits a tetrahedron better than a square, with $NH_2CH_2CH_2NHCH_2CH_2NHCH_2CH_2NH_2$ which fits a square better than a tetrahedron, we find that zinc prefers the former and copper the latter. A relative selectivity factor of four orders of magnitude can be achieved. Whereas sites can be designed to give strain in the metal coordination sphere they can also be designed to give selectivity of uptake. While noting such factors it should be remembered that biological systems can exert kinetic control of insertion of metals into sites as they do in the insertion of magnesium into chlorophyll.

Cobalt-containing enzymes

Vitamin B_{12} enzymes

It is now generally believed that cobalt(II) is an intermediate in coenzyme B_{12} catalysis in enzymic reactions[8,21]. Hence we must accept homolytic fission and the generation of diradicals in the catalytic sites of these enzymes. Such radicals have been seen as major species, up to 90 per cent of the total enzyme, in diol dehydratase, ethanolamine deaminase, ribonucleotide reductase and even in methyl-transfer enzymes. Apart from the nature of the radical centres, one of which is cobalt(II) and the other in some way associated with the ribose fragment which had been bound to the cobalt(III), the considerable separation of the two radicals as indicated by an examination of their e.p.r. spectra, is quite striking. It would appear that homolytic fission is followed by a separation of the radicals through some conformational change of the enzyme. As has been stressed, change of redox state will generate a change in conformation.

In an earlier section the reaction of P-450 has been described as involving heterolytic fission of a metal–nitrogen bond on binding of carbon monoxide or more particularly on the binding of molecular oxygen. This is another type of reaction which can be associated with large conformation changes in ligand and protein. In yet other cases a ligand may be completely rejected, e.g. on reduction of aquo-methemoglobin. a water molecule is rejected and there is a very large conformation change in the protein.

Redox changes in enzymes

Table 2 lists some of the most important changes of oxidation state in enzymes. Each change is expected to force some change of geometry at the

Table 2. Changes of redox state in enzymes

Metal	Redox states	Enzymes
Copper	Cu(I), Cu(II)	Most copper-containing enzymes
Iron	Fe(II), Fe(III)	Cytochromes
	Fe(III), Fe(IV), Fe(V)	Peroxidase, catalase
Molybdenum	Mo(III), to Mo(VI)	Nitrogenase
Cobalt	Co(I)?, Co(II), Co(III)	B_{12}-requiring enzymes
Manganese	Mn(II) to Mn(IV)?	Photosynthetic enzymes

active site, e.g. Cu(I) prefers tetrahedral to tetragonal, Cu(II), geometry. Again on change of redox state ligand partners may change. For example in peroxidase

$$\text{Fe(II)}\cdot\text{no sixth ligand}; \quad \text{Fe(III)}\cdot H_2O; \quad \text{Fe(IV)}\cdot O^{2-}$$

a similar sequence may apply in manganese and molybdenum enzymes.

In the case of cobalt in vitamin B_{12} coenzyme we expect the sequence to be

$$\text{Co(III)–carbanion plus imidazole; Co(II)–imidazole; Co(I)}\cdot\text{no ligands.}$$

These changes in partner must have profound effects on the geometry of the protein associated with the metal as we shall see. Even if there is no change in bond angle or of ligand partner bond lengths must alter. Such changes in stereochemistry introduce much of the opposing activation energy to homogeneous redox reactions. A good catalyst must reduce this energy and an enzyme may well do so by utilizing a large part of the protein to partition energy effectively at the active site. Similar observations apply to acid–base (substitution) reactions in the next section.

Zinc sites

Carbonic anhydrase

This enzyme has zinc at the active site. Some data on metal substitution are given in *Table 3*. Only cobalt(II) and zinc(II) are of high activity but the manganese(II) enzyme shows some activity. The physical properties indicate that 4(5) coordination is required for activity. Interest then centres on the function of the water molecule held in the coordination sphere for water is a *substrate* of this enzyme. It is still not known if the metal-bound water is activated or if the reactive water is in the second coordination sphere. Work on proton relaxation suggests that it is unlikely that coordinated water takes part in the reaction[22]. The mechanism discussed by Dennard and Williams[23] for model catalysts and invoked by Khalifa[24] for carbonic anhydrase seems plausible.

Table 3. The properties of different metallo-carbonic anhydrases

	Mn(II)	Co(II)	Ni(II)	Cu(II)	Zn(II)
Method of study	n.m.r.	e.p.r., or absorption spectra	—	e.p.r., or absorption spectra	crystal structure
Geometry observed	Four-coordinate	Four- (or five) coordinate (tetrahedral)	—	Four- (or five) coordinate (planar)	Four-coordinate (tetrahedral)
Activity	5%	75%	Very low	zero	100%

Note: None of the geometries are supposed to be regular but the observed differences between the metals are well beyond experimental error. The metals are open-sided with one or possibly two water molecules.

Carboxypeptidase

The zinc site of carboxypeptidase is still a difficult problem. *Table 4* gives a list of observed physical properties of different metal ion forms of the enzyme[1]. It is clear that the coordination number can vary from four to

Table 4. The properties of different metallo-carboxypeptidases

	Mn(II)	Co(II)	Ni(II)	Cu(II)	Zn(II)
Method of study	n.m.r. on $H_2O \cdot F^-$	e.p.r., Absorp. spectra	Absorp. spectra	e.p.r., Absorp. spectra	Crystal structure
Geometry observed	Five coordinate	Four- or five coordinate (tetrahedral)	Six coordinate	Four- or five coordinate (planar)	Four coordinate (tetrahedral)
Activity	4%	75%	50%	zero	100%

Note: None of the geometries are supposed to be regular but the observed differences are well beyond experimental error. It is noteworthy too that different metals discriminate in favour of particular types of substrates[1]. The metals are open-sided with one, two or three water molecules.

six and that here the activity of the enzyme cannot be simply related to the observed ground state geometry of the free enzyme or the enzyme bound to inhibitors. Thus the nickel enzyme described by Gray *et al.* apparently has an octahedral structure[25]. Curiously, its spectrum is almost identical with that of $[Ni(H_2O)_6]^{2+}$ while one expects it to be shifted to much higher energy if two nitrogen ligands are bound, *Table 5*. Thus although the symmetry is not unusual the bond lengths are strange if the same coordination groups are used as in the zinc enzyme.

The fact that there is so little change in activity for such a large geometry change and indeed a large change in the number of water molecules in the active site, means that the role of the metal is obscure. Much more experimental work will be necessary before this enzyme can be understood.

The entatic state

The general hypothesis presented by Vallee and Williams[2] was that metal ion sites in enzymes would be found to be of irregular coordination geometry. It is clear from the above and earlier data that this is the case. The general

Table 5. The absorption spectra of some nickel complexes

Nickel complex	Absorption maxima (nm)		
$Ni(H_2O)_6$	850	1350	2530
Ni in MgO	860	1380	2450
$Ni(H_2O)_4$ethylenediamine	980	1580	2680
$Ni(H_2O)_2$(glycinate)$_2$	1000	1600	2700
$Ni(H_2O)_2$(ethylenediamine)$_2$	1050	1790	2810
Ni · carboxypeptidase	900	1460	2427

Note: The closest parallel exists between nickel in carboxypeptidase and in water and nickel oxide. In complexes with nitrogen ligands the nickel bands move to considerably higher energy

advantage of such a situation in catalysis had been described earlier and was elaborated in diagrammatic form in later publications, see *Figure 5*. It is supposed that the catalyst is in an energized ground state prior to contact with the substrate. If this is the case, it should be possible to generate similar catalyst centres in other *organized* phases for the raising of the ground state is brought about by the overall packing of a large assembly. Away from the active site region the packing of the protein has stabilized the system as a whole and it is the conflict between the stereochemical demands of many centres which has elevated the catalytic site. In what follows we shall look at other *organized* states of matter before asking how we can find out more about them.

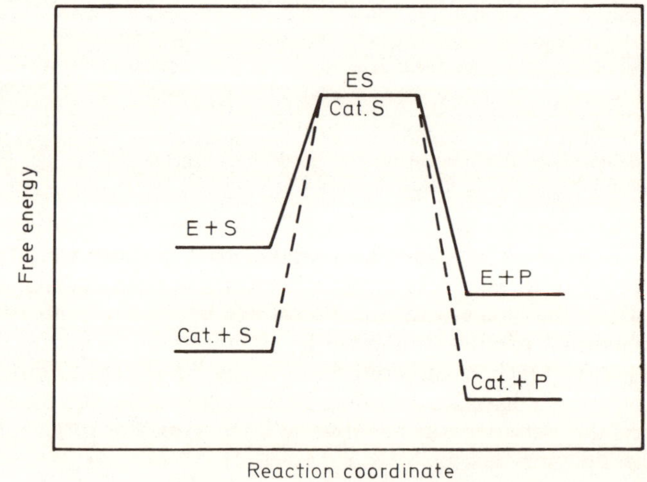

Figure 5. A diagrammatic representation of the way in which the ground state differs in an enzyme catalyst, E, and a homogeneous catalyst, Cat. A heterogeneous catalyst is to be likened to an enzyme in that it too has a promoted energy due to the nature of its highly structured matrix

SOLID STATE CATALYSTS

In an article by Boudart[26] it is pointed out there are many examples of *structure insensitive* bulk *metal* catalysts. In these cases it is of no consequence if the sizes of the metal particles are large or small. It has often been suggested that there is another class of metals for which the metal particle sizes are very important and that there are special catalytic sites in these metal catalysts. Boudart points out that the existence of such sites has never been proved. Turning to catalysts of the semi- or non-conducting metal-compound type a very different picture of catalysis has emerged. Very instructive examples are provided by various metal oxides. *Table 6*. Burwell and co-workers have

Table 6. Dependence of catalysis on surface states

Catalyst	Form used	Catalysis	Ref.
Cr_2O_3	Amorphous	High	1
	Microcrystalline	Low	
α-$TiCl_3$	0001 plane	Low	
	1010 plane	High	2
	Screw dislocation	High	
Zeolites	No long-range order	High	3
Al-doped SiO_2	γ-ray treated	High	4
	Untreated	Low	
MgO	Neutron or u.v. treated	High	5
	Untreated	Low	

[1] R. L. Burwell, G. L. Haller, K. C. Taylor and J. F. Read. *Adv. Catalysis* **20**, 1 (1969).
[2] E. J. Arlman, *J. Catalysis*, **3**, 89 (1964); M. Boudart, *Amer. Scientist*, (57), 1, 97 (1969).
[3] J. A. Rabo, C. L. Angell, P. H. Kasai and V. Schomaker, *Disc. Faraday Soc.* **41**, 328 (1966).
[4] H. W. Kohn, *J. Catalysis*, **2**, 208 (1963).
[5] J. H. Lunsford and T. W. Leland, *J. Phys. Chem.* **73**, 130 (1969).

shown that a Cr(III) oxide gel becomes active only on heating the gel to 400°C. The heating creates a number of *coordinatively unsaturated* acid–base pairs, $Cr^{3+} \cdot O^{2-}$. The availability of these sites is highly dependent upon the size and shape of the particles. In the case of MgO, *Table 6*, the catalytic sites are the imperfect corners of octahedra of MgO_6 (NaCl structure). These un-neutralized corners are in fact OH^- groups which undergo exchange. In yet other catalysts of this type it has been shown that trapped radicals (V-centres) are the source of the catalysis, compare B_{12} enzymes.

Generalizing from this limited range of results we see that catalysis owes itself to an especially energized centre. This energy can result from defects during crystallization, crystal growth along a screw axis, or the unavoidable appearance of corners. The first two cases are kinetically stable perhaps but the third is thermodynamically unavoidable. Further ways of generating the energy are by exposure of crystals to radiation at such a temperature that they cannot recover e.g. γ-rays, neutrons, or light. We have already postulated that biological systems can have similar states, arising from energy partition between the large protein and the smaller active site.

A further lesson can be learnt from the solid state. Thomas and his co-workers[27] have surveyed the literature and summarized their own experience on enhanced *intrinsic reactivity* at structural faults in solids. They state, 'It has long been recognized that emergent dislocations and related defects, function as centres of enhanced reactivity at the surfaces of a wide range of solids'. The implication here is that the excess *ground-state* energy of the defects causes them to decompose more readily. In the case where adsorption occurs at such a site (but now the reaction is that of an *absorbed* molecule) it is the *ground-state* energy of the defect which causes heightened catalysis. Elsewhere Vallee and Williams[2] have termed such a condition of a catalyst centre 'an entatic state'. Even in *Table 6* we note the elevated energy of the MgO_5OH unit of MgO catalysts and the V-centre radicals. Should they not be compared with the reactivity of zinc in carbonic anhydrase and of the enzyme radical in the iron-containing ribose reductase?

Homogeneous catalysis

If we consider homogeneous catalysis then it must suffer from one disadvantage—the inability to build-in relatively large 'strain' energy at one site through cooperative stability which gives only a small energy improvement per atom but a large one per assembly. Thus the strain must be put in by building ligands for the central atom which are rigidly fixed so that strain cannot be relieved. To some extent we see this principle applied in co-enzymes which contain metal ions. The hole at the centre of chlorophyll and porphyrin has so developed that it cannot be adjusted to fit high-spin iron(II) or magnesium(II). The general idea is to build the strain into covalent bonds which have little (kinetic) chance of reassembling in an alternative geometry. Unfortunately this has the disadvantage that there is no flexible matching of the four different periods in the reaction ground state, bound substrate, bound intermediate (or bound transition) states and bound product. Homogeneous systems are inherently at a disadvantage.

Enzymic catalysis

If our view of the nature of enzymic catalysis is correct then the energetics of catalysis could involve a relatively large fraction of a protein. The situation arises through the *cooperative* energies of a large number of interacting centres. Consider the case of the catalytic oxygen atom of MgO. This oxygen is reduced to —OH during the course of the reaction. Now replacing O^{2-} by OH^- means that all the oxygen and magnesium atoms near the site will readjust their bond distances (slightly). In the case of a metal (or nonmetal) atom which alters its state in a protein the adjustment is likely to be much greater and to be open to inspection at much greater distances from the metal for the cooperative interactions in the protein are much less 'structured' than those in MgO. How shall we discover if such changes do occur? What further advantages does a protein assembly gain from the additional flexibility it has relative to a heterogeneous catalyst? Moreover if we are correct in likening a protein to a solid state structure then we should make a much more general enquiry into the nature of cooperativity in phases.

COOPERATIVITY

From this discussion we see that the great distinction between biological systems and true homogeneous solutions lies in cooperativity. Cooperativity in chemistry is often thought to belong to the highly organized solid phase, e.g. the cooperative energy of a sodium chloride crystal. While this is not the case such simple systems give us insight into the nature of cooperative interactions. What striking properties are generated by cooperativity in a lattice? Clearly consideration of energy belongs to the whole lattice and not to individual atoms so that reactivities at certain points arise from interactions in the whole system. As a consequence, and although in a very simple phase we can usefully describe the energy change per mole, in a very complex solid phase the partitioning of energy to the individual atoms is impossibly difficult—so it is in a protein. Thus although a protein is not a cooperative unit of simple repeating units of almost indefinite extent, like NaCl, but is a complex assembly ABCDEFG ... etc (no repeats) of definite micro-volume, yet it is still a highly cooperative entity—it gives only one stable structure. Such cooperative microphases are of extreme interest for their physico-chemical properties are only somewhat like those of macrophases such as NaCl. They are in fact more similar to the surface layers of crystals which have not yet been satisfactorily treated by thermodynamics. The parallel is not only in the limited extent of the two phase systems but also in their irregular structures and in the mobilities of centres in the structures. It may be for these reasons that both give similar catalytic sites in enzymes and in heterogeneous surface catalysts.

Unlike a repeating lattice such as MgO where the effects of cooperativity can only be seen at the surface, because the interior cannot be reached, a protein is a more open structure and any part of it can have enhanced reactivity or much reduced reactivity. A simple consideration illustrates again how this can be understood at the structural level as seen in macrophases. When NaCl molecules enter a lattice there is a complete change in the structure. The sodium ions are surrounded by six chloride ions, the Na^+—Cl^- bond distance is totally different from that in the NaCl molecule. Its reactivity must obviously be different from NaCl molecules. Yet our development of considerations of reactivity of amino acid groups (and metal ions) in proteins has been from simple molecular models—solution chemistry. In so far as a protein is a *large* molecule the cooperative energy can be such as to alter bond angles and bond lengths in familiar chemical groups so as to make them hyper-reactive. Can we detect the structural features by physical measurements? The early part of this paper describes how we can see such factors through the use of many types of spectroscopy in certain metallo-enzymes. Can we detect them in proteins which have no metals? We return to this point later.

We have now seen how cooperativity adjusts the ground state energy and structure of a large group of atoms. A third feature of cooperativity is seen in the effect of temperature changes, which produce phase changes. A phase change of a large amount of a solid crystalline material occurs at a fixed temperature, all the material melts at once, e.g. NaCl. Now, gross conformation changes of proteins or of DNA (denaturations), unlike an equilibrium between two conformations of a small molecule, also occur over a very short temperature range, indicative of the fact that the heat and entropy change for a change of state of a protein molecule and a DNA molecule are large and intermediate between those of macroscopic crystals and isolated molecules.

The denaturation is a type of phase change. All these large molecular weight systems have cooperative structures. But as this is the case we must be able to construct other systems of any degree of cooperativity (molecular weight and structures), so that the melting of a solid is only one extreme and protein denaturation another. Amongst such systems lies a whole world of biological phases, such as membranes, protein–protein connections, protein–membrane interactions, and the whole realm of surface chemistry. A thorough re-analysis of the phase relationships of such bodies is therefore required. I believe that this will show that we must not restrict ourselves by discussing solids, liquids and vapours but must accept a vast range of partially organized phases between crystalline three-dimensional regular lattices and liquids. Phases can be of one or two dimensions, of small volume, and of a great variety of restricted mobilities. The properties of these phases will be different from our present expectation. They need not show *very* sharp phase changes, and the condition for a phase change will depend on the total amount of material in the system. They will show highly selective mutual solubility for all mutual solution must now be highly cooperative; there can be no relevance for ideal laws of solution. This may well be the explanation of the association of particular molecular species in well defined membranes in biology. Likewise the sensitivity of such structured cooperative systems to impurities and foreign chemicals could be quite unexpected, for the impurities could interact through the generation of phase changes, each impurity centre acting cooperatively with the next. Just as it is not to be expected that we shall be able to give a detailed atomic level description of protein energetics so these phases will not be open to discussion in the language of small gas molecules. A first need then is for an appropriate thermodynamic framework, see reference 28 for example.

I note in passing that all kinds of devices may be built into biological reactions through cooperativity, e.g. energy conservation can be action at a distance, memory can arise through the association of 'phase' changes with ion movements. After all memory in a computer is nothing more than a 'phase change' in magnetization. Memory may well be a property of the steady states achieved by action involving any large component of a biological cell system because of the nature of biological microassemblies—they are cooperative phases.

It follows too that there is no single critical point which differentiates solids from liquids. The idea of a melting point independent of the amount of material comes from lattices of infinite extension. We must expect phases in which the fluctuation of cooperative ordering makes us treat them as 'liquids' to some degree and 'solids' to another. In this light our recent findings[29] that molecules such as adenosine-5′-phosphate, (AMP), adenosine-5′-diphosphate (ADP) and adenosine-5′-triphosphate have '*structures*' in aqueous solution means that we must expect '*structures*' for growing nucleotide (and protein) chains, throwing light on a further problem in biology—the unfolding and folding of macromolecules. In solid state chemistry the idea of 'microdomains' has often been canvassed. Liquids can obviously have microdomains though their fluctuations may be quite rapid.

Returning to the problem of protein reactivity and supposing that we accept the idea that a protein phase is somewhat intermediate between a solid and a liquid with a high degree of cooperative energy and considerable fluctional mobility, it is very easy to see how the problem may have been misconstructed. Consider electron transfer in cytochrome-*c* or in any other electron-transfer protein[3, 13, 30, 31]. If the protein is anything like a crystal it is hard to conceive how electron transfer of low activation energy is possible. Electron transfer between the iron centres would be easy if these centres existed in free solution but the protein is obviously much more sterically-hindered than is true of simple small complex ions. If, however, the whole pocket in which the heme-group sits has considerable fluctional mobility then the expected barriers of the solid state to electron transfer disappear, for normally in a solid the cooperativity prevents easy electron transfer while in such an irregular cooperative phase as cytochrome-*c* it could well be that its structure aids electron transfer. Moreover there is evidence that cytochrome-*c*, like many proteins, has a high fluctional mobility for in both the oxidized and reduced states proton exchange with nearly all the peptide bonds is fast. Water can penetrate deeply into the protein matrix. Cytochrome-*c* is but one example of a highly cooperative phase with no long-range order which repeats. Our problem now becomes that of devising methods of sufficient sensitivity to study this type of phase, whether it be a protein, a membrane, or DNA.

NEW N.M.R. METHODS

We have published recently new procedures for the study of protein n.m.r. spectra[32]. They are two types of difference spectra and exemplify a great variety of difference spectra which are now possible in n.m.r. In the first case a difference is taken between the Fourier transform of the n.m.r. spectrum and the same Fourier transform multiplied by a line broadening function. Such subtraction makes for an increase in the effective resolution at the expense of signal strength. The enhanced spectrum of hen egg white lysozyme is shown in *Figure 6*. Utilizing a variety of relatively conventional procedures the peaks have been assigned and we can in fact assign nearly all aromatic and methyl residues. At this stage it is possible to follow the response of the methyl and aromatic proteins to changes in pH, to metal ion binding, or to inhibitor and (in principle) substrate binding. All of these experiments have been carried out. The titration of the carboxylates at residues 35 and 52, or their binding by Ca^{2+} or La^{3+} causes a movement in tryptophan 108 which we can follow readily. The binding of tri-*N*-acetylglucosamine, an inhibitor, has a different effect. However, it will now be possible to follow such effects deep into the lysozyme molecule, testing if this protein is somewhat flexible. The exact functional significance of the limited flexibility already proved cannot be assessed yet but it is tempting to suppose that the enzyme utilizes energy partition in the way described here.

In still larger assemblies this energy partition becomes the method for communication between different proteins so that cooperativity extends over a considerable volume. Here we stray from catalysis to the nature of life—another subject.

Before closing I wish to speculate about the possibility of utilizing the ideas discussed here in synthetic systems, made by man.

Figure 6(a)

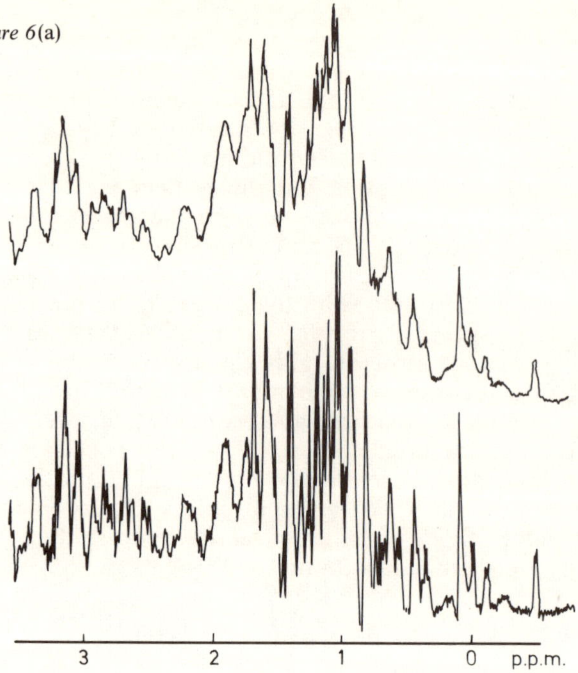

Figure 6(b)

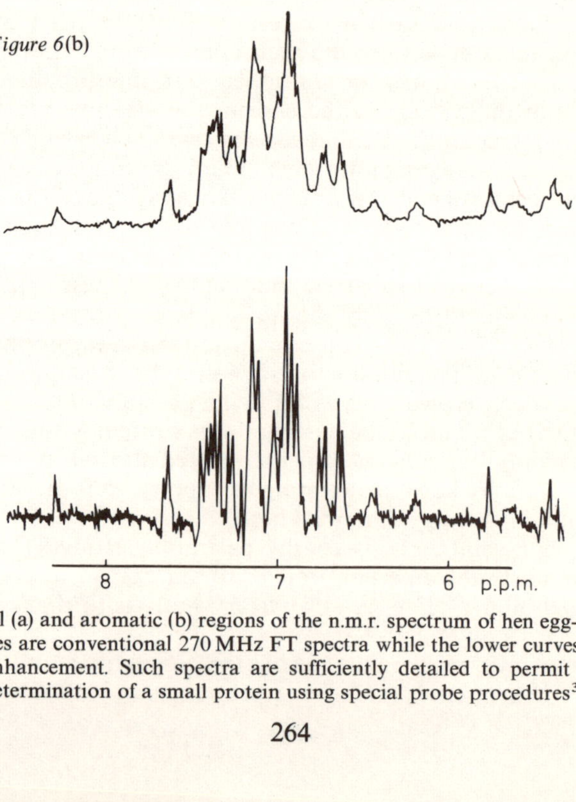

Figure 6. Methyl (a) and aromatic (b) regions of the n.m.r. spectrum of hen egg-white lysozyme. The upper curves are conventional 270 MHz FT spectra while the lower curves show the effect of resolution enhancement. Such spectra are sufficiently detailed to permit a full structure determination of a small protein using special probe procedures[32]

SYNTHETIC POLYMERS AS CATALYSTS

The above discussion leads to a suggestion for the structure frame which must be generated in synthetic polymers if we are to make them ourselves. Firstly they must be built so that (in our chosen solvent) they will have a maintained conformation overall. The ground rules for doing this are known from protein chemistry. There must be relatively strong interactions between at least some of the units in the polymer. In part this maintains solubility too. Secondly we do not need to have specially designed reactive centres as in homogeneous catalysis for the energy of the polymer conformation will generate these centres. It follows that it is no use using a polymer from a single monomer; we must copolymerize or even mix several monomers with acidic and basic sidechains. If we choose the units correctly we do not need to work at such low temperatures as biological systems so we do not need to have as highly activated groups as in those systems. A starting point would be the combination of leucine, glutamic acid, and lysine. Various polymers of molecular weight around 10000 could be made and their structures studied. Their activity as esterases at very different pH values could then be evaluated. This would be a hit-and-miss research effort but this is the way in which I believe biological catalysts have evolved.

I wish to record my indebtedness to the members of my own research group, and to my fellow-members in the Oxford Enzyme Group.

REFERENCES

1. B. L. Vallee and W. E. C. Wacker, *The Proteins*, Vol. 5, Ed. H. Neurath, Academic Press, New York (1970).
2. B. L. Vallee and R. J. P. Williams, *Proc. Nat. Acad. Sci. Wash.* **59**, 498 (1968).
3. R. J. P. Williams, *Biochem. Soc. Trans.* **1**, 1 (1973).
4. L. H. Jensen, L. C. Sieker, K. D. Watenpaugh, E. T. Adman and J. R. Herriott, *Biochem. Soc. Trans.* **1**, 27 (1973).
5. L. E. Mortensen, W. G. Zumft, T. C. Huang and G. Palmer, *Biochem. Soc. Trans.* **1**, 35 (1973); R. R. Eady, B. E. Smith, R. N. F. Thornley, D. A. Ware and J. R. Postgate, *Biochem. Soc. Trans.* **1**, 37 (1973).
6. M. Kelly and G. Lang, *Biochim. Biophys. Acta*, **223**, 86 (1970).
7. A. Ehrenberg and P. Reichard, *J. Biol. Chem.* **247**, 3485 (1972).
8. S. A. Cockle, H. A. O Hill, R. J. P. Williams, S. P. Davies and M. A. Foster, *J. Amer. Chem. Soc.* **94**, 275 (1972).
9. H. Beinert, N. R. Orme-Johnson and R. E. Hanson, *Biochem. Soc. Trans.* **1**, 32 (1973).
10. P. F. Hollenberg and L. P. Hager, *J. Biol. Chem.* **248**, 2630 (1973).
11. H. A. O. Hill, A. Roder and R. J. P. Williams, *Struct. Bonding (Berlin)*, **8**, 123 (1970).
12. B. Dunford and R. J. P. Williams, to be published.
13. J. R. Mckellar, J. A. Weightman and R. J. P. Williams, *Disc. Faraday Soc.* **51**, 179 (1971).
14. K. J. H. Van Buuren, P. F. Zuurendonk, B. F. Van Gelder and A. O. Muijsers, *Biochim. Biophys. Acta*, **256**, 243 (1972).
15. R. J. P. Williams, in *Molybdenum*, Chem. Soc. London (1974).
16. L. Morpurgo, C. Giovagnoli and G. Rotilio, *Biochim. Biophys. Acta*, **322**, 207 (1973).
17. See reference 32 for description of improved resolution of n.m.r. spectra by new techniques.
18. R. J. West and S. F. Lincoln, *Inorg. Chem.* **12**, 494 (1973).
19. W. Byers, G. Curzon, K. Garbett, B. E. Speyer, S. N. Young and R. J. P. Williams, *Biochim. Biophys. Acta*, **310**, 38 (1973).
20. R. J. P. Williams, *Biol. Revs. Cambridge Phil. Soc.* **28**, 381 (1953).
21. R. H. Abeles, *Biochem. Soc. Trans.* **1**, 39 (1973).

[22] A. Lanir, S. Gradsztajn and G. Navon, *FEBS. Letters*, **30**, 351 (1973).
[23] A. Dennard and R. J. P. Williams, in *Transition Metal Chemistry*, Vol. 2, p. 115. Ed. R. L. Carlin, Dekker, New York (1966).
[24] R. G. Khalifah, *J. Biol. Chem.* **246**, 2561 (1971).
[25] R. C. Rosenberg, C. A. Root, R. Wang, M. Cerdino and H. B. Gray, *Proc. Nat. Acad. Sci. Wash.* **70**, 161 (1973).
[26] M. Boudart, in Robert Welch Foundation Conference on Chemical Research, No. XIV, *Solid State Chemistry*, p. 299 (1973).
[27] J. M. Thomas, E. L. Evans and J. O. Williams, *Proc. Roy. Soc.* (*London*) **A331**, 417 (1972).
[28] G. Kemeny and I. M. Goklany, *J. Theoret. Biol.* **35**, **40**, 107 (1973).
[29] C. D. Barry, A. C. T. North, J. A. Glasel, R. J. P. Williams and A. Xavier, *Nature* (*London*), **232**, 236 (1971).
[30] M. V. Volkenstein, *J. Theoret. Biol.* **34**, 193 (1972).
[31] F. S. Dainton, *Ber. Deutsch. Chem. Ges.* **75**, 608 (1971).
[32] I. D. Campbell, C. M. Dobson, R. J. P. Williams and A. Xavier, *J. Magnet. Res.* **11**, 172 (1973).

KINETICS OF INTRAMOLECULAR AND INTERMOLECULAR REACTIONS INVOLVING TWO FUNCTIONAL GROUPS ATTACHED TO POLYMERS

H. MORAWETZ

Department of Chemistry, Polytechnic Institute of Brooklyn, Brooklyn, N.Y. 11201, USA

ABSTRACT

Three types of investigations are described. (a) When a polymer chain carries reactive substituents, the reaction rate may be increased by very large factors if these groups are sufficiently close to each other. More distant groups interact at rates governed by the chain flexibility. A technique is described for studying the effect of the excluded volume on the probability of cyclic chain conformations. (b) When reactive and catalytic substituents are attached to different polymer chains, their reaction should be hindered by the resistance of polymer coils to mutual interpenetration in good solvent media. Two theoretical approaches to this 'kinetic excluded volume effect' are described. Experimental data do not conform to the expected behaviour and the reason for this discrepancy is unknown. (c) Crosslinked polyacrylamide beads carrying nucleophilic substituents were evaluated as catalysts for the solvolysis of nitrophenyl esters in a solution phase. Under certain conditions, groups attached to the gel were much more efficient catalysts than their analogues in homogeneous solution. This effect was ascribed to a cooperative action of two catalytic groups.

INTRODUCTION

It is a widely held generalization that functional groups attached to macromolecules have the same reactivity as their low molecular weight analogues. This is a useful concept as a starting point for the analysis of kinetic data involving reactions of polymeric reagents with low molecular weight species. If the polymer behaves very differently from its analogue, this difference may be the result of several factors. (a) The sensitive bond may be attached very close to the backbone of the polymer and the resulting steric hindrance may then not be easily simulated in a molecule of low molecular weight. This would account, for instance, for the well-known difficulty of the solvolysis of poly (methacrylic acid) esters. (b) The 'effective solvent medium' in the neighbourhood of a polymer chain may be significantly different from the solvent in bulk. This explanation has been invoked, for instance, to explain the fact that poly(*N*-isopropylacrylamide) or poly-(*N*-vinylacetamide) undergo isotopic hydrogen exchange much more slowly than their low molecular weight analogues[1,2]. (c) If a low molecular weight reagent is drawn into the domain of a macromolecule by long-range Coulombic interactions[3,4] or by hydrophobic bonding[5,6], the rate of its reaction with a group attached to the polymer may be many orders of magnitude above that characterizing an analogous process between two small molecules.

In this report I shall be concerned with a different class of effects resulting from the macromolecular nature of a chemical reagent, that is, with effects observed when two interacting groups are both appended to macromolecules. Three cases will be discussed, (a) reactions involving two interacting groups attached to the same chain backbone, (b) reactions involving two groups attached to different macromolecules in highly dilute solutions and (c) co-operative catalytic effects of groups attached to a crosslinked gel interacting with a small molecule.

KINETICS OF REACTIONS INVOLVING INTERACTIONS OF TWO GROUPS ATTACHED TO THE SAME POLYMER CHAIN

Intramolecular reactions of two groups carried by the same chain molecule are of two types. In the first class, interaction of groups carried by substituents of two neighbouring monomer residues of a vinyl polymer is heavily favoured since it leads to a 5- or 6-membered cyclic structure in the transition state. A good example of a process of this type is the hydrolysis of acrylic acid copolymers with small concentrations of an aryl acrylate or methacrylate. This reaction proceeds by the mechanism

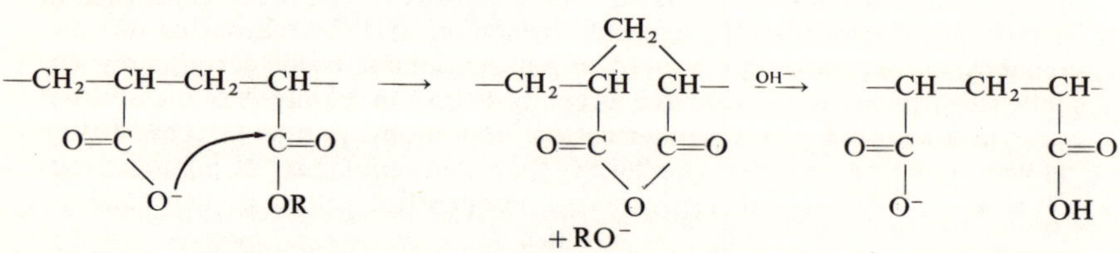

so that the solvolysis rate is exactly proportional to the fraction of the carboxyl groups which is ionized[7]. A comparison of this intramolecular process with the analogous bimolecular reaction, i.e. the attack of acetate on aryl esters leads to the following important observations:

(i) An intramolecular process corresponds to a very high 'effective local concentration', C_{eff}, of the carboxylate at the reaction site. For instance, the data of Gaetjens and Morawetz[7] show that the intramolecular process in phenyl succinate is 4500 times as fast as for phenyl acetate in 1M acetate solution, i.e. C_{eff} is 4500.

(ii) The probability of a neighbouring group attack is highly sensitive to the size of the cyclic structure in the transition state and to a variety of steric restraints. For instance, glutaric anhydride is formed two orders of magnitude more slowly from glutaric monoesters than succinic anhydride from succinic monoesters[7], and ester displacement from the monoester anion

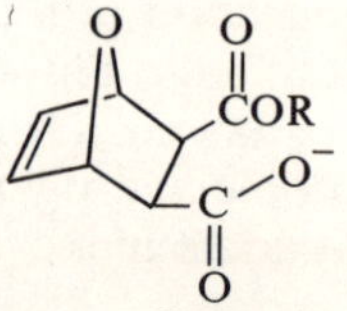

takes place 230 times as rapidly as for the corresponding succinic acid monoester[8].

(iii) The ease with which the carboxylate can attack the neighbouring ester group may be highly sensitive to the relative steric configuration of the carbons to which the carboxylate and ester groups are attached. Thus, the solvolysis of methacrylic acid copolymers carrying a small number of aryl ester groups deviates strongly from first-order kinetics, since neighbouring group interactions in the sequences

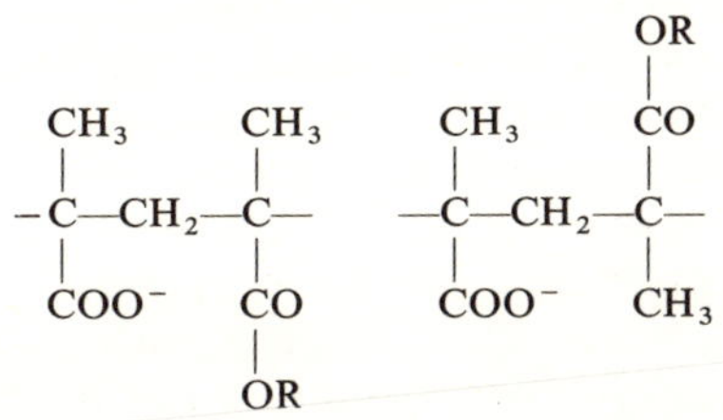

are characterized by rate constants differing by an order of magnitude[9].

A very different situation arises if the potentially interacting groups are at a larger distance from one another so that their interaction would lead to larger cyclic structures. It is well known that 7–12 membered ring formation involves large steric strain. For very large rings a theoretical estimate of the probability of a cyclic conformation may be obtained in the absence of an excluded volume effect. Using the model of a freely jointed chain with Z links of length b, this probability corresponds to an effective local concentration, C_{eff}, of the second chain end in the vicinity of the first chain end[10]

$$\begin{aligned} C_{\text{eff}} &= (\bar{N}_A/1000) \lim_{h \to 0} W(h)\, \mathrm{d}h/4\pi h^2\, \mathrm{d}h \\ &= (\bar{N}_A/1000)(3/2\pi\langle h^2\rangle)^{\frac{3}{2}} \end{aligned} \quad (1)$$

Here C_{eff} is expressed in units of mol l^{-1}, $\bar{N}_A$ is Avogadro's number, $W(h)$ is the probability distribution function of the end-to-end displacement h and $\langle h^2\rangle = Zb^2$ is the mean value of h^2. If two interacting groups are attached to the chain ends, the rate constant k_1 governing the intramolecular reaction will be

$$k_1 = (\bar{N}_A k_2/1000)(3/2\pi\langle h^2\rangle)^{\frac{3}{2}} \equiv C'Z^{-\frac{3}{2}} \quad (2)$$

where k_2 is the second-order rate constant characterizing the analogous intermolecular reaction†. For real polymer chains in θ-solvents where $\langle h^2\rangle = C_\infty Zb^2$ (with the 'characteristic ratio' C_∞ reflecting short-range interactions governing the chain flexibility[11]), equation 2 should be modified to

$$k_1 = C''Z^{-\frac{3}{2}}; \qquad C'' = C'/C_\infty^{\frac{3}{2}} \quad (3)$$

No corresponding analysis has been carried out for the rate of cyclization of chains if the excluded volume effect is operative. This will obviously

† This treatment assumes that the cyclization has a substantial activation energy, so that the rate of the process is not controlled by the *rate of conformational transitions* of the polymer chain.

decrease the probability of cyclic conformations and will lead to a steeper decline of k_1 with increasing Z. Computer generation of chains on a tetrahedral lattice has led to the estimate that k_1 should become proportional to Z^{-2} when the excluded volume effect is large and the chains are long[12] but no direct verification of the value of this 'chain closure exponent' exists.

Experimental studies designed to check the applicability of equation 1 for polymers in θ-solvents or the dependence of the cyclization rate on chain length in better solvent media encounter formidable difficulties. For straightforward interpretation of experimental data we should require not only monodisperse polymers, with each chain carrying the two interacting groups at its two ends, but the kinetics of the interaction of these groups would have to be observed at dilutions such that the intermolecular process becomes negligible. Since it appears at present impossible to meet these conditions, we have approached the problem in an indirect manner[13,14].

We prepared an acrylamide copolymer with a small proportion of a comonomer with a reactive side chain (a *p*-nitrophenyl ester) and a small proportion of a comonomer with a side chain containing a pyridine residue, which is known to catalyse the solvolysis of phenyl esters[15]. In analogy with equation 3 we would then expect the probability of solvolysis of a given ester group to be governed, in a θ-solvent, by a rate constant of the form

$$k_1 = C'' \sum_i{}' x_i^{-\frac{3}{2}} \tag{4}$$

where x_i is the number of bonds separating a given ester group from the ith catalytic pyridine residue attached to the same polymer chain. The summation excludes $x_i < x'$, where x' is the number of bonds in the smallest ring which can form without significant strain. Since the catalytic groups are distributed at random along the chain, each ester will be characterized by a different value of k_1 and the kinetics of the solvolysis will be given by

$$y(t) = \int W(k_1) \exp(-k_1 t)\, dk_1 \tag{5}$$

where $y(t)$ is the fraction of unreacted ester at time t and $W(k_1)$ is the probability distribution function of k_1.

For any given density of catalytic groups along the polymer chain, deviation from first-order kinetics should increase with the breadth of the distribution function $W(k_1)$. This, in turn, should increase with an increase in the 'chain closure exponent' which, as we have seen, has been estimated to increase from a value of $\frac{3}{2}$ for chains with zero volume to a value of 2 when the excluded volume becomes large. It was the main purpose of our study to compare these estimates with experimental evidence.

In analysing kinetic data obtained with polymers carrying substituents on a fraction ω of the monomer units we used a computer simulation technique. We generated 100 chains containing 1000 units with a probability ω that any one unit was catalytic. The reactive group was placed on the central unit of the chain and with the assumption that the catalytic and reactive groups could interact only if separated by ten or more monomer residues, the time dependence of the reaction corresponding to a ring closure exponent a was evaluated by

$$k_n = C\left[\sum_{j=1}^{j=490} p_{nj}(500-j)^{-a} + \sum_{j=510}^{j=1000} p_{nj}(j-500)^{-a}\right] \quad (6)$$

$$y(t) = (1/1000)\sum_n \exp(-k_n t) \quad (7)$$

Here $p_{nj} = 1$ or $p_{nj} = 0$ depending on whether the jth residue of the nth chain is, or is not, catalytic and the constant C is adjusted so as to make equation 7 yield the experimentally observed initial reaction rate. The procedure consists then in comparing the curvature of plots of log y against t when using data computed with different ring closure exponents and the curvature characterizing a similar plot of experimental data. The kinetic pattern observed in a good solvent medium was found to be well fitted by data computed with $a = 2$, that is, the ring closure exponent previously estimated for chains with a large excluded volume by generating chains on a tetrahedral lattice. On the other hand, when the solvent medium was adjusted to approach the θ-condition, the chain closure exponent seemed to be about 1.6, distinctly higher than the $\frac{3}{2}$ value predicted from the model of a random-flight chain with zero volume.

Another interesting conclusion from this study is obtained from $k_1^{\text{init}}/k_2\omega$ where k_1^{init} is the initial apparent first-order rate constant of the intramolecular process. This ratio may be identified as being equal to $C_{\text{eff}}^{\text{max}}$ the effective local concentration of catalytic groups at the site of a reactive group in a chain containing a single reactive residue and catalytic groups carried by all the other monomer units. We found $k_1^{\text{init}}/k_2\omega$ to be in the range 2.7–2.9M—a very substantial value. It is about three times as high as $C_{\text{eff}}^{\text{max}}$, values estimated from association equilibria involving two groups attached to a randomly coiled polymer chain[16].

THE SEARCH FOR A 'KINETIC EXCLUDED VOLUME EFFECT'

The large second virial coefficients characterizing the osmotic behaviour of flexible chain macromolecules in good solvent media have been interpreted as a consequence of a strong resistance of the molecular coils to mutual interpenetration. Some years ago, Morawetz and Song[17] suggested that the large excluded volume implied by the thermodynamic behaviour of dilute polymer solutions should also lead to pronounced kinetic effects. Specifically it was argued that the reaction rate in a highly dilute solution of two similar macromolecules carrying a small number of reactive or catalytic substituents, respectively, should be much smaller, in a good solvent medium, than the reaction of low molecular weight analogues of the interacting species. Qualitatively, such a result seemed plausible since the molecular chains were unlikely to penetrate each other, so that only the reactive and catalytic groups near the 'surface' of the coiled macromolecules at any given time would be accessible to each other. This concept also led to the expectation that the observed rate constant would decrease rapidly with the length of the chain to which the interacting groups are attached. Surprisingly, experimental results with systems containing mixtures of acrylamide copolymers carrying a small number of p-nitrophenyl ester groups or catalytic imidazole residues failed to reveal the expected 'kinetic excluded volume effect'.

We have now formulated a theory of this effect based on two models[18]:

(i) In the first model the two macromolecules carrying reactive and catalytic substituents, respectively, are assumed to have the same molecular volume v_2 and their randomly coiled chains are represented by spherical clouds with a uniform chain segment density and a volume v_e. Since the two polymers are very similar to each other, they are assigned the same polymer–solvent interaction parameter χ, while the polymer–polymer interaction parameter is taken as zero.

If a solution with a volume V contains one reactive and one catalytic polymer chain, the probability W_0 that a reactive group is, at any given time, within a volume v_j, jointly occupied by the two interpenetrating spheres is

$$W_0 = \frac{1}{v_e} \frac{\int_V v_j(r) \exp[-\Delta G(r)/kT]\, 4\pi r^2\, dr}{\int_V \exp[-\Delta G(r)/kT]\, 4\pi r^2\, dr} \tag{8}$$

where r is the distance between the centres of the spheres representing the polymer coils and $\Delta G(r)$ is the increase in free energy resulting from the interpenetration of the two spheres. If the system contains n_c catalytic chains and only binary interactions need to be considered, this probability becomes $W = n_c W_0$. Whenever a reactive group finds itself within a volume jointly occupied by the equivalent spheres of a reactive and a catalytic polymer, it is exposed to a local concentration c_l of catalytic groups which is related to their stoichiometric concentration c by $c_l = c(V/n_c v_e)$. The ratio of the second-order rate constants k_2 and k_2^0 characterizing a reaction between groups attached to the two polymer chains and a reaction between their low molecular weight analogues should then be

$$k_2/k_2^0 = c_l W/c = (V/v_e)W_0 \tag{9}$$

According to Flory[19], $\Delta G(r) = 2kTv_j(\frac{1}{2} - \chi)(v_2/v_e)^2/v_1$ where v_1 is the molecular volume of the solvent. Since $\Delta G(r) = 0$ for $r \geqslant 2R_e$, where R_e is the radius of the equivalent spheres, and the solution is assumed to be dilute, the denominator in equation 8 may be approximated by V. With $x = r/2R_e$, $v_j = v_e(x^3 - 3x + 2)/2$ for $x < 1$ and $v_j = 0$ for $x \geqslant 1$,

$$k_2/k_2^0 = 12\int_0^1 (x^5 - 3x^3 + 2x^2)\exp[-X(x^3 - 3x + 2)/2]\, dx$$

$$X = (v_2^2/v_1 v_e)(\tfrac{1}{2} - \chi) \tag{10}$$

This result leads to a rapid decay of the k_2/k_2^0 ratio with an increase in the parameter X. For instance, for $X = 1, 2, 3$ and 4 the computed values for k_2/k_2^0 are 0.56, 0.36, 0.25 and 0.19.

(ii) A second approach to an analysis of the 'kinetic excluded volume effect' is based on the transition state theory. Let us consider two polymer chains with a reactive and a catalytic group attached to the central chain segment. It is assumed that the mixing of the two chains is thermoneutral. If we then compare the activation parameters for the formation of the transition state from the two polymeric reagents and their low molecular weight analogues, we find that the only difference is due to the decrease in configurational entropy when the two linear chains combine to a cruciform transition state. Specifically, this treatment leads to

$$k_2/k_2^0 = \Omega^\ddagger/\Omega^2 \tag{11}$$

where $\Omega^\ddagger$ and Ω are the conformational partition functions for the transition state, in which the two macromolecules have their central segments connected to each other, and the linear reagent molecules, respectively.

To get a feeling for the behaviour of $\Omega^\ddagger/\Omega^2$, we generated linear chains and chains connected by links of various lengths by a computer on a tetrahedral lattice. We assumed then $\Omega^\ddagger/\Omega^2 = (f^\ddagger)^2/f^2$ where $f^\ddagger$ and f are the fractions of non-selfintersecting chains representing the bridged transition state and the linear reagent molecules, respectively. The conclusions may be summarized as follows: (a) For long chains, the dependence of k_2/k_2^0 on the number Z of links in the reagent chains is represented by $k_2/k_2^0 = 6.2Z^{-0.56}$ and $k_2/k_2^0 = 7.0Z^{-0.48}$ when the formation of the transition state involves 8-link and 16-link bridges between the reagent chains. (b) If the interacting groups are attached to the chain end, rather than to the central chain segment, the variation of k_2/k_2^0 with Z is much less pronounced, i.e. $k_2/k_2^0 = 1.2Z^{-0.15}$.

The model involving specifically the chain character of the reagents and the bridged transition state seems much more realistic than the spherical cloud model. It is particularly useful in revealing the importance of the location along the chain molecule to which the interacting groups are attached. However, the simplification required for computer simulation, i.e. the neglect of the effect of side chains, of polymer–solvent interaction and of differences in the energy of different chain conformations, makes it impossible to compare the results quantitatively with experimental data.

At any rate, a recent experimental study[19] did not bear out the theoretical predictions. Using a reactive acrylamide copolymer with $\overline{M} = 1.24 \times 10^5$ and catalytic copolymers with $\overline{M} = 1.3 \times 10^5$, 5×10^5 and 15.5×10^5, the k_2/k_2^0 ratios were found to be 1.20, 0.79, 0.79 in water and 0.40, 0.29, 0.18 in glycol. These results are disturbing on several counts. (a) It is not clear why the rate constant involving the reactive and the shortest catalytic polymer in water solution should actually be *higher* than for the reaction of their analogues. (b) The decay of k_2/k_2^0 with increasing chainlength of the catalytic polymer is much less than predicted by equation 10. (c) The k_2/k_2^0 values are lower in glycol than in water and they decay more rapidly with an increasing chainlength of the catalytic polymer, although the intrinsic viscosities are lower in glycol solution, implying a smaller excluded volume. We are at present unable to account for these discrepancies. We plan to carry out similar studies using polyions as carriers of the reactive and catalytic chain substituents, since the excluded volume can then be varied within wide limits by varying the concentration of simple electrolytes in the systems.

BEHAVIOUR OF CROSSLINKED GELS CARRYING CATALYTIC SUBSTITUENTS

In recent years, the use of crosslinked polymer gels as chemical reagents has assumed great importance, particularly for the synthesis of polypeptides[20, 21]. The work carried out in this area has generally employed commercially available resins as starting materials for the preparation of their reactive derivatives and little attention has been paid to the role which the structure of the crosslinked network may play in optimizing the reactivity of functional groups attached to the polymeric skeleton. The study described below was designed to make a contribution towards the clarification of this problem.

We used as a model reaction the solvolysis of *p*-nitrophenyl-acetate (NPA) and -butyrate (NPB) catalysed by crosslinked acrylamide copolymers containing 10 weight per cent of *N*-acrylyl-4-aminomethylpyridine. The resin was packed into a thermostated column and an aqueous solution of the ester was driven through the resin bed at various speeds. From the dependence of the ester conversion on the contact time a first-order rate constant was calculated and the 'catalytic efficiency' of the resin was defined as the ratio of this rate constant to the rate constant expected in homogeneous solution with the same volume concentration of the catalytic monomer. A preliminary report of the results obtained has been published[22]. Here I should only like to comment on two features of the data, the dependence of the catalytic efficiency on the degree of crosslinking of the resin, and the topology of the network structure.

Table 1. Dependence of the catalytic efficiency of acrylamide copolymers carrying pendant pyridine residues on the degree of crosslinking (DC). (All polymerizations carried out with a dilution ratio of 1 : 1 of monomer and water).

DC, %	Swelling ratio	Catalytic efficiency NPA	Catalytic efficiency NPB
0.5	17.0	0.33	0.59
1	12.4	0.74	1.5
2	7.1	1.00	4.5
3	5.6	0.66	2.8
4	5.0	0.21	0.91

Table 1 illustrates how the catalytic efficiency of resins prepared by polymerization of the mixed monomers diluted with an equal weight of water depends on the degree of crosslinking. We see that this efficiency passes through a maximum as the degree of crosslinking and the equilibrium swelling of the resin is being varied. The high values of these maxima are particularly remarkable in view of our observation that analogous soluble copolymers have no significant catalytic action on the ester solvolysis. The optimum catalytic efficiency of 4.5 observed in this series with PNB is particularly striking, since it is too large to be accounted for by the concentration of the ester due to partitioning between the aqueous and the resin phase. We believe that the high catalytic efficiency results from a cooperative attack of two pyridine residues on the ester bond. This effect will become less pronounced with increasing swelling of the resin; at very low swelling, on the other hand, the catalytic efficiency decreases due to an unfavourable change in the solvent medium.

However, it would be a mistake to believe that the swelling ratio of the resin determines uniquely the catalytic efficiency. *Table 2* lists results obtained with three resin samples which had swelling ratios in the narrow range of 11.5–13.1, but which differed in the network structure, since the dilution of the monomer during the polymerization was increased at the same time that the crosslink density was raised. There is a pronounced trend towards higher catalytic efficiencies for heavily crosslinked resins prepared at high monomer dilution.

Table 2. Dependence of catalytic efficiency of acrylamide copolymers carrying pendant pyridine residues on the network structure at similar swelling ratios.

DC, %	Monomer dilution ratio	Swelling ratio	Catalytic efficiency NPA	Catalytic efficiency NPB
1	1	12.4	0.74	1.5
2	2	13.1	1.8	4.3
3	3	11.5	2.6	7.2

CONCLUDING REMARKS

In the preceding discussion I have concentrated on work carried out in our laboratory and I should now like to remark on some related studies carried out elsewhere. Loucheux and Banderet[23] described an investigation of an intramolecular reaction involving two groups appended, at some distance from each other, to a flexible polymer chain. They treated their data as if the interacting groups could be represented as moving freely within the volume of a sphere hydrodynamically equivalent to the polymer coil. This treatment is clearly faulty, since for a series of polymer chains with a constant density of interacting groups the concentration of these groups within the equivalent sphere would decrease with an increasing polymer chainlength, thus leading to the implausible prediction of a decreasing reaction rate. On the other hand, it is not quite correct to say that the rate of intramolecular interaction of the ends of very long molecular chains cannot be observed. Wang and Davidson[24] studied the rate of cyclization of a double helical DNA with complementary single strands extending at the two ends of the molecule and they found that their result conformed to equation 2 although this ought to be valid only in the absence of an excluded volume effect. Apparently, these authors were unaware of the complication which an excluded volume (which ought to be substantial for a chain with a high density of ionic charges) might introduce into the theoretical analysis of the intramolecular process. I should also like to refer to the work of Sisido[25] who has shown that the kinetics of the intramolecular reaction involving reactive and catalytic substituents distributed at random along a chain molecule may be solved analytically, without recourse to computer simulation.

Vollmert and Stutz[26] studied solutions of two similar copolymers carrying groups which reacted with each other to form a covalent bond. They determined the limiting conversion of the interacting groups and tried to interpret such data in terms of the mutual interpenetration of the molecular coils. This treatment cannot be correct, since the formation of a gel structure introduces restraints on the mutual accessibility of the reactive groups which are unrelated to the extent to which the dissolved polymer coils can interpenetrate each other. However, in a study whose results are highly relevant to our investigation, Wetmur and Davidson[27] measured the kinetics of the renaturation of DNA and found that the rate constant governing the pairing-up of complementary nucleotide residues is subject to a 'kinetic excluded volume effect'. This effect was analysed in detail by Wetmur[28]. It would, of course, be expected that such an effect should be particularly large when the two interacting chain molecules are both highly charged.

A cooperative effect of two interacting catalytic groups, which we invoked to explain the high catalytic efficiency of gels carrying nucleophilic substituents, has been demonstrated on a *dissolved* catalytic polymer by Overberger *et al.*[29]. A similar mechanism may also have been operative in the study of Kiefer *et al.*[6], who reported that imidazole residues appended to polyethylene imine which also carried hydrophobic substituents were 10^{12} times more efficient in catalysing the hydrolysis of 4-nitrocatechol sulphate than free imidazole. Even if all the ester molecules are drawn into the polymer domain by hydrophobic interactions, the local concentration of imidazole would not be sufficient to explain this astounding acceleration factor, so that some changes in the reaction mechanism must have been involved.

ACKNOWLEDGEMENT

I am grateful for financial support of this work by Grant GM-05811 of the National Institutes of Health.

REFERENCES

1. J. S. Scarpa, D. D. Mueller and I. M. Klotz, *J. Amer. Chem. Soc.*, **89**, 6024 (1967).
2. A. Hvidt and R. Corett, *J. Amer. Chem. Soc.*, **92**, 5546 (1970).
3. H. Ladenheim, E. M. Loebl and H. Morawetz, *J. Amer. Chem. Soc.*, **81**, 4860 (1959).
4. R. L. Letsinger and T. J. Savereide, *J. Amer. Chem. Soc.*, **84**, 3127 (1962).
5. C. G. Overberger, M. Morimoto, I. Cho and J. C. Salamone, *J. Amer. Chem. Soc.*, **93**, 3228 (1971).
6. H. C. Kiefer, N. I. Congdon, I. S. Scarpa and I. M. Klotz, *Proc. Natl. Acad. Sci., U.S.*, **69**, 2155 (1971).
7. E. Gaetjens and H. Morawetz, *J. Amer. Chem. Soc.*, **82**, 5328 (1960).
8. T. C. Bruice and U. K. Pandit, *J. Amer. Chem. Soc.*, **82**, 5858 (1960).
9. E. Gaetjens and H. Morawetz, *J. Amer. Chem. Soc.*, **83** 1738 (1961).
10. W. Kuhn, *Kolloid-Z*, **68**, 2 (1934).
11. P. J. Flory, *Statistical Mechanics of Chain Molecules*, Interscience Publ., New York, N.Y. (1969).
12. M. E. Fisher, *J. Chem. Phys.*, **45**, 1469 (1966).
13. N. Goodman and H. Morawetz, *J. Polymer Sci., Pt C*, **31**, 177 (1970).
14. N. Goodman and H. Morawetz, *J. Polymer Sci., Pt A-2*, **9**, 1657 (1971).
15. W. P. Jencks and J. Carriuolo, *J. Amer. Chem. Soc.*, **82**, 1778 (1960).
16. H. Morawetz, *Macromolecules in Solution*, Interscience Publ., New York, N.Y., (1965), p. 433.
17. H. Morawetz and W. R. Song, *J. Amer. Chem. Soc.*, **88**, 5714 (1966).
18. H. Morawetz. J.-R. Cho and P. Gans, *Macromolecules*, **6**, 624 (1973).
19. J.-R. Cho and H. Morawetz, *Macromolecules*, in press.
20. R. B. Merrifield, *Advan. Enzymol.*, **52**, 221 (1969).

[21] M. Fridkin. A. Patchornik and E. Katchalski, *J. Amer. Chem. Soc.*, **90**, 2953 (1968).
[22] J. I. Kau and H. Morawetz, *Amer. Chem. Soc., Div. Polymer Chem., Polymer Preprints*, **13**, 819 (1972).
[23] M. H. Loucheux and M. Banderet, *J. Polymer Sci.*, **48**, 405 (1960); *Bull. Soc. Chim. France*, 1964, 1220.
[24] J. C. Wang and N. Davidson, *J. Mol. Biol.*, **14**, 469 (1960).
[25] M. Sisido, *Polymer J.*, **3**, 84 (1972).
[26] B. Vollmert and H. Stutz, *Angew. Makromol. Chem.*, **3**, 182 (1968).
[27] J. G. Wetmur and N. Davidson, *J. Mol. Biol.*, **31**, 349 (1968).
[28] J. G. Wetmur, *Biopolymers*, **10**, 601 (1971).
[29] C. G. Overberger, T. St. Pierre, N. Vorchheimer, J. Lee and S. Yaroslovsky, *J. Amer. Chem. Soc.*, **87**, 296 (1965).